Gender Ironies of Nationalism

The interplay between nation, gender and sexuality pressures people to negotiate their identities in complex ways. The empowerment of one nation, one gender or one sexuality invariably occurs at the expense of another.

In *Gender Ironies of Nationalism*, international case studies offer new insights into the compound intimacies and multiple identities that result from these negotiations – calling into profound question assumptions about nationalism as monolithic, much less gender neutral. The contributors conclude that control over access to benefits of belonging to the nation is invariably gendered. Nationalism frequently becomes the language through which sexual control and repression are justified and through which masculine prowess is expressed and strategically exercised.

By exploring the ways in which nations are comprised of sexed bodies, and the central role of sexuality in nation-building and in the construction of national identity, the contributors expose a fundamental set of "gender ironies." Despite its rhetoric of equality for all who partake in the "national project," globally *nation* remains the property of men. Yet, while it is men who claim the prerogatives of nation and of national authority it is, for the most part, women who actually accept the obligations of nation-building and of sustaining national identity.

Finally, if both "nation" and "gender" help to construct a fiction of "innateness," the fierceness with which sexuality is wielded in defense of national bonds reveals the fragile, fragmented – and strained – status of nationalism itself.

Tamar Mayer is Professor of Geography at Middlebury College, Middlebury, Vermont, USA.

Gender Ironies of Nationalism

Sexing the nation

Edited by Tamar Mayer

London and New York

First published 2000
by Routledge
11 New Fetter Lane, London EC4P 4EE

Simultaneously published in the USA and Canada
by Routledge
29 West 35th Street, New York, NY 10001

Routledge is an imprint of the Taylor & Francis Group

Typeset in Galliard by The Florence Group, Stoodleigh, Devon
Printed and bound in Great Britain by
TJ International Ltd, Padstow, Cornwall

British Library Cataloguing in Publication Data
A catalogue record for this book is available from the British Library

Library of Congress Cataloging in Publication Data
Gender ironies of nationalism : sexing the nation / edited by Tamar Mayer.
p. cm.
1. Sex role–Political aspects. 2. Nationalism.
3. Nationalism and feminism. I. Mayer, Tamar, 1952–
HQ1075.G4643 1999
305.3–dc21 98–52189
CIP

ISBN 0–415–16254–8 (hbk)
ISBN 0–415–16255–6 (pbk)

In memory of my father, Artur Mayer,
For my mother, Shoshana Mayer,
And in honor of their great hopes

Contents

Plates

Contributors

Cihan Ahmetbeyzade is a graduate student in Anthropology. She is a Turkish woman who has studied extensively the changing nature of production of tribal Kurds and the contribution of peasant Kurdish women to their household economy in eastern Turkey.

Holly Allen is an Assistant Professor of American Studies at Middlebury College. She received her Ph.D. from Yale University in 1996 and is currently working on a study of gender, citizenship and U.S. national identity entitled *Fallen Women and Forgotten Men: Gendered Concepts of Community, Home, and Nation, 1932–1945.*

William Chaloupka is a Professor of Environmental Studies at the University of Montana in Missoula, where he also teaches courses in Political Science. He is the author of *Knowing Nukes: Politics and Culture of the Atom* (University of Minnesota Press, 1992) and co-editor of *In the Nature of Things: Language, Politics, and the Environment* (with Jane Bennett) (University of Minnesota Press, 1993) and of *Jean Baudrillard: The Disappearance of Art and Politics* (with William Stearns) (Macmillan and St. Martin's, 1992). He is currently writing a book about cynicism in American politics and culture.

Steve Derné is an Assistant Professor of Sociology at State University of New York – Geneseo. He is the author of *Culture in Action: Family Life, Emotion, and Male Dominance in Banaras, India* (SUNY Press, 1995) and is currently completing an ethnographic study of filmgoing in India, *Movies, Masculinity, and "Modernity"* which explores Indian men's identity conflicts about "masculinity" and "modern" life.

Leslie K. Dwyer is a Ph.D. candidate in Anthropology at Princeton University. She has conducted fieldwork on Islam, modernity, sexuality and bio-medicine in Java, Indonesia, and is completing her dissertation entitled "Making modern Muslims: gender, sexuality and embodied politics in urban Java, Indonesia." Her research interests include the politics of bodies, visuality and tourism; she is involved in an ethnographic documentary of Islam in Indonesia.

Tamara Hamlish is an Assistant Professor of Anthropology at Beloit College. She has published several articles on the anthropology of Chinese art and art exhibition, and is currently completing a book on representations of nationalism in the Chinese National Palace Museum.

Linden Lewis is an Associate Professor of Sociology at Bucknell University. He was co-director of the Race/Gender Resource Center at Bucknell from 1994 to 1996. His research area is the Caribbean, with special interests in issues of labor, popular culture, race, gender, nationalism, globalization and the state. He has published extensively in these areas and is currently editing a book on gender, sexuality and popular culture in the Caribbean.

Andrew Light is an Assistant Professor of Philosophy and Environmental Studies at the State University of New York at Binghamton. In addition to publishing over twenty articles on environmental philosophy, philosophy of technology and political theory, he has edited or co-edited five anthologies including *Environmental Pragmatism* (with Eric Katz) (Routledge, 1996). He is co-editor of the journal *Philosophy and Geography* and managing editor of the *Radical Philosophy Review.*

Jeanne Marecek is a Professor of Psychology and Women's Studies at Swarthmore College. She is co-editor of *Making a Difference: Psychology and the Construction of Gender* (with Rachel Hare-Mustin) (Yale University Press, 1994), series co-editor of *New Directions in Theory and Psychology* (with Rachel Hare-Mustin) (Westview Press), series co-editor of *Qualitative Studies in Psychology* (with Michelle Fine) (New York University Press) and the guest editor of a *Psychology of Women Quarterly* special issue, "Theory and Methodology in Feminist Psychology" (1989, 13, 4).

Angela K. Martin is a Ph.D. candidate at the University of Kentucky. She is currently a visiting Assistant Professor of Anthropology at Michigan State University, where she works on the Detroit Metropolitan Congregational Project, the first large-scale ethnographic study of contemporary religious practice and belief in the United States. Her research interests include religious and national identity, health, the body and embodiment, and gender. She has conducted ethnographic fieldwork in the Republic of Ireland and in the United States.

Tamar Mayer is Professor and Chair of the Geography Department at Middlebury College, and the editor of *Women and the Israeli Occupation: The Politics of Change* (Routledge, 1994). Her research interests focus on the interplay among nationalism, gender and sexuality, particularly in the Middle East, and on the relationships among nationalism, landscape and memory.

Mary H. Moran is an Associate Professor of Anthropology at Colgate University. She is the author of *Civilized Women: Gender and Prestige in Southeastern Liberia* (Cornell University Press, 1990) and a number of articles on Liberia. She is currently the editor of the American Ethnological Society's Monograph Series and is the general editor of the series *Women and Change in the Developing World* (Lynn Reinner).

Julie Mostov is an Associate Professor of History and Politics at Drexel University. Her research and interest areas are modern political thought, and democratic theory and ethics. She is the author of *Power, Process, and Popular Sovereignty* (Temple University Press, 1992) and is currently working on another book entitled *Guardians, Warriors, and Traitors – What Place for Citizens? The Politics of National Identity in Eastern Europe.*

Elizabeth A. Povinelli is an Associate Professor of Anthropology at the University of Chicago. She is the author of *Labor's Lot: The Culture, History and Power of Aboriginal Actions* (University of Chicago Press, 1994) and is currently completing a book on Australian multiculturalism and the crisis of indigenous citizenship.

Acknowledgments

I wish to thank many people who have been important to this project, who have pushed me to think about the connections among nationalism, gender and sexuality, and who have helped me see its completion. Tristan Palmer, formerly of Routledge, encouraged me to imagine this project while I was still working on *Women and the Israeli Occupation*. His enthusiasm was almost contagious and his suggestions were valuable. While on academic leave from Middlebury College I was fortunate to be a Research Associate at the Five College Women's Studies Research Center, at Mount Holyoke College. It was at the Center that the idea of the book took shape. I was part of a reading group on sexuality in which Renée Romano, Liana Borghi, Bat-Ami Bar On and Martha Verburgge helped me formulate connections between gender and sexuality and between these two constructed categories and nationalism.

Over the years, I have shared with Joanne Jacobson my fascination with the construction of nationalism, especially the construction of Jewish nationalism. Her critical eye, her support and our excellent discussions on the gender ironies of nationalism show throughout this book. From the early days of this project Nancy Shumate insisted on my understanding that nationalism, like so many other modern phenomena, has roots and parallels in antiquity; that if I looked at the past I would better understand the present. Her incisiveness, originality and sense of irony were helpful at the different stages of this project. And Tahl Mayer, who is not an academic, at least not yet, has shown sophistication about nationalism and gender not generally found among his grade-school peers. His patience with my work schedule, which often kept me away from him, made it possible to complete the book in a timely fashion.

I dedicate this book to my parents, Shoshana and Artur Mayer, who taught me through example the value of national identity. I am deeply saddened that my father's sudden death – within a few months' sight of its publication – kept him from reading the book and from participating in the conversation that I hope it will provoke.

1 Gender ironies of nationalism

Setting the stage

Tamar Mayer

The nation is a process of becoming
(Bauer 1996 [1924])

In his famous 1924 essay "The nation," Otto Bauer asserted that "national character is changeable" (1996 [1924]: 40), and that the idea of *nation* is bound up with ego (1996 [1924]: 63). He suggested that "if someone slights the nation they slight me too . . . [F]or the nation is nowhere but in me and *my kind*" (ibid., emphasis added). The ideology which members of the community, those who are of the *same kind*, share – through which they identify with the nation and express their national loyalty – is what we call nationalism. Hence nationalism is the exercise of internal hegemony, the exclusive empowerment of those who share a sense of belonging to the same "imagined community" (Anderson 1991). This empowerment is clearly intertwined with what Bauer called "ego." But what kind of ego is at stake in the case of the "nation"? The chapters in this volume argue that the national ego is intertwined with male and female ego, that it is inseparable from gender and sexuality. They further argue that nationalism becomes the language through which sexual control and repression (specifically, but not exclusively, of women and homosexuals) is justified, and masculine prowess is expressed and exercised.

Because nationalism, gender and sexuality are all socially and culturally constructed, they frequently play an important role in constructing one another – by invoking and helping to construct the "us" versus "them" distinction and the exclusion of the Other. The empowerment of one gender, one nation or one sexuality virtually always occurs at the expense and disempowerment of another. But because people have multiple identities, the interplay among nation, gender and sexuality often pressures people to negotiate their identities in complex ways.

The title of this book, *Gender Ironies of Nationalism*, is meant to convey the idea that the links between "gender" and "nation" tell us about some of the more profound ironies of modern social life. Despite its rhetoric of equality for all who partake in the "national project," *nation* remains, like other feminized entities – emphatically, historically and globally – the property of men. At the same time, if it is gendered, *nation* remains – quite like gender and sexuality – a construction that speaks to the conflicted urges of human community. For both "nation" and "gender" help construct a fiction of "innateness" in the name of bonds whose fragile, endangered status is evidenced in the fierceness with which they are defended – and in the fierceness with which the role of the imagination in the construction of transcendent categories and the urge to reify those categories are both, at once, revealed and denied. The subtitle *Sexing the Nation* emphasizes, further, that when sexed bodies comprise the nation we can no longer think of the nation as sexless. Rather, by exploring the gender ironies of nationalism we expose the fact that sexuality plays a key role in nation-building and in sustaining national identity.

The chapters in this volume demonstrate the many complex intimacies between gender and nation and sexuality. They show, in particular, that control over access to the benefits of belonging to the nation is virtually always gendered; that through control over reproduction, sexuality and the means of representation the authority to define the nation lies mainly with men. Finally, these chapters emphatically establish the relationship between gender boundaries and the nation: for they demonstrate that while it is men who claim the prerogatives of nation and nation-building it is for the most part women who actually tend to accept the obligation of nation and nation-building.

Definitions

Two sets of categories – nation and state, and gender and sexuality – are the bare bones of each of the chapters in this volume. Although nation is not to state what gender is to sexuality – because nation could be conceived without state but gender and sexuality remain inevitably connected – there are parallels across these sets of categories: all of these categories are socially or culturally constructed in opposition (or sometimes in relationship which is not binary) to the Other, and all of them involve power relationships. But before we turn to discussion of how nation, state, gender and sexuality intersect globally, it is important to define and explore each of these categories separately. Although *nation* and *state* are often used interchangeably, they are emphatically not synonymous. A state is a sovereign political unit which has tangible boundaries, abides by international law and is recognized by the international community. But while it may have tangible characteristics (Connor 1972) and is always self-defined, a nation is not tangible.

A nation "is a soul, a spiritual principle" (Renan 1990: 19), a "moral consciousness" (Renan 1990: 20) which its members believe must be maintained at all times and at all costs. The nation is a glorified ethnic group whose members are often attached to a specific territory (Smith 1981, Connor 1978) over which they strive for sovereignty or at least the ability to manage their own affairs. Members of the nation believe in their common origins and in the uniqueness of their common history, and they hope for a shared destiny (Smith 1986). They amplify the past and keep memories of communal sufferings alive. They share national symbols like customs, language and religion, and are often blind to the fact that their national narrative is based on myths and on what Etienne Balibar (1991) calls "fictive ethnicity."[1] Myth remains in fact essential to the life of the nation, for it is by embracing myths about the nation's creation that members perpetuate not only national myths but also the nation itself.

The nation is sustained as well through both reactive and proactive measures. Nationalistic ideology can serve as "emotional glue" – by *othering* the nation when it occupies minority status (Calhoun 1997, Hechter 1975, Deutsch 1953) – and when there is no threat from outside or when threat does not appear imminent, through regular, even repetitive, exercises of solidarity which become accepted by members of the nation as "natural." As many of the chapters here demonstrate, cultural, religious and political ceremonies – along with education (Chatterjee 1993, Anderson 1991, Hobsbawm 1990, Gellner 1983), exploitation of national media and museums and control of the national "moral code" – keep national consciousness alive and the nation "real."

For the sake of maintaining parallels with the other set of categories discussed in this volume – gender and sexuality – and of better understanding the distinction between nation and state grounding the volume, it is important to develop here more fully some aspects of this complex relationship.[2] While a nation can live without a state, a state usually does not exist without a nation: we know of many stateless nations (some of whose national consciousness has been raised because of the state system).[3] While there are many multinational states, there are no nation-less states.[4] Furthermore, even though state is often perceived as the political extension of nation (Connor 1978), it must be viewed as a separate entity, because rarely do we find a pure nation-state that constructs a 100 percent fit between a nation and the state territory that it occupies. More often than not, instead, we find states which house many nations, leading to a hierarchy among these nations and creating a competition among them over control of resources and the exercise of power as a means to achieve national hegemony within the state.

As important as these discussions about the nation and the state are, they omit an essential discussion about gender and sexuality. Since the mid-1980s scholars have begun to demonstrate that we cannot understand nation and nationalism without understanding that gender and sexuality are integral to both.[5] These scholars have shown that power, control and hegemony exist not only in the relationships between nation and state but also in the relationships between gender and sexuality, and between nation and state and gender and sexuality.

The distinction between gender and sexuality remains considerably less sharp and more complicated than the distinction between nation and state, because our understanding of these categories varies historically and geographically and because our definitions of them are still being debated. But for the sake of setting the stage for discussion and providing common language with which to read this volume I shall, nevertheless, offer some working definitions of these categories. First, what it means to be "male" or "female," "masculine" or "feminine," "man" or "woman" is inevitably, socially constructed – for culture gives gender and sexuality meanings that are particular to time and space, and provides the arena within which a given subject is positioned. Within this arena, gender serves as the cultural marker of biological sex (Vance 1984: 9) and sexuality serves as the cultural marking of desire (Foucault 1978).

More specifically, gender is the "dichotomized social production and reproduction of male and female identities and behaviors" (Sedgwick 1990: 27). Social re-production (re)produces gender through daily repetition of acts/performances – or what Judith Butler (1990) calls "performativity." In other words, what we perform repeatedly – based on norms that predate us – is what we become, regardless of biological chromosomes. In this sense, gender is divorced from sex (biology) and, therefore, "masculinity" does not necessarily have to be the domain of a biological "male" or "femininity" the domain of the biological "female." As the chapters by Derné, Mayer, Martin, Allen and Lewis in this volume illustrate, at this time in the life of the nation "masculine" and "feminine" identities do seem to be fixed, with "masculinity" the domain of (biological) men and "femininity" the domain of the (biological) females. But because nation, gender and sexuality are always in the process of becoming, because they evolve continuously, associating "masculinity" with men and "femininity" with women in a national context could eventually change if either the discourse of the nation or that of gender and sexuality changes. Dwyer (Chapter 2), Moran (Chapter 5), Povinelli (Chapter 7) and Ahmetbeyzade (Chapter 8) show that in Indonesia, Liberia, Australia and Turkey these identities have already begun to change.

Sexuality, too, is not fixed in time and space. It too is a cultural construction, which refers both to an individual's sexed desire and to an individual's sexed being, embracing ideas about "pleasure *and* physiology, fantasy *and* anatomy" (Bristow 1997: 1, original emphasis). But without understanding that sexuality is also "a domain of restriction, repression, . . . danger . . . and agency" (Vance 1984: 1) and is an "actively contested political and symbolic terrain in which groups struggle to implement sexual programs and alter sexual arrangements and ideologies" (Vance 1995: 41), we cannot fully understand the importance of sexuality to gender and to nation. Like gender, sexuality (and, as the chapters in the volume show, also nation) is organized into systems of power "which reward and encourage some individuals and activities while punishing and suppressing others" (Rubin 1984: 309). As the chapters in this volume amply demonstrate, throughout the contemporary world these power systems generally reward heterosexual males and often punish women and gays.

To complicate these definitions even more, we need to recognize that neither gender nor sexuality ought to be discussed in the singular. Rather, because both gender and sexuality vary geographically, across lines of age, class, ethnicity, sexual orientation and religion, and because both are articulated through a variety of positions, languages and institutions, we witness a multiplicity of gender identities and sexualities (see Lancaster and Leonardo 1997, McClintock *et al.* 1997, Duncan 1996, Berger *et al.* 1995, Brittan 1989, Vance 1984). Therefore neither gender nor sexuality is a "fixed" category: each is always implicated in the other; each is always ambivalent, always complicated, always a product of individual and institutional power.

Nation, gender, sexuality: liaison of/over bodies

The nation is comprised of sexed subjects whose "performativity" constructs not only their own gender identity but the identity of the entire nation as well. Through repetition of accepted norms and behaviors – control over reproduction, militarism and heroism, and heterosexuality – members help to construct the privileged nation; equally, the repetitive performance of these acts in the name of the nation helps to construct gender and sexuality. Moreover, because nation, gender and sexuality are all constructed in opposition, or at least in relation to, an(O)ther, they are all part of culturally constructed hierarchies, and all of them involve power. One nation, one gender and one particular sexuality is always favored by the social, political and cultural institutions which it helps to construct and which it benefits from – and thus each seeks to occupy the most favored position in the hierarchy (of nation, gender and sexuality); each tries to achieve hegemony; and each in the process becomes a contested territory, even the arena of battle among nations, genders and sexualities.

Until recently the literature on nationalism has been gender blind. But feminist scholarship's identification of gender as a category of analysis has led to the exploration of the relationship among nation and gender/sexuality. Feminist research has steadily revealed that men and women participate differently in the national project (Yuval-Davis 1997, McClintock 1995, Kondo 1990, Enloe 1989, Yuval-Davis and Anthias 1989, Jayawardena 1986).[6] Much of this scholarship has focused on women's marginality *vis-à-vis* the construction of nation, and as a result these discussions have, for the most part, neglected to analyze *men* as an equally constructed category.[7] This imbalance has arisen, I believe, from Women's Studies' tendency until recently to concentrate on recovering women's experience, without necessarily positioning it in the larger context of gender construction, and from the unmarked status of masculinity within the nation and in nationalist discourse. However, as gender and its connection to sexuality continue to be explored, scholarship about nationalism has come to involve, more explicitly, analysis of both men's and women's relationship to the construction of the nation and of the ways in which national discourse constructs man and woman.[8] It is this discussion to which this volume contributes.

When we examine the intersections among nation, gender and sexuality, we become aware that Otto Bauer (1996 [1924]) might have been ahead of his time when he observed that "the idea of the nation is bound up with 'ego'." Although Bauer's reference to the nation's "ego" does not even mention gender (which is understandable, given the time he wrote), our understanding of this intersection is improved if we understand that who *I* am is connected to who the nation is; that my "ego" is often inseparable from the "ego" of the nation to which *I* belong and which helps to define *my* identity. Because the nation was produced as a heterosexual male construct its "ego" is intimately connected to patriarchal hierarchies and norms. These enable men and nation to achieve superiority over women and a different Other by controlling them. As a result, the intersection of nation, gender and sexuality is a discourse about a moral code, which mobilizes men (and sometimes women) to become its sole protectors and women its biological and symbolic reproducers.[9]

But I do not want to imply that the relationships which either the nation's ego or male's ego reproduce are monolithic, where men are active and women are passive; rather that it is important to recognize that women, too, participate culturally in reproducing the nation, defending the "moral code" and partaking in controlling the Other (e.g. De Grazia 1992, Koonz 1987). Therefore, they too sometimes contribute to the nation's ego. For it is usually women, Yuval-Davis (1997: 37) argues, "especially older women who are given the roles of the cultural reproduction of 'the nation' and who are empowered to rule what is 'appropriate' behavior and appearance and what is not and to exert control over other women who might be constructed as 'deviant'."

Gender control over nation and sexuality

The nation has largely been constructed as a hetero-male project, and imagined as a brotherhood (Anderson 1991: 16) which has typically sprung, as Enloe (1989: 44) suggests, "from masculinized memory, masculinized humiliation and masculinized hope." Nation, therefore, regardless of location, largely remains the domain of men. But because not all men, and certainly not all people, are created equal, this "horizontal comradeship" (Anderson 1991) remains gender, sexuality, race and class specific.[10] Furthermore, because nationalism is about difference – and imagined community can therefore not be inclusive (Chatterjee 1996) – internal hierarchies often occur along lines of gender, race, class and sexuality, despite the national discourse of internal unity. It is men who are generally expected to defend the "moral consciousness" and the "ego" of the nation. Men tend to assume this role because their identity is so often intertwined with that of the nation that it translates into a "personalized image of the nation" (Hroch 1996:

90–91). Because men "regard the nation – that is themselves – as a single body" (ibid.), their own "ego" becomes at stake in national conflicts, and they frequently seek to sustain control over reproduction and representation of both sexuality and nation and over the boundaries of the nation, through defining who is included in, or excluded from it.

Reproduction

Much of the literature on gender and nation has focused on the centrality of women to the national project. Yuval-Davis and Anthias (1989) have suggested that women's national importance is based on their reproductive roles, which include biological and ideological reproduction, reproduction of ethnic or national boundaries, transmission of culture and participation in national struggles. Their centrality is also based on women's symbolic status, connected to their reproductive roles, as representatives of purity. Only pure and modest women can re-produce the pure nation; without purity in biological reproduction the nation clearly cannot survive. The chapters in this volume by Dwyer, Martin, Mostov and Marecek (Chapters 2, 3, 4 and 6) use examples from Indonesia, the Republic of Ireland, the former Yugoslavia and Sri Lanka to illustrate the fact that reproduction is culturally constructed, and that fertility is frequently hailed by the nation's subjects and their leaders as a sign of both national prosperity and virility. It follows in all these cases that men control fertility and reproduction. Furthermore, these chapters show that when the nation is faced with internal and external pressures it polices and employs coercive means to control sexuality. These means can often be seen, as well, as racist.

Examining the politics of family planning in Indonesia, Dwyer (Chapter 2) shows that women there have little choice about their reproductive behavior; that they are not at liberty to define the size of their family; and that the reproductive choices which they do have remain cultural constructs. For the nation's sake, in the years before Western development, women were encouraged to have more children; but since the World Bank and the United States Agency for International Development (U.S.A.I.D.) have begun their activities in Indonesia and development has become the cornerstone of Indonesian nationalism women have been encouraged, by all means, to limit their fertility. Ultimately, through family planning the state has orchestrated control over both women's sexuality and the public articulation of nationalism. And the Indonesian situation is far from unique. In Ireland, for example, as Martin argues in Chapter 3, the state also controls women's sexuality and reproduction through its judicial system; and there, too, the state maps out the contours of national identity. By using the twinned logic of religion and nation to prohibit abortions (including travel to neighboring states for abortions, even for adolescent girls), the state interferes with women's reproductive choices and in effect sets the discursive relationship among state, nation, and reproduction.

Abortion is also constructed as the enemy of the nation in the former Yugoslavia, according to Mostov (Chapter 4), who shows that Bosnian, Croat and Serbian women are encouraged by religious and national leaders to have more children in the name of nationalism. Women who have abortions are figured as "moral enemies of the state" – but reproduction is celebrated only if it is consummated with men of that nation. The case of the former Yugoslavia demonstrates how in a multinational state, especially one that has experienced major international wars, the "us"/"them" construction remains especially strong.

In both Liberia and Sri Lanka, as Moran and Marecek (Chapters 5 and 6) show, the reproduction of the traditional nation is carried out by women who are extolled as custodians of the national cultural heritage – if they conform to the ideals of traditional womanhood, and if they do not they come under attack by men of these nations. In both these countries, as in Indonesia, Western ideals about womanhood have been adopted, and thus reproduction of the nation in the late twentieth century means something quite different than a few decades ago. Furthermore, by legalizing a "moral code" concerning marriage and family, the state in both Liberia and Sri Lanka actively controls both sexuality and the nation.

Western ideals and technologies also participate in reproduction of the nation, especially in the non-Western world. While nationalism has become equated in Indonesia and Liberia and for the Sinhala in Sri Lanka with "modernity" and modeled on Western ideals about family and progress, in Sri Lanka the nation is at the same time reproduced in resistance against Western dress and lifestyle. As Western ideas such as family planning, the nuclear family, monogamous relationships and "civilized" behavior replace traditional practices, the nation is thus forced to negotiate its identity in complex ways: for while it may aspire the approval and resources of the West, the only way the nation can be fully reproduced is if it remains "traditional" and its women remain modest and pure. One way to ensure this type of reproduction, as Marecek, Martin, Mostov and Moran all suggest, is by mandating women's confinement to the home, to the private sphere, where they remain under the watchful eye of their husbands.

These national battles over reproduction, representation and control over sexuality are inevitably complicated by social and political hierarchies: not only are there gender, class and sexual hierarchies within every nation, but hierarchies separate different nations as well. Therefore in Asia, Africa and Latin America biological reproduction is negotiated not only by husband and wife but also by the nation's elites, whose interests frequently coincide with the interests of Western developers and politicians. Reproducing the nation has become in non-Western nations, then, paradoxically, in significant part the domain of the West and its white populations.

In addition to biological reproduction, the nation is reproduced culturally, socially and symbolically through the performativity of its members. This is also the way that norms of gender and sexuality are reproduced when they intersect with nation. Narratives about the creation of the nation, which posit the proper behavior of women as mothers and defenders of culture and national values, are discussed by Mostov for the case of Yugoslavia, by Martin for the case of Ireland, by Marecek for the case of Sinhala in Sri Lanka, by Moran for the case of Liberia, by Ahmetbeyzade for the case of the Kurds in Turkey (Chapter 8) and by Derné for the case of Hindus in India (Chapter 10). Narratives which construct the role of men as defenders of the nation and its traditions are discussed by Mostov for Yugoslavia, Derné for India, Lewis for the Caribbean (Chapter 11), Mayer for Jewish Israel (Chapter 12) and Allen for the U.S. (Chapter 13). Their gender roles are reproduced in the name of the nation and in the process they reproduce the nation itself. Furthermore, as Povinelli suggests in Chapter 7, the nation may also be reproduced through sexual "performativity," as in Australia, through the violent ritual sex of Aborigines. Although until the early part of the twentieth century these acts and their significance to Australian identity were misunderstood and challenged hegemonic understandings of Australian nationalism, ritual sex has more recently prompted the Australian public to consider a new, more heterogeneous and tolerant, foundation for its national narrative.

Cultural ritual is also central to the reproduction of both nationhood and gender roles, as Hamlish argues in Chapter 9, about China. The traditional practice of calligraphy is an important medium through which the nation is reproduced and gender is sexualized. Because the practice of calligraphy has not changed much over many years and remains almost exclusively the domain of men, women in China – unlike in other parts of the world – remain marginal to the cultural reproduction of the nation. Because calligraphy embodies traditional – gendered – values and beliefs, Hamlish argues, it remains an obstacle in the way of progress, actively helping to sustain "traditional" reproduction of both nation and gender roles.

In each of these cases, culture and ritual are central strategies through which the nation projects itself inwardly, as well as to the outside world, and through which it mobilizes its members. The nation's self-representation always involves myths about the nation's creation and about its members. As established by Connor (1990, 1978), Hobsbawm (1990), Smith (1986), Gellner (1983) and others, myth is such a crucial element in the life of the nation that without it the nation cannot survive. But because myth, by definition, does not necessarily represent with historical accuracy the nation's past, the "reality" that is constructed intends to represent both the nation and its members in a way that will continue both to benefit the unity of the nation and to sustain the myth.

Representation

National narratives construct the ideal image of the nation. This discourse is a way for the nation to present itself to multiple audiences: to the national community (regardless of gender, sexuality, race and class) and to the international community. In order to survive and to justify its existence, the nation must preserve its uniqueness; it does so by constructing myths about national creation and by defining "proper behaviors" for members of the nation and for the nation itself. Because elites play a major role in constructing the nation and its narratives, the nation is generally represented so that it serves the aspirations of the elite (Anderson 1991, Hobsbawm 1990, Connor 1987, Mosse 1985, Brass 1979). In these narratives the nation is virtually always feminized and characterized as in need of protection; women are figured as the biological and cultural reproducers of the nation and as "pure" and "modest," and men defend the national image and protect the nation's territory, women's "purity" and "modesty," and the "moral code." Thus women are represented as the nation's social and biological womb and the men as its protectors: "women [are] sedate rather than dynamic . . . [t]hey [stand] for immutability rather than progress, providing the backdrop against which men determine[d] the fate of the nation" (Mosse 1985: 23). Although in reality these prescribed identities are often challenged, rarely do we find in the national rhetoric ambivalence over any of these identities.

"Purity," "modesty" and "chastity" are common themes in national narratives of gender, nation and sexuality (e.g. Chatterjee 1993, Katrak 1992, Kandiyoti 1991) and they are discussed extensively in this volume. Dwyer, Martin, Mostov, Moran, Marecek and Derné all show that when a nation is constructed in opposition to the Other there emerges a profound distinction not only between us and them but also more pointedly, between our women and theirs. Our women are always "pure" and "moral" while their women are "deviant" and "immoral." As these contributors illustrate, representing women in this way guarantees women's inferiority, for the favored members of the nation – the loyal sons – must defend *our* women's "purity," as well as the "moral code" of the nation. These men praise traditional roles for women but embrace for themselves practices which are based on modernity.

In the case of Indonesia, for example, as Dwyer shows (Chapter 2), even as the ideal of nationhood is becoming intimately connected with the path of modern progress, representation of the "ideal" woman – who restricts her fertility for the sake of the nation, who is "modest" and who is committed to her nuclear family – is becoming more important in public life. Religious sermons, family planning propaganda in schools and youth movements, and commercials on television and billboards all hail the benefits of contraception to small and "happy" families in their representations of women and the nation.

The "ideal" nation and its "model" members are represented in arts, literature and the media, in public speeches and in the writings of the nation's leaders – in every medium through which the nation is mobilized. Other media through which the relationships among gender, sexuality and nation are represented and which are discussed in the volume include newspaper cartoons in Liberia, calligraphy in China, the rhetoric of remembrance in Israel and the political debates surrounding the participation of women and gays in the U.S. military. Moran argues in Chapter 5 that because newspapers are one important medium through which African social life is constructed, given meaning and revised, and where the nation constitutes itself as "civilized," women's representation in Liberian newspaper cartoons exemplifies national tensions between becoming modern, "civilized" and remaining "native." Although these cartoons offer different visions of "civilized" womanhood and female citizenship, their frequent portrayal of modern women as "predators" and "aggressors" – as enemies of national development – has, according to Moran, contributed to the negative representation of women in contemporary Liberian nationalism. Hamlish argues in Chapter 9 that because in China the images of calligraphy embody a timelessness that transcends the particularities of any given historical moment, appropriating calligraphy as an instance of national heritage contributes to representations of the nation as a singular, unified community. And because it is largely men who have participated in the calligraphic tradition, it is men's vision through which the nation has been reproduced and represented.

A common theme in the literature concerning gender and nation is the feminization of the motherland and the call of the nation's sons to defend her. Mostov and Mayer argue, in Chapters 4 and 12, that in the cases of both the former Yugoslavia and Jewish Israel it is the nation's men who are made into heroes, and it is through imagery of men that the nation represents itself. Such representation is sustained, Mayer argues, through the rhetoric of remembrance and through public embrace of a "cult of toughness" that represents both the Jewish nation and its Israeli men. Representation of the nation through its military is a pattern which is also explored by Allen (Chapter 13), who asserts that the U.S. military performs the most important representative function in American life. After examining the rhetoric of threat to the national fabric in U.S. Congressional debates about inclusion of women and gays in the military, Allen argues that the military has become a leading defender of heterosexual national (and "family") values.

But representation of the American nation is not only the domain of mainstream groups. Light and Chaloupka argue in Chapter 14 that for that part of the American public which they call "angry white men" the formulation of an American sense of nationhood depends greatly on creating myths about white male supremacy. It is through the white supremacy discourse of the right, through major anti-government challenges mounted in places like Waco, Texas, in Ruby Ridge, Idaho and Oklahoma City, that the extreme right's vision of an American nation is constructed and played out. Light and Chaloupka also suggest that it is because mainstream representations of the nation are limited and exclusive that fringe groups like white supremacists construct their own vision of the nation and fight for it.

Inclusion/exclusion – whose nation is it, anyway?

The nation, as Benedict Anderson (1991) has framed it, is an "imagined community" whose members conceive it to be united, exclusive and worthy of sacrifice (Breuilly 1996). While it may feel central when the nation is constructed *vis-à-vis* the Other, this discourse of "unity" is often challenged when the nation's inner workings are examined, especially in relation to gender and to sexuality.

Because the nation is often constructed by elites who have the power to define the nation in ways that further their own interests, the same elites are also able to define who is central and who is marginal to the national project. In the intersection of nation, gender and sexuality the nation is constructed to respect a "moral code" which is often based on masculinity and heterosexuality. This is the reason why the leaders of the nation may try to represent their nation as "modest" – and in turn speak in terms of the ideals of the nation in imposing on women a traditional moral code (see Mostov, Martin, Moran, Ahmetbeyzade and Derné).

Allen's analysis of Congressional debates over the inclusion of gays in the U.S. military and of women in combat units (Chapter 13) suggests that the military has been an important vehicle for American national imagining. Opponents in Congress have, in effect, tried to exclude women and gays from fully participating in the nation; for it is precisely because of the military's significance to the nation and the importance of military service to membership in U.S. national community, Allen argues, that female combatants and gay soldiers seek access to a military role. The military has also been central, Mayer argues (Chapter 12), in defining who is part of the Jewish nation and who is not. Jewish national imagining was possible only through creating the *New Jew*, the *Muscle Jew*, an exclusively male figure who became a fighter; the pioneer redeemer of the biblical homeland, the crown jewel of Zionism. Although women were clearly important to the Jewish national project (at the very least, as biological reproducers), because so much about Jewish nationalism has since the beginning of the twentieth century revolved around militarism and defense, men have been elevated to much more central roles – while women have in many ways remained marginal to the Jewish national discourse. Furthermore, as the national memory remains so closely linked to acts of heroism and to what Mayer calls the "pantheon of male heroes," women continue for the most part to remain excluded from the Jewish national project and its imagining.

Chinese women experience a different sort of exclusion. Hamlish (Chapter 9) argues that because the nation's symbolic reproduction is closely linked to calligraphy and because only a small cohort of ruling elite – mostly men – actually practiced this art during the nationally formative imperial period, women in China have for centuries been excluded from the national discourse and from participation in its symbolic reproduction. Furthermore, even though more Chinese women have recently become calligraphers they have continued to remain marginalized – and their art invisible, distant from the sphere of national symbol-making, as they are restricted to a space defined by gender.

Exclusion from the national discourse, Ahmetbeyzade argues (Chapter 8), has also been the experience of Kurdish peasant women in Turkey. Although within their own communities Kurdish peasant women have access to power (they are often heads of households, they organize peasant networks and they often represent peasant communities to the Turkish state), they remain marginalized in the larger context of the Turkish nation. Through social policies, on the one hand, and military repression, on the other, the Turkish state excludes Kurds in general and Kurdish women in particular.

The discourse of national inclusion and exclusion is in fact central throughout the world, as three more chapters in the volume evidence. Light and Chaloupka argue in Chapter 14 that it is directly in response to the discourse of nationalism articulated by the left that the far right's formulation of itself around themes of national self-identity has emerged in the U.S. White male identity politics have focused on the formation of a new, exclusionary American national identity – one which is based on white racial pride and on a desire to return to the ideal of white supremacy associated with earlier American culture. Moran and Povinelli (Chapters 5 and 7) suggest that in nations which include both modern and native populations, access to membership in the nation involves complex cultural negotiation. In the case of Australia, Povinelli argues, sexual behavior has been an important marker of who is included in or excluded from the "imagined nation." Aborigines' ritual sex practices – some of which were violent and all of which were public – revealed to white Australians the fact the national "imagined community" that they had conceived as unified was in actuality fragmented, prompting an effort to forge a more tolerant and inclusive notion of national identity. As Moran shows in the case of Liberia, when the population is comprised of both "civilized" and "native," modern and rural, it is the "civilized" who are elevated to the highest positions within the national hierarchy while "natives," especially if they are women, and even if they have achieved some measure of "civilized" status, remain excluded from the imagined community. In other words, social transformation does not necessarily enable the members of the indigenous population to overcome their position in society and to become part of the nation as defined by the elite.

As these examples show, in determining who belongs to the nation and who does not, elites construct a code of "proper behavior" for members of the nation which becomes a sort of national boundary. In each case, the code which the elite promotes as essential to the continuation of the nation also furthers the elite's own interests; thus in the life of the nation one gender, one sexuality and one national narrative tends to rule. Even as groups of women and men, straight and gay, have begun to challenge these models and gender identities have become more fluid, the hegemony of one gender and one sexuality within the nation remains relatively unchanged all over the world.

Gender/sexual boundaries and nation

The nation has been constructed as the hegemonic domain of both masculinity and heterosexuality, and thus has been a major site for the institutionalization of gender differences (McClintock 1995). Because the nation has been symbolically figured as a family (McClintock *et al.* 1997) – and as such has acquired a patriarchal hierarchy within which members are assigned distinct roles in accordance with their gender – as in the patriarchal family, for the nation to sustain itself it needs both masculinity and femininity. For without masculinity, femininity cannot exist; and without these twin constructions the nation as we know it would not exist either.

Mosse argues in *Nationalism and Sexuality* (1985) that manliness has been the idea on which the nation is built and the arena where a passive femininity is constructed. Even when the binarism of hetero-patriarchal norms is challenged, it virtually always remains the case that it is men who claim the authority to define the nation and its boundaries; to define the process of nation-building; and to articulate what masculinities and femininities are appropriate to the nation. The chapters in this volume discuss the relationship between nation-building and both masculinity and femininity, and assert that these categories were constructed at the same time as the nation, and that very often they play an active part in defining the boundaries of the nation through a relationship to the body.

Masculinity and the nation

In her essay "How to build a man," Ann Fausto-Sterling (1995: 127) tells us that "men are made not born" and that we "construct masculinity through social discourse." Male behavior depends on existing social relations and on the social code that predetermines these relations. Therefore the expression of masculinity will depend on the image that men have of themselves (Brittan 1989) relative to women, community, society and the nation.

The five chapters in this volume which explicitly discuss the dynamics of interaction between masculinity and the nation suggest that the construction of the nation was simultaneous to the construction of masculinity in India, the Caribbean and Israel, and that male bonding or *Männerbund*, in Mosse's words, is central to the perpetuation of the nation in the cases of Israel and the United States.

Both Indian and Caribbean nationalism developed in reaction to British imperialism and to imperialism's feminization and infantilization both of the colonies themselves and of indigenous men.[11] In reaction to the powerlessness which they experienced during colonialism, Derné argues in

Chapter 10, Indian men developed sharper consciousness of their nation and their bodies. And as the British challenged their masculinity, Indian men emphasized both control over their own bodies and control over Indian women's bodies – through body-building and celibacy, and through controlling Indian women's sexuality. Many Indian men's sense of masculinity has, increasingly, come to depend on preserving women's femininity, modesty and religiosity; because the nationalist discourse, built around the intersection between nation and masculinity, has focused on "protecting" women and, especially, their sexuality from assaults by foreigners.

British imperialism has also sought to feminize the Anglophone Caribbean, but the intersection between nationalism and masculinity took a different route there. In the Caribbean, Lewis argues in Chapter 11, nationalism's connection to masculinity developed simultaneously in two different arenas: in trade unions, where men who held central positions articulated ideas about a Caribbean-based self-determination; and in the Black communities, where a Black consciousness was growing and a Black national pride was evolving. Because much of the Anglophone Caribbean nationalist project was formulated by men as part of projects of colonial resistance, the nationalism that developed there was masculinist in nature – and therefore reproduced many of the male-dominated political and social institutions of the colonial era. Men's goals, which were embraced both by women and men as if they were to benefit the entire nation, defined the nationalist project in the colonial era and continue to do so in the post-colonial era as well.

Reaction to being feminized has also led to the intersection of masculinity and nationalism among Jews. Mayer shows in Chapter 12 that we cannot conceive of Jewish nationalism without understanding how masculine a project it has been. From its inception, the idea which stood behind Zionism (Jewish nationalism) was the transformation of the social, political, economic and psychological profile of the Jews of Europe, the creation of a *New Jew*, a *Muscle Jew*, who would be the antithesis of the pejoratively "feminized" Diaspora Jew. And, in turn, the political and economic transformation which led to the creation of a Jewish homeland in Palestine required the construction of a physically fit *New Jew*, who upon arrival in Palestine would take up arms to protect himself, his communities and what he believed was his land. Because Jewish history in Palestine has been burdened by a continuous struggle for survival, a militarized notion of Jewish nationhood developed which further shaped Jewish nationalism in Palestine (and later in Israel) as masculine. Mayer argues that because national myths of creation and survival – of wars and heroic, even miraculous, saving episodes – have been integral to the daily Jewish experience in Palestine and central to formal and informal Hebrew education there, a militarized nationalism and an almost exclusively male cult of heroism has developed there. The homosocial experiences that the militarized setting has offered and the male bonding experiences that have occurred in military units have also helped to build the intimate connection between masculinity and Jewish nationalism.

Attempts to protect homosociality in the military have been important as well in the U.S., where they have been at the center of Congressional debates. Allen argues in Chapter 13 that efforts to prevent women from becoming combatants and gays from participating in the military are a way to preserve the U.S. military as a perpetual military fraternity, an heterosexual masculine zone. Because the military in the U.S. serves to defend political ideals as well as gender and sexual ideals (such as heterosexual masculinity and the model of male-headed households) which are at the heart of the mainstream American notion of nation, an inclusive military force could, according to opponents, threaten the national (heterosexual male) fabric. At the heart of the debates in Congress about inclusion or exclusion of women and gays is the dilemma of how to protect, uninterrupted, the existing connection between nationalism and masculinity.

Light and Chaloupka take up in Chapter 14 yet another aspect of the relationship between masculinity and nationalism, as they demonstrate the connection between leftist versions of identity politics and right-wing nationalism. They argue that right-wing nationalism is a reactive nationalism, formulated by white men whose idea of the nation is intertwined with their race and gender supremacy, in reaction to liberal formulations of the American nation. Because they are committed to the idea that their formulation of nationalism is the only correct one, right-wing nationalists see themselves as "saving" the American nation from its current government, and present such activities as the bombing of the Alfred P. Murrah Federal Building in Oklahoma City and the siege in Ruby Ridge as justifiable acts of defense of the American people. These acts of saving the nation from its own government is the ultimate masculine task.

In all these cases, the connection between masculinity and nationalism remains strong: men take the liberty to define the nation and the nation-building process, while women for the most part accept their obligation to reproduce the nation biologically and symbolically. Although some of these roles have begun to be challenged, we can still generalize that masculinity and femininity remain fixed categories when they interact with the nation.

Femininity

Because the national project was initially defined by men and almost immediately became a masculinist project, femininity has in the national context been constructed in relation to men, to nation and, later, to state policies. As many of the chapters in this volume illustrate, femininity is generally produced as a means of supporting the nation's construction, through symbolic, moral and biological reproduction; in turn, it is precisely because it is a masculine project that nation becomes feminized and figured in service to male needs.

In Indonesia, for example, as Dwyer shows in Chapter 2, women are encouraged to control their fertility so that they can participate in the process of nation-building by becoming guardians of family morality and national development. Because women are the primary users of contraception, femininity has become associated in Indonesia not only with reproduction but also with the control of reproduction. Membership in contraception "acceptor clubs" offers Indonesian women a privileged and approved means to participate in national identity and nation-building.

Martin argues in Chapter 3 that in Ireland, where women have historically been charged with the labor of representing or embodying the nation, femininity is ascribed through a religious discourse. Not only are Irish women constructed as equivalents to home and motherhood, but also their femininity and their relation to the nation are structured around the Virgin Mary so that, according to Martin, femininity is inscribed and embodied as a product of the everyday discursive practices that comprise the devotion to Mary. Women are encouraged to represent and manifest the ideal of Mary in their own "essence" – in their behavior, their motherhood and their relationships with others. In other words, it is through their mimetic performance of Mary's model that individual Irish women come to embody femininity and, by extension, the Irish nation.

And in Liberia, where the nation is divided along a civilized/native binarism, femininity is defined by "civilized" men. "Civilized" women are defined by the fact that they do not participate in the physical difficulties of farm labor; ideally they are to be dependent housewives who are fully occupied with the care of the home and the children. Through these domestic practices, women produce and reproduce the honored status of the entire household as well as the next generation of "civilized" people. Yet if, therefore, "civilized" women more than "native" women participate in the nation-building process, both their "feminine" and their "civilized" status can be maintained only as long as these women do not challenge the "proper codes of behavior" or engage in activities that may threaten their respectability. If they do so, their civilized status will be stripped and their femininity questioned.

Femininity as constructed through patriarchal relations and women's obligation to the nation has in fact begun to be challenged – by women themselves. Ahmetbeyzade argues in Chapter 8 that because so many Kurdish peasant mothers and wives are heads of households and have access to power and resources within their own communities they are able to negotiate their own positions by resisting familial, tribal and even state patriarchies. No longer are these women just biological and symbolic reproducers of the nation. Rather, they have accepted a new obligation to nation-building, one which is based on their own ability to control communication among Kurdish peasant villages and on their own growing political consciousness.

The body as boundary

As several of the chapters in this volume demonstrate, when nation, gender and sexuality intersect, the body becomes an important marker – even a boundary – for the nation. In the hierarchical relationship between masculinity and femininity, when men (and sometimes older women) control the "proper behavior" of women, in effect they control women's bodies and sexuality. And because women's bodies represent the "purity" of the nation and thus are guarded heavily by men, an attack on these bodies becomes an attack on the nation's men.

Dwyer, Martin and Mostov (Chapters 2, 3 and 4) show different ways in which women's bodies have become the nation's boundary. In Indonesia, Dwyer argues, sexuality has become a primary idiom through which national identity is articulated: encouragement (and coercion) of women into participation in family planning programs and use of contraceptive technology becomes a means of controlling their fertility and, as a result, altering their bodies. Because family planning is such an important part of Indonesian nationalism, women and their bodies thus become the nation's marker.

In Ireland the body plays a different role in marking the boundaries of the nation. Martin analyzes the case of Miss X, a 14 year old who became pregnant as a result of a rape, and who subsequently made plans to go to England for an abortion because abortions are illegal in Ireland. The debates surrounding Miss X's possible travel to England for abortion, Martin argues, show how Miss X's pregnant body became synonymous with the ideal image of the Irish nation; how for many in Ireland the death of Miss X's baby came to correspond mimetically to the death of the Irish nation. Even though her rape was a personal act of violence, through restricting Miss X's travel to obtain an abortion in England and, in effect, prohibiting her from leaving Ireland altogether, the boundaries of her female body and the boundaries of the Irish nation became conflated.

A third case of how the body serves as a boundary for the nation is discussed by Mostov in Chapter 4. In the former Yugoslavia, where women's bodies have been important to the collective national body in their reproductive capacity, violent personal acts like rape have acquired national significance. Mass rapes such as in Bosnia, Mostov asserts, are about the invasion of the Other's boundaries, the occupation of the Other's symbolic space, property and territory: rape of women becomes an attack on the nation, figuring as a violation of national boundaries, a violation of national autonomy and national sovereignty.

Ultimately in the interplay of nation, gender and sexuality and in the mutual roles that they play in constructing each other, power becomes the most important narrative – because power, more than any other discourse, determines the hierarchical relations within each of these discourses and among them. And because the national project has been imagined by men and has been designed as a masculine construct, patriarchal hierarchies have become the foundation of the nation as much as the foundation of both gender and sexuality. As nation, gender and sexuality interact with one another, one nation, one gender and one sexuality come to dominate; and therefore what the nation is, its "ego," becomes imbedded in what men are and what women are assigned to be. The nation and men so often seem to mirror one another and be each other's extension, therefore, as Bauer put it, in his 1924 essay: "if someone slights the nation they slight me too." However, as social, political and economic conditions in each nation are never static the hegemony of the-male-nation has begun to be challenged. And as the nation is always in the process of "becoming" so are gender and sexuality. Challenges to heteronormativity are likely, therefore, to yield changes to the nation which will no doubt become the grounds of discussion of – and tension over – power conflicts in the years to come.

Notes

1 For general discussion of the role of myth in nationalism see Gellner (1996), Connor (1990) and Hutchinson (1987).
2 For an important contribution to this discussion see Herb and Kaplan (1998) and see, in particular, G. Herb's essay "National identity and territory" (1998) where he provides a lucid framework for the discussion of nation, state and territory. Also central to the discussion are Calhoun (1997), Breuilly (1993), Hobsbawm (1990) and Connor (1978).
3 A few examples of stateless nations are the Palestinians, the Kurds and the Druze, the Basques, Quebecois, Sami people, Zapotecs, the Berbers, the Moros, the Sikhs, Flemings, Bretons and Catalans.
4 The list of multinational states is very long but some examples which we may want to think about are Belgium, Canada, China, Cyprus, Czechoslovakia, Ethiopia, France, Germany, India, Indonesia, Iraq, Italy, Lebanon, Malaysia, Nigeria, South Africa, Soviet Union, Spain, Sri Lanka, Sudan, Switzerland, Uganda and Zimbabwe.
5 See in particular the works of Yuval-Davis (1997), Radcliffe and Westwood (1996), McClintock (1995), Mayer (1994), De Grazia (1992), Parker *et al.* (1992), Enloe (1989), Yuval-Davis and Anthias (1989), Koonz (1987) and Mosse (1985), who all discuss explicitly the importance of gender or sexuality to our understanding of the nation and of nationalism.
6 Because many of these books do not distinguish between nation and state they discuss women's "citizenship" rather than women and nationalism, even though, as discussed in the early part of this introduction, the two are not the same and should not be conflated.
7 Notable exceptions are Herzfeld (1997), Pickering-Lazzi (1995), Parker *et al.* (1992), the two volumes by Theweleit (1987, 1989) and Mosse (1985).
8 Two important examples are Radcliffe and Westwood (1996), specifically Chapter 6, and Sharp's (1996) analysis.
9 In particular see Anthias' (1989) work on Greek-Cypriot nationalism and Kandiyoti's (1991) analysis of Turkish nationalism.
10 For an important discussion about who can do the "imagining" of the *nation* during colonial times and who can and should do it in a post-colonial era, see Chatterjee (1996).
11 For an excellent discussion of the impact of British imperialism on the feminization of the colonies see McClintock (1995).

References cited

Anderson, B. (1991) *Imagined Communities: Reflections on the Origin and Spread of Nationalism*, London: Verso.
Anthias, F. (1989) "Women and nationalism in Cyprus," in N. Yuval-Davis and F. Anthias (eds.) *Woman–Nation–Gender*, London: Macmillan, pp. 150–167.
Balibar, E. (1991) "The nation form: history and ideology," in E. Balibar and I. Wallerstein, *Race, Nation, Class: Ambiguous Identities*, London: Verso, pp. 86–106.
Bauer, O. (1996 [1924]) "The nation," in G. Balakrishnan (ed.) *Mapping the Nation*, London: Verso, pp. 39–77. (The original was published in 1924 as Chapter 1 of *Die Nationalitätenfrage und die Sozialdemokratie*, Vienna.)
Berger, M., B. Wallis and S. Watson (eds.) (1995) *Constructing Masculinity*, New York and London: Routledge.
Bhabha, H. (ed.) (1990) *Nation and Narration*, New York and London: Routledge.
Brass, P. (1979) "Elite groups, symbol manipulation and ethnic identity among Muslims of South Asia," in D. Taylor and M. Yapp (eds.) *Political Identity in South Asia*, London: Curzon Press, pp. 35–43.
Breuilly, J. (1993) *Nationalism and the State*, Chicago: University of Chicago Press.
—— (1996) "Approaches to nationalism," in G. Balakrishnan (ed.) *Mapping the Nation*, London: Verso, pp. 146–174.
Bristow, J. (1997) *Sexuality*, London and New York: Routledge.
Brittan, A. (1989) *Masculinity and Power*, Oxford and New York: Basil Blackwell.
Butler, J. (1990) *Gender Trouble*, New York and London: Routledge.
Calhoun, C. (1997) *Nationalism*, Minneapolis: University of Minnesota Press.
Chatterjee, P. (1993) *The Nation and its Fragments: Colonial and Postcolonial Histories*, Princeton, N.J.: Princeton University Press.
—— (1996) "Whose imagined community?" in G. Balakrishnan (ed.) *Mapping the Nation*, London: Verso, pp. 214–225.
Connor, W. (1972) "Nation building or nation destroying?" *World Politics* 24: 319–355.
—— (1978) "A nation is a nation, is a state, is an ethnic group, is a . . . ," *Ethnic and Racial Studies* 1: 377–400.
—— (1987) "Ethnonationalism," in M. Weiner and S. Huntington (eds.) *Understanding Political Development: An Analytic Study*, Boston, MA: Little, Brown, pp. 196–220.
—— (1990) "When is a nation?" *Ethnic and Racial Studies* 13: 92–100.
De Grazia, V. (1992) *How Fascism Ruled Women: Italy 1922–1945*, Berkeley: University of California Press.
Deutsch, K. (1953) *Nationalism and Social Communication: An Inquiry into the Foundations of Nationality*, Cambridge, MA: Technology Press of MIT.
Duncan, N. (ed.) (1996) *Bodyspace: Destabilizing Geographies of Gender and Sexuality*, London and New York: Routledge.
Enloe, C. (1989) *Bananas, Beaches, and Bases : Making Feminist Sense of International Politics*, Berkeley, CA: University of California Press.
Fausto-Sterling, A. (1995) "How to build a man," in M. Berger, B. Wallis and S. Watson (eds.) *Constructing Masculinity*, New York and London: Routledge, pp. 127–134.
Foucault, M. (1978) *History of Sexuality*, vol. 1, New York: Random House.
Gellner, E. (1983) *Nations and Nationalism*, Oxford: Basil Blackwell.

—— (1996) "The coming of nationalism and its interpretation: the myths of nation and class," in G. Balakrishnan (ed.) *Mapping the Nation*, London: Verso, pp. 98–145.

Hechter, M. (1975) *Internal Colonialism: The Celtic Fringe in British National Development, 1536–1966*, Berkeley and Los Angeles: University of California Press.

Herb, G. (1998) "National identity and territory," in G. Herb and D. Kaplan (eds.) *Nested Identities: Nationalism, Territory, and State*, Lanham, MD: Rowman & Littlefield.

Herb, G. and D. Kaplan (eds.) (1998) *Nested Identities: Nationalism, Territory, and State*, Lanham, MD: Rowman & Littlefield.

Herzfeld, M. (1997) *Cultural Intimacy: Social Poetics in the Nation-State*, New York: Routledge.

Hobsbawm, E. (1990) *Nations and Nationalism since 1780: Programme, Myth, Reality*, Cambridge: Cambridge University Press.

Hroch, M. (1996) "From national movement to the fully-formed nation: the nation-building process in Europe," in G. Balakrishnan (ed.) *Mapping the Nation*, London: Verso, pp. 78–97.

Hutchinson, J. (1987) *The Dynamics of Cultural Nationalism*, London: Allen & Unwin.

Jayawardena, K. (1986) *Feminism and Nationalism in the Third World*, London: Zed.

Kandiyoti, D. (ed.) (1991) *Women, Islam and the State*, London: Macmillan.

Katrak, K. (1992) "Indian nationalism, Gandhian 'satyagraha,' and representation of female sexuality," in A. Parker, M. Russo, D. Sommer and P. Yaeger (eds.) *Nationalisms and Sexualities*, London: Routledge, pp. 395–406.

Kondo, D. (1990) *Crafting Self: Power, Gender, Discourses of Identity in a Japanese Work Place*, Chicago: University of Chicago Press.

Koonz, C. (1987) *Mothers in the Fatherland*, New York: St. Martin's Press.

Lancaster, R. and M. Leonardo (eds.) (1997) *The Gender/Sexuality Reader*, New York and London: Routledge.

McClintock, A. (1995) *Imperial Leather: Race, Gender, and Sexuality in the Colonial Contest*, New York and London: Routledge.

McClintock, A., A. Mufti and E. Shohat (eds.) (1997) *Dangerous Liaisons: Gender, Nation and Postcolonial Perspectives*, Minneapolis: University of Minnesota Press.

Mayer, T. (1994) "Heightened Palestinian nationalism: military occupation, repression, difference and gender," in T. Mayer (ed.) *Women and the Israeli Occupation: The Politics of Change*, London: Routledge.

Mosse, G. (1985) *Nationalism and Sexuality: Middle-Class Morality and Sexual Norms in Modern Europe*, Madison, WI: University of Wisconsin Press.

Parker, A., M. Russo, D. Sommer and P. Yaeger (eds.) (1992) *Nationalisms and Sexualities*, London: Routledge.

Pickering-Lazzi, R. (1995) *Mothers of Invention: Women, Italian Fascism, and Culture*, Minneapolis: University of Minnesota Press.

Radcliffe, S. and S. Westwood (1996) *Remaking the Nation: Place, Identity, and Politics in Latin America*, London: Routledge.

Renan, E. (1990) "What is a nation?" in H. Bhabha (ed.) *Nation and Narration*, New York and London: Routledge, pp. 8–22.

Rubin, G. (1984) "Thinking sex: notes for a radical theory of the politics of sexuality," in C. Vance (ed.) *Pleasure and Danger: Exploring Female Sexuality*, Boston and London: Routledge & Kegan Paul, pp. 267–319.

Sedgwick, E. (1990) *Epistemology of the Closet*, Berkeley and Los Angeles: University of California Press.

Sharp, J. (1996) “Gendering nationhood: a feminist engagement with national identity,” in N. Duncan (ed.) *BodySpace: Destabilizing Geographies of Gender and Sexuality*, London: Routledge, pp. 97–108.

Smith, A. (1981) “State and homelands: the social and geopolitical implications of national territory,” *Millennium: Journal of International Studies* 10(3): 187–202.

—— (1986) *The Ethnic Origins of Nations*, Oxford: Blackwell.

Theweleit, K. (1987, 1989) *Male Fantasies*, vols 1 and 2, Minneapolis: University of Minnesota Press.

Vance, C. (1984) “Pleasure and danger: toward a politics of sexuality,” in C. Vance (ed.) *Pleasure and Danger: Exploring Female Sexuality*, Boston, MA and London: Routledge & Kegan Paul, pp. 1–28.

—— (1995) “Social construction theory and sexuality,” in M. Berger, B. Wallis and S. Watson (eds.) *Constructing Masculinity*, New York and London: Routledge, pp. 37–48.

Yuval-Davis, N. (1997) *Gender and Nation*, London: Sage.

Yuval-Davis, N. and F. Anthias (eds.) (1989) *Woman–Nation–Gender*, London: Macmillan.

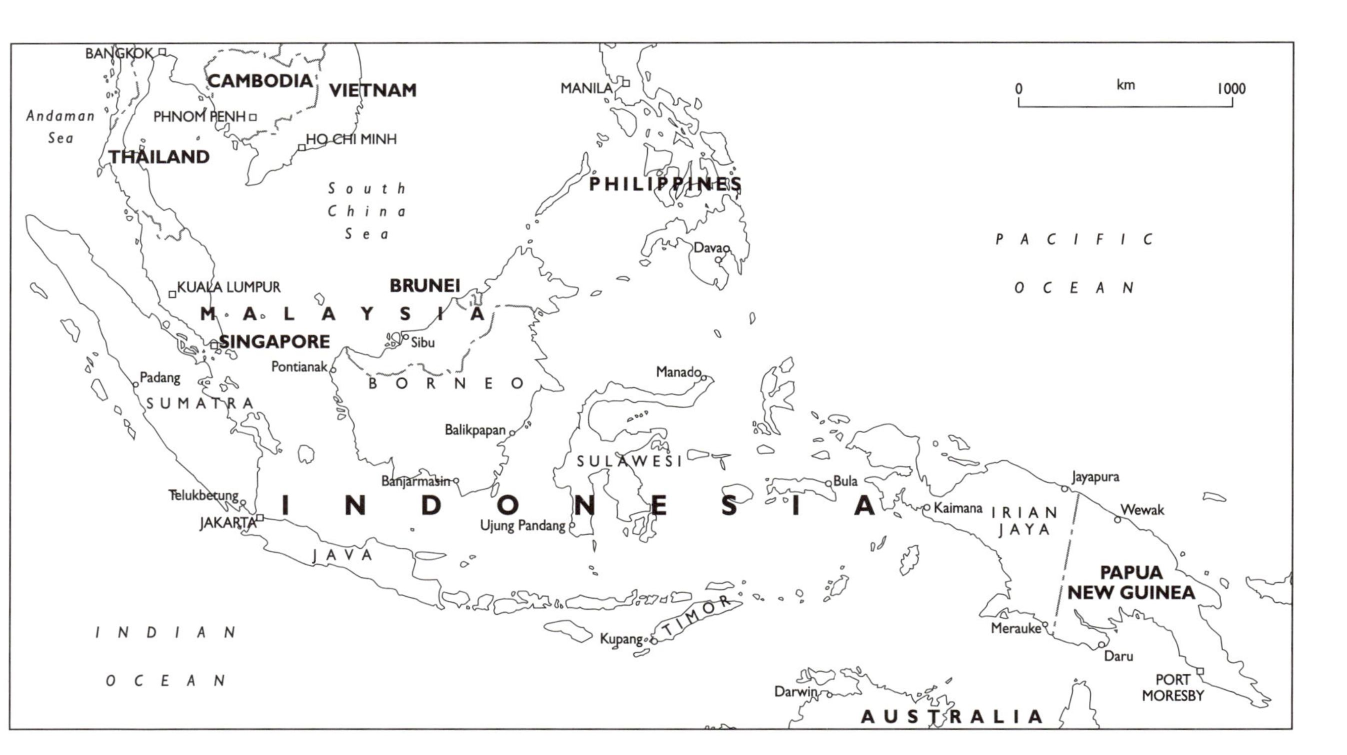
BANGKOK
CAMBODIA
VIETNAM
MANILA
0
km
1000
Andaman
Sea
PHNOM PENH
HO CHI MINH
THAILAND
PHILIPPINES
South
China
Sea
PACIFIC
OCEAN
Davao
KUALA LUMPUR
BRUNEI
MALAYSIA
SINGAPORE
Sibu
Pontianak
BORNEO
Manado
Padang
SUMATRA
Balikpapan
SULAWESI
Banjarmasin
Bula
Jayapura
Telukbetung
INDONESIA
Kaimana
IRIAN
JAYA
Wewak
JAKARTA
Ujung Pandang
JAVA
PAPUA
NEW GUINEA
TIMOR
INDIAN
OCEAN
Kupang
Merauke
Daru
PORT
MORESBY
Darwin
AUSTRALIA

Indonesia

2 Spectacular sexuality

Nationalism, development and the politics of family planning in Indonesia

Leslie K. Dwyer

A dusty August breeze winds its way through the narrow alleys of a Jakarta neighborhood, carrying with it the pungent smells of street food and the sharp alcoholic odors of disinfectant. Vendors push their carts over cracked pavement, trying to make some extra *rupiah* from the crowd that has gathered in this run-down section of town. At one end of a blocked-off intersection, a stage that was set up for performers and speechmakers is now occupied by a group of transvestite singers crooning an amplified pop song about a love that was never meant to be. At the other end is parked a long, gleaming white bus, its doors swung open to display a sparkling collection of video monitors and medical equipment. Scattered throughout the square are a number of booths, hung with bright banners identifying them as the condom booth, the Norplant booth, the Depo-Provera booth, the I.U.D. (intrauterine device)-insertion booth, the birth control pill dispensing booth, the immunization booth, the circumcision booth and the booth where men and women can sign up for surgical sterilization. Milling about laughing, talking, paying sporadic attention to the scene around them are groups of older women in faded batik *sarongs*, younger women in matching school skirts, plastic shoes and pressed white blouses, and young boys with stonewashed jeans, slicked-back hair and t-shirts. There are babies and grandfathers, squatters and day laborers, market traders and university students, *hajjis* and Christians. At once spectators, consumers and targets, this diverse group of "urban poor" has been demographically selected by the Indonesian government to be the state's "guests" at a family planning "party."

Since 1970, when family planning first gained state approval, many such events have been staged in Indonesia. As part of its plan to lower population growth rates by increasing numbers of contraceptive users, the government has injected an array of pills and parties, advertisements and incentive plans, surveys and statistics, persuasions and oftentimes coercions into Indonesian public culture. But this Jakarta neighborhood, like similarly saturated places throughout the archipelago, is more than just a bull's-eye on a demographer's dartboard. It is a nodal point in a network of discourses of nationalism and development, sexuality and citizenship, local culture and transnational movements whose trajectories extend far beyond the state's official borders. This neighborhood is home not only to potential participants in a ritual of citizenship meant to ensure the progress of the Indonesian nation on the road to a zero-growth-rate modernity, but also to social actors whose cultural, religious, political and sexual practices may challenge, divert or dismiss the state's call for their bodies and minds. It is a theater for nationalist spectacles meant to evoke an "imagined community" (Anderson 1991) of planned families and selectively fertile bodies, but the symbols employed here may be used in ways unscripted by the state to create alternative versions of "nationalism."

Given these complexities, how can we approach this scene and others like it to begin to analyze the relationships between nationalism and sexuality, culture and power operating in contemporary Indonesia? And what might we learn from the Indonesian situation that could illuminate these issues in other areas? Using ethnographic and historical data on Indonesia and its population control program, this chapter will attempt to draw some general conclusions about the crucial roles reproductive policies and politics may play in the circulation of national and sexual meanings.

Sexuality and the national

> The "locality" of national culture is neither unified nor unitary in relation to itself, nor must it be seen simply as "other" in relation to what is outside or beyond it. The boundary is Janus-faced and the problem of outside/inside must always itself be a process of hybridity, incorporating new "people" in relation to the body politic, generating other sites of meaning and, inevitably, in the political process, producing unmanned sites of political antagonism and unpredictable forces for political representation.
>
> (Homi Bhabha, *Nation and Narration*, 1990)

> We are worshippers at the shrines of Terminus, god of boundary stones.
>
> (Marshall Sahlins, *Islands of History*, 1985)

In recent years interest among social scientists in the historical and contemporary significance of nationalisms has grown considerably.[1] With the rise of transnational movements in an era of multinational corporations, "flexible accumulation" (Harvey 1990) and international "scapes" (Appadurai 1990) of media, capital, commodities, images and persons, it has seemed puzzling to many that identities continue to be framed in the idiom of the nation. Yet despite scholarly attempts to unlink notions of "place" and "culture" to account for these global flows (Gupta and Ferguson 1992), claims to nationness continue to be contested, often violently, especially in the post-colonial world.[2]

Attempting to understand this seeming paradox, many scholars have turned attention to the means by which the nation may be made to seem "natural" to its inhabitants. They have pointed to a number of ways in which this "naturalization" might be accomplished, including communal ritual practice (e.g. Kaplan 1995, Kelly 1995); educational programs and disciplinary technologies designed to instill a normalized "nationalist morality" in a regulated citizenry (e.g. Chatterjee 1993, Corrigan and Sayer 1985); shared commodity use (e.g. Kemper 1993); national media (e.g. Abu-Lughod 1993, Dominguez 1993, Handelman and Shamgar-Handelman 1993, Mankekar 1993, Anderson 1991); museums and cultural exhibits (e.g. Pemberton 1994, Duncan 1991, Errington 1989a); and the promotion of "national histories" (Anderson 1991, Dirks 1990, Alonso 1988, Wright 1985, Hobsbawm 1983). But other scholars have sought to understand how the perceived or asserted "naturalness" of particular sexualities or of the body may play a role of its own in the creation and maintenance of nationalisms. Nationalist constructs may naturalize themselves by reference to certain ideas of sexuality and gender, and in turn sexuality and gender may be reified as essential, non-negotiable attributes of national identity.

Many of these analyses have highlighted attempts by national elites to repress unapproved forms of sexuality in order to fix and employ a privileged pattern for nation-making purposes. Whether framed in terms of a strict code of sexual "respectability" required to counteract the perceived fragmentation of modernity (Mosse 1985), in terms of a closely guarded "inner domain" of female virtue required to mark off a space of sovereignty in the face of colonial incursions (Chatterjee 1993), an image of heterosexual male virility and female passivity required to keep the suppressed erotics of homosocially bonded males hidden (Parker *et al.* 1992), or a narrative of shared "national culture" that hides the sexual difference and violence operating within it (Bhattacharjee 1992, Enloe 1989), nationalist discourses have often been theorized as making images and practices of sexuality the malleable means of reproducing homogeneous and bounded communities. In that social reproduction requires biological reproduction, the goals of nations have been understood to involve a purposeful conflation of the two domains – social and biological – in order to bring recalcitrant bodies under control. Many analysts have described modern techniques of "bio-power" (Foucault 1978) that expand the domain of civil society to encompass the family, and the health, hygiene, bodily discipline and moral education of the national public (e.g. Horn 1994, Rabinow 1989, Donzelot 1979). Meanwhile, in line with Douglas's (1973) pioneering analysis of the body as a "natural symbol" used to express social and cultural categories, other scholars have explored the ways in which the cultural qualities attributed to the "body social" are metaphorically transferred to the "body individual," requiring the policing of personal bodily boundaries in order to secure the nation's integrity (e.g. Martin 1990, Bland and Mort 1984). With its borders guarded and its sexual transgressors identified through new techniques of knowledge and excluded through new techniques of power, the nation can, so the argument goes, consolidate a stronger sense of community.

Certainly, the repression and policing of sexualities labeled as aberrant have played critical roles in nationalist identity politics. As documentation of exclusionary or violent practices directed against those who do not adhere to prevailing notions of "virtue" have shown,[3] nationalist images often contain normative prescriptions for gender and sexual roles, defined against examples of "deviance" that purportedly threaten the social or symbolic reproduction of the nation. Especially in cases where the nation is anthropomorphized as possessing certain gendered or sexualized qualities – e.g. the virtuous and nurturing mother, the virile and militaristic male – deviations from this ideal embodiment may be seen as challenging the well-being, or even the very existence, of the nation (Koester 1995).

Yet analyses of nationalism and sexuality that foreground themes of category transgression and the repression of alternative sexualities may not be entirely sufficient for understanding the complexities of the contemporary post-colonial world. The issue of "boundaries" and their policing has perhaps seemed especially compelling to nationalism studies where, following Anderson's (1991) influential work, the nation has been theorized as a *limited* "imagined community," a modern enclosed and homogeneous territory rather than a pre-modern fading of influence out from diverse centers of kingly rule. But such theoretical framings may become problematic for understanding situations where national or cultural boundaries are permeable, contested or ambiguous as people with different interests in or kinds of access to transnational media, commodities, tourism or employment debate the meanings of national or sexual modernity and the relevance of cross-cultural connections in their lives.[4] We should also, I suggest, be cautious that our theories do not unreflexively replicate the claims of nationalisms themselves. In stressing the hegemonic power of nationalisms to subordinate sexuality we may run the risk of mistaking the desires of national elites for ethnographic actuality. Not only do we then lose much of the critical power of our analyses, but also we may end up reproducing entrenched Western cultural constructs that position mind over body and "planning" over passion in assuming sexuality to be easily amenable to conscious, strategic control. We anthropomorphize "nationalism" as an agent acting *upon* sexuality when we fail to allow for situations where the possibilities for nationalist discourse might themselves be limited by sexualities.

In this chapter I argue for two shifts in approach toward issues of nationalism and sexuality, which I believe permit us to better understand the situation of post-colonial nations like Indonesia. First, I argue that to focus primarily on nationalism as an agent of sexual control or repression is to lose sight of the possibility that the everyday encounter between sexuality and nationalism may be more complex. For instance, when sexuality is marginalized at one level of public culture, it may simultaneously be used as a grammar or idiom for articulating contestations, or as a nexus around which political debates or resistances cohere.[5] Indeed, the very techniques through which the erasure of non-approved sexuality may be attempted, such as public debate and denunciation, can have the unintended effect of broadening the range of available sexual alternatives.[6] Such an awareness of the complexities of national discursive space encourages us to view nationalism less as a coherent "thing" (Fox 1990) or as a state agent with unitary intentions and a consolidated apparatus of control than as a dialogic process.[7] This lets us, by extension, move from polarities of power and resistance as counterweights on an ideally balanceable scale of justice to see nationalism as always producing and requiring resistances or hybrid forms of knowledge and practice.[8]

Second, I suggest that our analyses not overlook the important fact that nationalisms, especially post-colonial nationalisms, are positioned not only "inward" toward control and containment of the sexuality of their own populations and the maintenance of their distinct categories and boundaries, but also "outward" to display, define and accommodate themselves toward multiple audiences. As Parker *et al.* (1992) point out, nations, like all symbolic constructs, exist only as identities-in-difference, in relation to other nations and other ways of imagining community.[9] But nations also attempt to position themselves relationally so as to channel and dam global flows of ideas and financial or political incentives. With the expanding role of multinational corporations in third world countries and the political and economic benefits of "development" aid for post-colonial elites, states must produce images of "national culture" – e.g. culturally "stable" enough to protect investments, culturally "flexible" enough to adapt to shifting production requirements[10] – that will localize capital and influence within their necessarily permeable boundaries. It is not simply "global culture" and "local culture" that interact in these encounters, for the multiple discourses of nation also work to mediate transnational negotiations.[11]

Taken together, these two shifts in focus allow us to better engage with the fact that one of the most critical "sexual issues" influencing post-colonial nationalisms today is the pressure to implement programs of "population control." In line with international ideologies that assert fertility control to be a prerequisite to national economic development and the conservation of ecological resources, and directly encouraged by aid organizations such as U.S.A.I.D. and the World Bank that have tied development funds to the implementation of family planning programs (Hartmann 1995), many third world countries like Indonesia have instituted policies designed to bring the fertility of their citizens under state influence. To understand the relationships between nationalism and sexuality being created and contested in Indonesia, then, we must be attuned to Indonesia's positioning within global networks of knowledge and power, and the multiple orientations of Indonesian nationalism that this entails. We must also, I argue, recognize that in implementing population control programs states rely not only on repressions of sexuality or proclamations of propriety but also on practices that may incite possibilities for desire and debate outside the hegemonic frame of nationalist narrative. In Indonesia, family planning programs work not only to "control" births and bodies, but also to create spaces for competing articulations and experiences of sexuality that in turn foster alternative nationalisms.

Before I narrow my focus to the Indonesian case, however, I want to step back and ask why family planning might constitute an important site for an investigation into nationalism and sexuality. After all, scholarship in this field has tended to ignore the tremendous role that population control policies have played in the construction and contestation of these notions in the post-colonial world, despite the fact that state promotion of contraceptive devices may be viewed as one of the most intimate inscriptions of nationalism upon gendered and sexualized bodies. But this neglect may be due, I suggest, not to the irrelevance of population control politics to issues of nation and sexuality, but to the fact that the very connections between family planning and sexuality may seem to inhabitants of Western cultural worlds to be at once self-evident and invisible. Confronted with the question of what family planning might have to do with sexuality, the answer seems to be: at once everything and nearly nothing. Before, then, I turn to a discussion of the Indonesian case, I want to briefly address the issue of how challenging our own culturally embedded expectations surrounding "sexuality" and "planning" might lead us to a better understanding of nationalisms in cultural spaces like Indonesia.

Planning and passion/desire and development

> Modern life is based on control and science: We control the speed of our automobile. We control machines. We endeavor to control disease and death. Let us control the size of our family to ensure health and happiness.
>
> (1940s U.S. family planning poster)[12]

> . . . on being confronted with a complete machine made up of six stones in the right-hand pocket of my coat (the pocket that serves as the source of the stones), five stones in the right-hand pocket of my trousers, and five in the left-hand pocket (transmission pockets), with the remaining pocket of my coat receiving the stones that have already been handled, as each of the stones moves forward one pocket, how can we determine the effect of this circuit of distribution in which the mouth, too, plays a role as a stone-sucking machine? Where in this entire circuit do we find the production of sexual pleasure?
>
> (Gilles Deleuze and Félix Guattari, *Anti-Oedipus: Capitalism and Schizophrenia*, 1983)

In his 1798 *An Essay on the Principle of Population*, Thomas Malthus offered a "postulatum" identifying sexual desire as the human universal at the root of population increase. The "passion between the sexes," Malthus had come to believe, was so strong that only "moral propriety" could halt the "geometric" expansion of humanity.[13] Sexual desire was even, he argued, responsible for the differences in population growth rates between industrialized and non-industrialized societies. In the latter, he wrote, where women were subjected to "the constant and unremitting drudgery of preparing every thing for the reception of their tyrannic lords" (1976 [1798]: 28), these "vicious customs" acted to suppress the natural passions that would have led to a level of reproduction on par with the civilized societies of Europe. Of course, Malthus wrote long before contraception became the population control means of choice, before anxieties about population expansion shifted from urban Europe to focus on the "dark, teeming masses" of the third world, and before more sophisticated and less racist analyses of the dynamics of reproductive behavior appeared. Still, notions of "sexuality" or "desire," much less "passion," have been strikingly lacking from many recent debates over reproductive interventions.

Contemporary bio-medical technologies devoted to the control and spacing of births are designed to work on humans presumed to be bifurcated into rational minds and irrational yet controllable bodily impulses.[14] Premised upon the possibility of sexuality's containment under the powerful aegis of "planning," birth control seems to erase much of desire's disruptive potential. These purportedly value-neutral and universally effective technologies simultaneously appear to dissolve the relevance of cultural differences in sexual or reproductive practices and politics.[15] Discussions of "population control" tend to follow this dualistic logic even further, using statistical techniques to render humans quantifiable as disembodied and decultured units of (under-)productive space or consumer (over-)demand that can then be targeted for policies aimed at their reduction or redistribution.[16] Sexuality is acknowledged merely as the "common-sense cause" of overpopulation, as an intractable and thus uninteresting inevitability, presumed stable in form and force from culture to culture. In these discourses, reproduction is severed from sexuality and sexual or gendered politics to be discussed as an effect of conscious "strategies" that can be redesigned to meet changing political or economic demands. Even feminist and postcolonial critiques of population policy tend to articulate their oppositions in terms of rational "choice" and individual or national freedom or sovereignty,[17] rather than acknowledging or questioning the presence of pleasure or desire. While feminist scholars have begun to examine what Ginsburg and Rapp (1995) call the "euphemistic violence" that becomes embedded in culture "when a woman's identity as a mother (or nonmother) is split off from her sexuality and broader social relations" (1995: 4), our analyses have been slower to realize that by evaluating population control practices mainly in terms of unfettered access to contraception and the presence or absence of "coercion" or reproductive "freedom" we are subjecting women to a similar splitting, and perpetrating a similar violence by failing to account for sexuality.[18]

If, as Foucault has argued, modern Western culture has come to view sexuality as the site of deeply buried truths about human nature, as an essential core that lies underneath rational consciousness yet may hold the power to shape its form and thereby construct our identities,[19] it is perhaps not surprising that family planning is generally seen as decidedly "unsexy," in the widest sense of the word. Contraception is experienced by many as a somewhat distasteful necessity threatening the pleasure of sex that, fundamentally, is not even "about" sex or sexuality at all. The ideal situation, according to this cultural logic (and the manufacturers of contraceptives such as Norplant and Depo-Provera who profit from it), is the one in which family planning and sexuality are held so far apart as to seem virtually unconnected. Locating "planning" in the sterile space of the doctor's office and pleasure in the intimate privacy of the bedroom offers the reassurance that the "truth" of sexuality can remain uncontaminated by the practical necessities of planning; that mind and body, conscious and unconscious, can remain distinct and separate.

But is this wishful tendency to assert the separateness of family planning and sexuality or desire and power a universal phenomenon? And can our analyses of family planning technologies, and their relations to post-colonial nationalisms, remain entrenched in the Western liberal discourses of individual "choice" (cf. Dumont 1986) and objectivist assumptions about the possibilities of "planning" bodily practices (cf. Bourdieu 1977)[20] found in calls to gain for women "control over their own bodies," or in arguments that locate "liberation" in shifting the balance of power *over* sexuality? How can we understand places like Indonesia, where reproductive technologies are not embedded in market idioms of commodity selection and "sold" to "consumers," but are instead implicated in fields of power that link gender and sexuality to nationalism?

Anthropologists studying the cultural construction of sexuality have documented the wide variety of forms that sexual experience can take, thus challenging the naturalness perceived to reside in our own categories. Examples of the power of culture to give meaning to sexuality without holding sexuality to be a core of "identity" or an underlying, essential "truth" might lead us to expect that the relations between sexuality and family planning might likewise be organized differently elsewhere.[21] As a number of scholars have shown, contraception practices are situated within cultural experiences of sexuality, gender and inequality, not within a disembodied and hypothetically neutral "rationality."[22] We might also expect that given a late capitalist international division of labor and the particular geopolitical positioning of post-colonial nations as recipients of development aid from the West, issues of "choice" or "planning" might take on particular meanings for third world people. Indeed family planning does seem to occupy a very different place in the lives of inhabitants of post-colonial countries, especially women, than it does for Euro-Americans. As Morsy writes:

> Within the framework of the contemporary global political economy, political and economic power relations influence the very types of reproductive technologies targeted at women in different parts of the world,

> and different classes in the same part. Whereas the new reproductive technology is expected to reverse the state of *infertility*, and insure the production of "state of the art," "quality-controlled" infants in the North, this expectation does not extend to the female "breeders" of the Southern hemisphere. . . . While control over procreation is increasingly transferred to the biomedical domain world wide, for the late comers to "modernization," the professional management of biological reproduction is likely to be part of population control programs.
>
> (Morsy n.d.: 3, original emphasis)

For third world women, who are the primary targets of population control programs rather than consumers of commodified convenience or perfected products, their status as users ("acceptors") or non-users (rejectors? resistors?) of family planning positions them within networks of highly contested and volatile discourses of nationalism, progress and sexual politics. For these women, their sexuality and how they do or do not control it is deeply implicated in practices of citizenship and transnational power.

How, then, can we analyze family planning interventions and their implanted (often literally in the case of Indonesia, which had as of 1991 absorbed 75 percent of the world's supply of Norplant: Cohen 1991) yet contested presence in the lives of citizens of post-colonial nations? I suggest turning to theories that do not simply replicate our own cultural valorizations of liberal individualism, rational "choice" and dichotomies between conscious minds and unconscious bodies. By examining the case of Indonesia, the extent to which the production of desires and the exercise of power is inherent in attempts to manage women's fertility becomes apparent. In Indonesia the connections between family planning and sexuality become clear when the presence of contraceptive technologies creates spaces for competing articulations of sexuality, linking passion and power, planning and desire together in ways that are often unintended by the Western producers and state distributors of these devices. It is in these discursive spaces, as much as in the proclamations of nationalist elites, that we can locate the meanings of nationalism and sexuality in Indonesia.

Genealogies of population control

> "Having a room of one's own" is a desire but also a control. Inversely, a regulatory mechanism is haunted by everything that overruns it and already causes it to split apart from within.
>
> (Gilles Deleuze, "The birth of the social," 1979)

I first became intrigued by Indonesia's population control program during a trip to Java in 1993. One afternoon I decided to visit Borobudur, the massive monument built in Central Java by Mahayana Buddhists in the ninth century and dug out from under volcanic debris and vegetation by the British colonial governor Raffles a thousand years later. In the late 1970s the Indonesian government, aided by Unesco funds for historic preservation, renovated the monument into a nationalist paean to "cultural heritage." At Borobudur I joined a group of Indonesians for the guided tour of the temple, led by a Muslim university student from Yogyakarta, Java's center of modernist Islamic learning 40 kilometers to the south. Above the inattentive chatter of schoolchildren, liberated from the classroom in order to learn "their history," our guide ponderously read to us from a dog-eared pamphlet on Buddhist philosophy available, he noted, in the official gift shop. "Life in all its manifestations is suffering, caused by desire and lust," he intoned. "But suffering can be annihilated, and the Buddha says he will show you the way of annihilation." Climbing from the lower levels of the monument, whose walls depict the sensuous pleasures of the flesh and the tortures accruing to bad actions, to the upper strata displaying the Buddha's progressive enlightenment, our tour-guide waxed more and more morose. But as we came upon a level crowned with statues of meditating Buddhas his face suddenly became animated. Pointing to a statue of the Buddha with its right arm extended and its fingers drawn into a "V," he called for our notice. "In America," he said, looking pointedly at me, "they call that the peace sign. Or the 'V' of victory. But in Indonesia we have our own sign for victory. When someone makes that sign in Indonesia, he is showing that he knows two children are enough." *Dua anak cukup*, the Indonesian government's slogan for its family planning program, could be found, he implied, victoriously resplendent here in this monument to "national culture."

The ironies of a ninth-century ascetic Buddha, its presence mediated by colonial reconstruction and U.N. renovation, preaching birth control in a twentieth-century stronghold of Muslim piety in Indonesia to an American anthropologist/tourist were not, I realized later, as jarring as they first appeared. Indeed, such juxtapositions, bound up in ambivalent desires and dominations, accommodations and appropriations, are what have characterized Indonesia's "population miracle." As the hybrid offspring of colonial, nationalist, Muslim and international development histories of culture and control, contemporary Indonesian population control constitutes less a smooth "syncretism,"[23] than an unstable space filled with shifting and competing meanings.[24]

According to international development experts and state planners,[25] there are two kinds of birth control in Indonesia now: modern, technological methods of contraception developed in Western laboratories according to culturally neutral scientific criteria; and "indigenous" methods of birth spacing.[26] These "indigenous" methods – a label which itself ignores the tremendous diversity to be found in multi-ethnic, multilingual Indonesia – are, in population control logic, tied to (at best) "tradition" and (at worst) "ignorance." Culture, power and history are seen as immaterial to biomedical contraception, at the same time they are conversely viewed as either things to be overcome or as "factors" to be manipulated in the case of "indigenous" fertility regulation. As the World Bank, in a 1994 report entitled *Strategies for Family Planning Promotion*, advises, a

> program implies a long-term commitment to the family planning program client. It must recognize that *behavior change is evolutionary* and affected by many determinants, facilitated or hindered by *factors in the client's environment.* Interventions must go beyond persuasive messages *to directly address the barriers to behavior change.*
>
> (Piotrow *et al.* 1994: v, added emphases)

In these kinds of narratives, local meanings of sexuality, bodies, reproduction, health, gender or development itself, are reduced to environment: a static, timeless state that can be overcome only by an "evolutionary" advance to an acultural and apolitical bio-medicine. Yet despite these attempts to divide fertility practices into clean categories of "Western" versus "indigenous," it is clear that reproductive and sexual practices of all kinds are culturally conditioned, historically situated, and implicated in translocal relations of power. Birth control in the West was developed and promoted in a milieu of feminist, eugenic and racial debates (Gordon 1977) and continues to be evoked in heated controversies on these issues; while Indonesian "indigenous methods" of fertility control, far from constituting timeless tradition, have been shaped by long histories of cultural and political contestation. Furthermore, the use of bio-medical contraception in Indonesia itself occurs within a highly charged cultural and political field.

Much research still needs to be done on the history of reproduction and sexuality in Indonesia. But while the contours of this past are fuzzy, it is clear that Dutch colonialism had a profound impact on reproduction and sexuality in the East Indies colonies. Demographers and social historians have noted that beginning in the nineteenth century the population of Java, the most densely settled of the islands, underwent a rapid rise, increasing from estimates of 4.5 million in 1815 to 9.4 million in 1845 to 28.4 million at the close of the century (Alexander 1984, White 1973).[27] Reproduction became an overtly politicized issue as the Dutch sought first to explain this growth as a result of colonialism's civilizing influence (Alexander 1984, McNicoll and Singarimbun 1982), and then to relocate "excess" bodies away from Java to plantations in more sparsely populated areas such as Sumatra (Stoler 1985). Especially as sexual relations across racial lines became an issue of intense anxiety for both the colonizers and those under their control, population, reproduction and sexuality became objects of increasing concern. Not only the brute *quantity* of bodies but also their *composition* as ethnicized, racialized or class-inflected entities became the subject of new regimes of classification and control. Indeed, Stoler argues, Dutch colonial power was based in large part on creating and managing racial categories through the surveillance of sex:

> Colonial control was predicated on identifying who was "white," who was "native," and which children could become citizens rather than subjects, designating who were legitimate progeny and who were not. ... Social and legal standing derived from the cultural prism through which color was viewed, from the silences, acknowledgments, and denials of the social circumstances in which one's parents had sex.
>
> (Stoler 1991: 53)

Before the turn of the twentieth century, when white women from Europe were encouraged to settle in the colonies as wives, male Dutch colonial employees were expected to take "native" women as concubines who could provide sexual and domestic services and ease the transition from European to colonial culture. Through these contacts and the desires and repulsions they involved, the sexuality of the colonies' indigenous inhabitants became typified in Dutch colonial thinking. Sexuality was conceptualized, however, in ambivalent and contradictory ways. Colonized women were seen simultaneously as perilous temptresses using their sexual wiles to maximize personal gain (Gouda 1993) and as less demanding and more nurturing than European women; while colonized men, especially Javanese, were alternately seen as sexually passive (Stoler 1985) and as sexually degenerate (Anderson 1990).[28] Because of the dangers believed to result from alliances between the colonizers and the colonized – including the probability of moral contagion, the potential for the blurring of boundaries of colonial control, and the possibility of symbolic anxiety over the ambivalent representations of sexuality produced – moral re-education was offered for "mixed race" children so they would not reproduce the sexual dangers of their parents (Stoler 1991). Despite these attempts to domesticate desire, however, the *nyai*, or native concubine, remained a central and controversial figure in a great many of the literary texts of the era, fascinating critics and defenders of colonialism, Dutch and Indonesians alike.[29]

By the time the Republic of Indonesia proclaimed its independence in 1945, emergent nationalism had shifted the meanings attached to population, reproduction and sexuality. Worldwide economic depression, the Japanese occupation of Indonesia from 1942 to 1945 and the struggles against returning Dutch forces destroyed lives and livelihoods and resulted in population growth rates that were stable or negative (McNicholl and Singarimbun 1982). Indonesia's first President, Sukarno, one of the founders of the Non-Aligned Movement and a resolute critic of Western imperialism, believed that a firmly pro-natalist stance could help recoup Indonesia's losses and develop the new nation's strength through sheer numerical expansion. "Population growth" was not, Sukarno argued, the problem. Rather, imbalances in the distribution of persons and resources were what needed to be corrected. *Transmigrasi* programs that relocated people away from overcrowded Java and Bali, and policies that rearranged land distribution, would, Sukarno claimed, alleviate poverty without threatening Indonesia's demographic advantage. Citizens of the new nation were encouraged to be not simply productive but reproductive.

The sexual politics that imbued the nationalism of the Sukarno era were, like those of the colonial period, complex and oftentimes contradictory. Rather than creating and codifying detailed degrees of sexual heterogeneity, as the Dutch had, nationalists stressed the generic "Indonesianness" of the inhabitants of the former East Indies. Sukarno's slogan of NASAKOM – nationalism, religion (*agama*) and communism – represented an attempt to meld diverse currents of Indonesian thought into a coherent national unity, despite differences within these social strands regarding sexuality, gender, reproduction or other cultural characteristics.[30] According to Anderson (1990), Sukarno's strategy for managing these potential tensions was to draw upon a Javanese concept of "Power" that linked personal and political charisma to mystical powers, including the ability to unify opposites and amass sexual potency.[31] Elaborating Javanese court histories that connected sexuality to spiritual and political power – histories that, as Pemberton (1994) argues, were themselves in many ways reified artifacts of the Javanese rulers' cultural struggles against Dutch dominance – Sukarno sought to arouse his followers and ensure their devotion by portraying himself as successor to the virile Arjuna, hero of the Hindu epic drama *Mahabharata*, who, when engaging in sex in a pool with a heavenly nymph, impregnates the goddesses swimming nearby through the force of his potency (Anderson 1990). Flagrant in his sexual dalliances, Sukarno ordered the Indonesian National Monument in Jakarta's Freedom Square built in the form of a Javanese *lingga-yoni*, a phallic display which, Anderson (1990: 175) reports, Sukarno joked "testified both to his own and to Indonesia's inexhaustible virility." The sexual potency of the head of state was, in this nationalist narrative, the visible sign of his inner power, and of the strength and legitimacy of his rule. The fertility of the people was a sign of the prosperity of the nation, and their allegiance in turn represented and augmented the power of their president.

Sexual power was not, however, to be wantonly deployed. According to Anderson (1990), the traditional Javanese concept of "Power" that Sukarno drew upon involved a combination of creativity and harmony, fertility and order:

> Power is the ability to give life. Power is also the ability to maintain a smooth tautness and to act like a magnet that aligns scattered iron filings in a patterned field of force. Conversely, the signs of a lessening in the tautness of a ruler's Power and of a diffusion of his strength are seen equally in the manifestations of disorder in the natural world – floods, eruptions, and plagues – and in inappropriate modes of social behavior – theft, greed, and murder.
>
> (Anderson 1990: 33)

Sexual excess and uncontrolled lust not only were signs of a misalignment in the social order and a lack of the control that comprised real "Power,"[32] but also were associated with the greed and moral degeneracy of imperialism and neo-imperialism. The Japanese and the Dutch were portrayed in the popular press of this period as avaricious rapists violating the purity of the Indonesian nation. In the last years of the Sukarno era, as American culture grew more and more to symbolize the external enemy against which Indonesia needed to define and protect itself, many American films were banned from Indonesian screens. The reasons were twofold: either they portrayed the people of Asia as primitive and underdeveloped, or they contained excessive amounts of sex (Sen 1994). And when Sukarno was removed from power following the abortive 1965 coup attempt attributed by the new government to communist forces, the bloody backlash against the Indonesian Communist Party (P.K.I.) and the left-wing women's group Gerwani gave rise to circulating stories of sexual degeneracy among Gerwani women, and the slogan popularized by posters and graffiti: GERWANI WHORES (Tiwon 1996, Sen 1994, Wieringa 1992). Sexuality had become, in multiple ways, a primary idiom through which national identity was articulated, intra-national divisions were stated or smoothed, and international conflicts were defined and waged.

Although Sukarno's government was opposed to family planning, biomedical contraception arrived in Indonesia in 1957 when an affiliate of the International Planned Parenthood Foundation, the Indonesian Planned Parenthood Association (I.P.P.A.), began work in Java. The International Planned Parenthood Foundation was one of a number of organizations founded in a post-World War II climate where the funding of "development" programs was seen by Western nations as a key means of ensuring world "political stability" and the access to the raw materials and resources of the "third world" that such an international order could provide (Escobar 1995). In contrast to the claim made by the Dutch that Javanese population growth was a positive sign of the beneficial effects of European rule, or the arguments of Sukarno that an increasing population signalled the expanding power of Indonesia, "overpopulation" was identified in development thought as a primary cause of poverty and political turmoil. As the masses of the non-West bred, development experts claimed, so would global disorder. Hartmann (1995) cites a 1967 advertisement placed in U.S. newspapers with funding from a former president of the World Bank and a founding member of the Population Council that typified such a position:

> A world with mass starvation in underdeveloped countries will be a world of chaos, riots and war. And a perfect breeding ground for Communism. . . . We cannot afford a half dozen Vietnams or even one more. . . . Our own national interest demands that we go all out to help the underdeveloped countries control their population.
>
> (Hartmann 1995: 106)

Although the I.P.P.A. was able to introduce birth control only on a limited basis, it initiated a pattern in which the Indonesian state would turn to international aid organizations for the technologies and resources to promote development at the site of women's bodies.

Under the post-1965 "New Order" government led by President Suharto, development has become the cornerstone of Indonesian nationalism, and family planning has been embraced both as one of development's crowning glories and as one of its most concrete, measurable manifestations. The Indonesian state's role in managing development has been offered as the primary justification for a system in which the government creates and controls articulations of nationalism, and alternative narratives risk being branded as "political" and thereby subversive of "national consensus." "Development" in this state-sponsored national discourse has come to refer not only to changes in socio-economic indicators, but also to realignments of moral and cultural values. As Hobart (1993) notes, out of several available words for "development" in the national language, Bahasa Indonesia, the current government has stressed the word *pembangunan*, from the root word whose meanings include "to get up," "to grow up," "to wake up" or "to build," implying that development involves a progressive awakening and growth that needs the careful cultivation of state planners and development experts to achieve fruition. Rather than employing a nationalist rhetoric stressing the power of the Indonesian people to ward off outside interference, as Sukarno did, in promoting *pembangunan* Suharto's government has articulated an ideal national culture in which people must be guided along the path to progress by experts armed with selectively chosen Western technologies. The results of this "development" – greater political "stability" and more visible signs of wealth, especially among an emergent middle class – can be read, according to New Order rhetoric, as a sign of Suharto's possession of the attributes of the Javanese man of power. But the creative charisma of roguish *Bung Karno* ("older brother Sukarno") appears to have been replaced in nationalist discourse by the paternal politics of *Pak Harto* ("Father Suharto") and *Bu Tien* ("Mother Tien" Suharto, the late first lady). Such an ideological stress on the first family as national parents has been replicated in social policy, as the government attempts to create nuclear families out of the fluid array of social ties that comprised kinship – especially Javanese kinship – in an earlier era.[33] This shift is reflected, in turn, in legislation intended to codify and increase the responsibilities of spouses toward one another, to prevent unmarried couples from cohabiting, and to foster a national image of the Indonesian woman as a virtuous housewife and mother, despite the fact that in Javanese culture divorce was formerly frequent and unremarkable, family ties were flexible and relevant only in particular situations, and women moved confidently through public space as key economic actors and decision-makers. If the Sukarno era was primarily devoted to the generative and creative aspect of a Javanese idea of power, Suharto's government has appeared to stress the power of order, control and material well-being.[34]

Under Suharto, Indonesia has implemented a population control program that international aid organizations have hailed as one of the most successful in the world. Indonesia, development experts have repeatedly claimed, has both dramatically increased numbers of birth control "acceptors" and effectively combined family planning with local cultural, especially Muslim, values. Indeed, major development agencies have taken Indonesia's program as a model, supporting its exportation and implementation in dozens of other "third world" nations. Since 1970, Indonesia has trained thousands of delegates from over 80 countries in its methods (Crossette 1994). In the words of the World Bank, Indonesia has undergone "one of the most impressive demographic transitions within the developing world" (International Bank for Reconstruction and Development (I.B.R.D.) 1990: ix), due to what, according to U.S.A.I.D., is a "success story unrivaled in family planning history" (U.S.A.I.D. 1979, cited in Hartmann 1995).[35] In this environment, population control has become both a politically charged icon of what it means for a nation to be "modern," and an extremely direct writing of state-sponsored nationalism on the bodies of Indonesian citizens.

Critics of Indonesian population control policies have also stressed themes of control in their analyses of New Order family planning, documenting instances in which coercive methods have been used to increase numbers of contraceptive users and to maximize influence over which methods they employ. These critics have drawn attention to practices such as the aptly named "safaris" in which National Family Planning Coordinating Board (B.K.K.B.N.) workers have descended on rural or isolated villages, military or police escorts in tow, presenting lectures on the benefits of contraception and offering free insertion of long-term, non-user-dependent devices such as Norplant or I.U.D.s,[36] and to situations where needed agricultural loans, access to high-yield rice varieties, hybrid coconut seedlings or school scholarships are given only to those who practice family planning (Hartmann 1995, Robinson 1989). Stressing the lack of options available to women and men in the face of a military presence in their village or a potential economic loss to their more educated, better equipped neighbors, these analysts have argued for the implementation of family planning programs that offer women more "choice" of whether, when and how to use birth control. As one critic argues, population programs in Indonesia need to move away from repressive manipulations, and instead identify "women's needs and desires" and recognize women's "rights to the control of their own fertility" (Smyth 1991).

The Indonesian state has in recent years outwardly appeared to respond to some of these critiques. In keeping with a shift in international development rhetoric away from the language of population control toward the language of women's health (Smyth 1991) and "community-based initiatives" (Escobar 1995), the state has renamed the safaris "service operations" (*operasi bakti*: Cohen 1991), trained local midwives in contraceptive insertion and prescription,[37] and instituted "cafeteria plans" that, working within a consumer idiom of "choice," purportedly make a wider variety of birth control options available, especially to those who pay for services at private clinics.[38] As the State Minister for Population claims, "we don't treat our culture roughly . . . it is like giving flowers to your wife before asking for a kiss. Eventually you get what you want" (Crossette 1994). Indonesia has also followed the suggestions of development agencies in implementing what the World Bank calls "enter-education," a combination of "education" and "entertainment" suggested "because it holds the audience's attention, evokes strong emotional responses and provides role models for behavior change" (Piotrow *et al.* 1994: 3). Who, after all, could object to the state throwing a family planning "party" with the Jakarta poor as invited "guests," especially one that combined needed health services such as immunizations and sensitivity to cultural forms such as traditional entertainers with information on family planning possibilities?

But given these kinds of practices it becomes, I argue, critical to develop an understanding of other kinds of powers besides state coercion at work in creating Indonesian family planning, and to construct critiques that are based upon logics other than "repression" versus "choice." "Women's needs and desires" are cultural constructs, not natural facts, and as such they are susceptible to techniques of power that attempt to work through culture and nationalist discourse. While the Indonesian family planning program has frequently been effected by coercion, it has more often operated by means of campaigns designed to produce desire for a "planned" family by managing the visibility of sexuality in national public culture. By creating spectacles – public cultural displays of images designed to simultaneously engage viewers and to inculcate them with certain emotional and ideological dispositions – the Indonesian state at once offers a compelling version of nationalism and sexuality and positions itself as the primary author of national representation. By staging these spectacles, the state works to instill a regime where people monitor themselves and their sexual practices to accommodate to public displays of sexual "normalcy." Displays of sexuality, and of the state's power to stage them, evoke – despite the apparent disjuncture between Old and New Order nationalisms – the power *both* of control *and* of creation, portraying the pleasures of sexuality as well as the powers of planning.

Examples of the Indonesian state's staging of such spectacles are numerous: encouraged by the government, Muslim leaders deliver sermons on family planning and female anatomy and sexuality in mosques; discussions about contraceptives and sexual morality are held in religious groups, state schools and Boy and Girl Scout meetings; and billboards and television commercials blare the benefits of condoms, birth control pills, or "small, happy and prosperous families." "Model couples" who have been using contraceptives long enough are paraded before national media audiences: those who have been "acceptors" for five years are offered a commemorative plaque or batik shirts decorated with the spiral pattern of an I.U.D., and after ten years become eligible to win an all-expenses-paid trip to Saudi Arabia for the *hajj* or to be brought to Jakarta to have tea with President Suharto flanked by lights and cameras. Children's television shows present songs and skits for preschoolers telling how "two children are enough," while puppeteers who present shadow plays are recruited by the government to add contraception to the sexual scenes of the dramas they present. Sexuality is also made visible when state university students perform mandatory "community service" as "family planning motivators" in rural villages; when the state conducts a national family planning survey that attempts to assess the contraceptive knowledge and behavior of every home in the country; or when some civil servants are delivered their paychecks on different days of the week depending on what contraceptive method they or their spouses use. The power relations created by manipulating the visibility of sexuality are made even more explicit when government census takers mandate the posting of signs on the doors of houses indicating the contraceptive "acceptor" status of the inhabitants,[39] or when village maps are posted in many Indonesian "town halls" indicating birth-control users, color-coded by contraceptive method (Cohen 1993) – the very same sorts of village maps that Clifford Geertz (1995) describes being used to show the location of former communist households in the years following the New Order's ascent to power.

These nationalist sexual spectacles encode a gender politics as well. Through these displays, Indonesian nationalism works to repress gender "deviance," both by channeling fluid cultural forms into newly rigid binary categories through the use of legislation or coercion, and by positioning women as both spectators and primary targets of family planning. Even as women's reproductive, sexual and gender-related behavior is placed under surveillance by the state, it is also rendered spectacular through the repeated exhibition of idealized female figures. These may include the regional leaders of *Dharma Wanita*, the mandatory organization for the wives of civil servants that assigns women a rank based on that of her husband; or, in the pages of a glossy women's magazine, the smiling, modest housewife standing in front of her Westernized house beside her suited husband and two happy children. Closer to home, villagers may be confronted with the woman who, by becoming a participant in *KB Mandiri* – independent, private-sector family planning – has received a plaque announcing her status, to be placed on the outside of her house. By being positioned as consumers of these spectacles, Indonesian women are, in Althusserian terms, "hailed" with a potential identity as sexual citizens. They are offered a place as participants in the process of nation-building by becoming guardians of family morality and national development through the acquisition of fertility-controlled bodies. Although Indonesia is certainly not unusual in its identification of women as the primary users of contraception – the majority of research and funding for contraception world-wide is for birth control methods that work on female bodies[40] – what is instructive in this case is the extent to which the pleasures of sexuality, modernity and citizenship are intertwined and made to seem interdependent in these nationalist spectacles. Not only may "acceptor clubs" be mandatory in Indonesian villages, but also such activities and the representations surrounding them offer a privileged and approved means for women to participate in national identity and obtain the benefits that accrue to those modelling their practices of "citizenship" on state ideology.

The sexual and gender politics of the Indonesian state are not, however, complete and coherent in their agency and effects. "Resistances" take many forms in Indonesia, although identifying and analyzing them offers theoretical and methodological challenges. What to a Western observer might seem to be agreement may, from a local standpoint, be viewed as an alternative to or protest against dominant modes of ordering experience. And what might appear from one perspective as refusal or challenge might not be articulated – or indeed articulable – in such oppositional terms.

Explicit protests against family planning policies or hegemonic state versions of nationalism and sexuality are rare in Indonesia. In large part due to state willingness, backed by military might, to regularly and rigorously denounce alternatives as "political," and thus anti-national, Indonesian women rarely are able to publicly critique family planning as a system of practices or representations. This does not mean, however, that women simply acquiesce either to state policy or to state rhetoric. Quite often women will refuse to use particular forms of birth control, especially the more invasive hormonal methods that the state favors, on grounds of "health risk," coopting a state language of bio-medical modernity in their assertions of the right to reject these "advanced" methods. And, more frequently, women rework state discourses through their participation in family planning, to create new ways of representing sexual and nationalist experience.

Looking at the ways women interpret and transform the meanings the state attributes to their bodies and sexualities, we find that, despite the ease with which Indonesians generally discuss family planning – it is not, for example, considered impolite to ask a relative stranger what kind of contraceptive she uses – women frequently attempt to reposition their own sexuality outside the public domain of *nationalist* discourse. Rural women may, for instance, refuse to use condoms since they must be requested from a state-appointed family planning official – in villages typically the headman's wife – who would learn and possibly pass on to others the frequency of their sexual activity. In the cities, where condoms can be bought in pharmacies only after showing a government identification card or a marriage certificate, women are similarly wary of exposing their sexuality to state control and neighborhood knowledge. Through such refusals Indonesian women assert a connection between family planning and sexuality, in the process making visible the state's attempts to claim its citizens' bodies as material grounds for the work of nationalist ideology. Women also attempt to place their desires outside of the state's control by using contraceptive "acceptor clubs" and religious group meetings officially "about" family planning to discuss other issues of concern to them, including lack of sexual fulfillment; methods of keeping a husband from taking another wife; or ways to gain a prospective partner's interest. For example, at one government-sponsored meeting

I attended, the ostensible purpose of which was to disseminate information about the benefits of Norplant, the divisions between nationalist ideology and local concerns became blatantly apparent. While the chair of the meeting lectured the women attendees about how they could participate in the development of Indonesia as a modern nation by switching to a more "modern" contraceptive technology, most of the women were chatting and laughing with each other along the edges of the room. Only when the chair asked for questions did anyone seem to pay much attention. After a long silence one younger woman called out, "I have read in a magazine about orgasm. Could you tell me what it is and how I can get one?" After the meeting, the chair complained to me that the major problem with the family planning program these days was that women were getting information from each other, rather than from the appropriate experts. "They come to the meetings and gossip, and then after they leave they gossip again." While the state's intention may be to promote "planning," these women can, to an important extent, fill the discursive space created by the population control program with their own experiences of pleasures and pains.

"Participation" in family planning also has other unintended effects. In Java, by representing and positioning women as the primary "planners" of their families, able to control their sexuality through rational calculation, the state may ironically be undermining its own social policies by invoking the cultural values many Javanese attribute to ascetic control. As Brenner (1995) argues, Javanese women consider themselves to be less susceptible than men to the vicissitudes of passion, a restraint they say justifies their participation in economic activities and community decision-making.[41] The state's attributing to women an ability to rationally domesticate sexuality through family planning may ironically galvanize women's understandings of their own power, countering government rhetoric that seeks to confine women's influence to the home. Women frequently stated that just as women were more suited to be "planners" of their families by using contraception, they were also more able to "plan" other activities ranging from economic decisions and purchases to smoothly run social activities and community organizations.

The Indonesian state has also been challenged in its attempts to create hegemonic national sexual ideologies and practices by its need to work with local cultural understandings of sexuality in order to implement its plans for reproducing itself at the desired rate of growth. Probably the most important of these negotiations have been with pious Muslims, both because of Muslims' increasingly influential role in Indonesian society, and because success at deflecting potential Islamic opposition to family planning has been high on the agenda of international development organizations interested in using the Indonesian case as a model for other post-colonial countries.[42]

The relations between nationalism and Islam in Indonesia are complex, given that approximately 85–90 percent of Indonesia's population identifies itself as Muslim but Indonesia is not an Islamic state. Traditionally, Indonesian

Islam has been described as "syncretic," mixing local culture with Muslim and Hindu–Buddhist elements into a unique system of belief and practice. But more orthodox Islamic elements have long existed in Indonesia, and their potential strength has been looked upon with suspicion, first by the Dutch and then the Indonesian state. Fearful that a substantial proportion of the population might inhabit an "imagined community" orienting them not only toward national development but toward the global Islamic *ummat*, or community of believers, including more radical influences from the Middle East, tensions between Islam and the state have ebbed and flowed throughout the history of Indonesia. Thus one of the first priorities of Indonesia's family planning program was to enlist Muslim leaders to help promote the state's goals.[43] Muslim "participation" in family planning, however, is quite often directed toward goals other than those articulated by the state, and has created a number of complex effects that offer alternative versions of nationalism and sexuality.

Muslim groups, especially Indonesia's major modernist organizations *Muhammadiyah* and *Aisyiyah*, run Islamic family planning clinics, hospitals and schools to train women in midwifery and family planning medicine. In so doing they have ironically challenged the gender politics of the New Order by expanding opportunities for women as trained professionals. Indeed, a number of women medical students told me that it was not merely religiously permissible but religiously mandated (*wajib*) for them to use their intelligence and skills to become physicians devoted to the care of women. And although Muslim organizations officially support the family planning program, they do so using logic very different from that of the state or international development agencies. For instance, the Legal Affairs Committee of *Muhammadiyah* decided in 1968 that although contraception violates Islamic doctrine, it is permissible "in case of emergency" with agreement by both husband and wife and respect for Islamic values; for example, abortion and permanent sterilization are prohibited and I.U.D.s should be inserted by female doctors or nurses (B.K.K.B.N. 1993). "Emergency" has been very broadly defined by *Muhammadiyah* to respond both to the state's neo-Malthusian declaration of an Indonesian "population emergency," and to fulfill the need for Muslims to have small, economically stable families that will "create healthy, intelligent, and religious" children (B.K.K.B.N. 1993) able to compete in the national arena with non-Muslims, especially the ethnic Chinese whom many perceive to be controlling much of Indonesia's economic resources and political power. *Muhammadiyah* has been careful to distance its support for family planning from Western "individualist" attempts to *limit* births for personal convenience. Its leaders have defined "planning" as an attempt to *produce* particular kinds of children, in the process improving Muslim economic and social standing, as expressed in the insistence on speaking neither of birth control (*pembatasan kelahiran*) nor of family planning (*keluarga berencana*) but family welfare (*keluarga sejahtera*) when discussing contraception (Sodhy *et al.* 1980).

Sentiments countering state development discourse were echoed in my discussions with individual Muslim women. One 19-year-old university student who had recently decided to try to live a more orthodox Islamic lifestyle told me:

> Your body does not belong to you. It does not belong to the government. It belongs to God. Therefore you have to take care of it, continue along with it the way it was given to you. I don't believe in putting things in your body that are not natural, that do not come from God. You can't use family planning like the spiral [I.U.D.] or the injection [Depo-Provera]. You can use the method where you check what days it is safe to have sex, or if your husband agrees you can use condoms. Yes, you can follow family planning, but not because you are afraid of becoming poor if you have many children. You have to trust God that you will be able to take care of your children. But you can follow family planning so that the children that you have will be able to be more educated and better taken care of. If you only have two children the attention that you are able to give them will be more concentrated. It will be stronger. The Muslim community will then become stronger.

While by no means all Muslim women share such opinions, at least a third of those I asked told me that they believed only the rhythm method to be permitted by Islam, even though they were aware of government statements to the contrary. On another occasion, at the wedding of two Muslim student activists, guests were asked to offer congratulations and advice to the couple on videotape. Many of the younger women only half-jokingly urged fruitfulness on the newly married couple. "Don't just have two children!" one exhorted. "Have many children who can go out and protest with us!" In these kinds of narratives, the benefits of family planning accrue neither to the individual engaged in rational "planning" in order to create a maximally convenient lifestyle, nor to a nation as an abstract, united entity. Family planning ideally creates community, but a kind of community identified by particular moral and religious values that indeed may be imagined as existing in tension with state or national-level priorities. What is articulated as "natural" here is neither the linear move toward the kind of "progress" or "modernity" articulated in state nationalism; nor a notion of the body as a unit in a population count. Bodies are sites for the deployment of ethical and political values, and "development" is articulated in moral and cultural terms, not in assertions of technical virtuosity.

Women also challenge the state's version of planning by participating in family planning in formal ways without accepting the ideological premises of state programs. Orthodox Muslim women frequently claimed that, in the end, it did not really matter if one used birth control or not, because God is more powerful than both contraceptive technology and, by extension, state control. One woman told me that her mother, who had been one of the first people in her village to try birth control, had given birth to six children despite taking the pill for fifteen years. "God wanted her to have a large family, and God is stronger than medicine," she said. Another woman told me that she had tried almost every kind of contraceptive, with no effect:

> The I.U.D., the injection, the pill, nothing could stop those children. There was the sign on my door saying that I was following family planning, but I told them [the village officials] when they asked me why I kept having children, what I want or what you want is not what God wants.

Alternative versions of sexuality are also offered by Muslims, and incorporated into family planning discourse. For instance, Indonesian Muslims are, in increasing numbers, articulating beliefs about sexuality that characterize the body not as a neutral instrument of rationality but as a heterogeneous space marked with areas that are endowed with an uncontrollable and irresistible sexual attractiveness (*aurat*). These beliefs have led to the creation of special clinics where women can have contraceptive devices inserted by other women, outside the vision of men and the realm of sexuality that men's presence implies. These practices challenge state characterizations of the family planning clinic as a domain of desexualized rational "planning," critiquing the notion that either the gynecological exam or the discussion of reproductive behavior is a politically or sexually neutral activity.[44] Muslim leaders have also encouraged national leaders, most recently in 1994 at the United Nations' International Conference on Population and Development in Cairo, to challenge their international sponsors, to denounce the practices of abortion, permanent sterilization and what they consider to be the sexually immoral practices encouraged by the new international development rhetoric of "choice."

These examples illustrate both the multiplicity of meanings present in nationalist discursive space and the multiple audiences toward which Indonesian nationalism is directed and to which it responds. State power is channeled and circumscribed by cultural values and practices as well as by the demands of international development. Here the conditions of possibility for state agency and the forms that state programs can take are limited by the cultural milieu that power attempts to work through, and are susceptible to the unpredictable and multi-voiced outcomes such interactions can produce. Indeed, the diverse orientations and agendas of the state and the multiple narratives of nationalism that this diversity entails produce rifts between Indonesia's version of "development" and that of the agencies that underwrite it. The World Bank has been pushing Indonesia to pursue sterilization as a form of birth control (I.B.R.D. 1990), placing the state in a situation in which its desire to advertise the Indonesian nation as a modern world leader in family planning conflicts with its need to accommodate Muslims who are opposed to such methods. The World Bank has also demonstrated its doubts about Indonesia's development by challenging the government's enthusiasm for staging family planning training programs for representatives from other countries. Through these programs Indonesia positions itself as an innovator in family planning, as a nation that can teach others how to achieve "development." But the World Bank argues that these activities might "distract" Indonesia from its immediate task of controlling its own growth rate (ibid.). We might interpret this conflict as a result of the reproduction of "underdevelopment" or "dependency" that Escobar (1995) argues is essential to an international development apparatus, or we might attribute it to what Bhabha describes as a colonial form of power that encourages the colonized to be "almost the same but not quite . . . almost the same but not white" (1984: 130). Yet whatever argument we choose to explain this situation, its effects demonstrate that even the state's versions of nationalism and sexuality are inherently unstable.

Family planning as national spectacle

> The spectacle is not a collection of images; rather, it is a social relationship between people that is mediated by images.
>
> (Guy Debord, *The Society of the Spectacle*, 1967)

Surveillance as a technique of power in modern societies has, since Foucault (1977), been the subject of sustained discussion. In the area of reproduction, Anagnost (1995) describes how China's "one-child" policy works to keep people regulating their own fertility as well as scrutinizing that of their neighbors.[45] Although the Chinese state does employ officials who monitor reproductive excesses and track and punish resistances (Greenhalgh 1994), for the most part it relies on a situation of "self-normalizing" power where the rule of law is internalized and made to seem like individual desire. But in the Indonesian case, in the absence of regulations absolutely restricting the number of children one can bear, power relies less on surveillance of behavior than on the production of spectacle. The Indonesian state attempts to bring sexuality out of a domain that might be construed as private into a domain of nationalist discourse not by extending surveillance into the bedroom but by, in effect, bringing the bedroom into the town square. By staging displays of sexuality in national public culture, the state offers representations that can ideally become models for people to follow in becoming sexual citizens, committed to the reproduction of Indonesianness.

In presenting these nationalist sexual spectacles, the Indonesian state attempts to offer particular representations of nationalism, as well as to display its own power to stage and manage such displays. It is both nationalist "content" and an apparatus for the production of nationalism that are made visible in these exhibitions. While film theorists have stressed the importance of attending both to cinematic narratives and to the technologies and cultural dispositions that enable people to receive representations that they feel are realistic and thus powerful, in the case of Indonesian family planning the spectacular apparatus is displayed as powerful and thus as real or compelling in people's lives. It is no coincidence, I would argue, that approximately 40 percent of the World Bank's lending to the Indonesian Family Planning Coordinating Board (B.K.K.B.N.) has been allocated for the construction of over 300 offices around the country for the B.K.K.B.N. (I.B.R.D. 1992). The World Bank justifies such an expenditure as follows:

> The buildings put in place are impressive – significantly better than other nearby government buildings, particularly in the countryside. Had Bank financing not been available, it is probable that the Government would have provided BKKBN with similar buildings, but they would have been built more slowly and been few, smaller, and more spartan. The impact of the Bank's inputs, therefore, was to speed up the expansion of the program and to help it achieve status in the eyes of the public and its own staff. . . . These are not inconsequential achievements, particularly for a new program that must change social mores and overcome doubts to acquire a clientele.
>
> (I.B.R.D. 1992: 29)

The World Bank offers no evidence for its claim that "impressive" family planning offices lead to a rise in the "status" of family planning or to an increase in "clientele." Indeed, many Indonesian women told me they would feel ill at ease in such places. I would argue, however, that such projects do have other cultural and political effects. Such monumental architectural encodings of the power of the state to provide and perform family planning, and to situate, infiltrate and represent women's bodies, work to elide "nation" and "state" in Indonesia. Rather than following the Foucauldian model of the panopticon where subjectivities are shaped by the uncertainty of whether the eye of power might be watching, or a model of a cinematic apparatus that produces "realism" precisely because its intervention remains hidden, these efforts allow the Indonesian family planning program to make visible its potential to generate and manage sexual spectacles, as well as to effect certain "surveillances." The shining facades of B.K.K.B.N. buildings do not hide state power behind their walls; rather they signal the ability and willingness of the state to create equally spectacular displays of nationalism and sexuality that can reach into the corners of everyday Indonesian life.

Spectacles of any sort are not, however, displays that inert individuals simply watch and internalize. They are staged in social settings and experienced in cultural terms. Within film studies feminists, especially, have begun to develop theories of spectatorship that highlight viewers' active engagements with film apparatuses and texts.[46] A number of other theorists, ranging from Bakhtin to De Certeau, support such arguments, describing how the reading of texts or the interpretation of symbols is as active a production as their writing. Questioning the passivity of spectatorship is especially appropriate to the

Javanese setting where performances – theater, shadow-plays, lectures – are not set apart from an audience and reverently attended to but are incorporated into a social milieu of backtalk, commentary and diversions of interest (Keeler 1987, Peacock 1987, Siegel 1986). As Siegel describes comedic theater performances in Surakarta, Central Java:

> The actors never ignore what the audience says, or pretend that their theatrical reality is somehow enclosed and thus precludes direct communication with the audience. And the audience response does not depend on a notion of their own invisibility. The assumption is that there is a single linguistic universe that allows anyone to enter the conversation.
> (Siegel 1986: 107–108)

Alternative discourses flow around spectacles, interrupting them and fragmenting them, incorporating them into local cultural concerns. As the example of the beleaguered government family planning lecturer who felt herself to be mired in a field of "gossip" shows, nationalist pronouncements lead to situations of narrative multiplicity that may challenge the position of the state as the sole author and owner of nationalist and sexual meaning. Again, this is not to pose a situation of binary "power" and "resistance." Most of the women I worked with in Java willingly attend state-sponsored women's groups and family planning meetings, even though they rarely "pay attention" in the way one would be expected to in such a setting in the West. And outside of these official spaces, family planning is discussed freely and frequently. To highlight the complexity of Javanese spectator practices is not to dismiss the power of state spectacles to absorb or entrance, but rather to "question the hyphen" in the nation-state (Kelly 1995, Appadurai 1993) by describing nationalism as a heteroglossic discursive space.

What the Indonesian case demonstrates, I believe, is that the relationship between nationalism and sexuality cannot be theorized simply as one of repression or erasure of "deviant" sexuality. In order to encourage its family planning program, the Indonesian state exhibits sexuality in the arena of public nationalist discourse, creating exemplary representations of desire and development that link pleasure and citizenship. But as Appadurai and Breckenridge (1988: 6) argue, "public culture is a zone of cultural debate." Power's possibilities are shaped by its objects, and vice versa. Nations narrate sexuality, and they evoke the potential for pleasure through their displays. But the culturally embedded meanings of desire and power, and indeed discourse itself, in turn impact on nationalism – fragmenting it, displacing it, replacing it with more and less partial alternatives, and providing it with both the matter and means of articulation.

By examining the case of Indonesia's population control program it becomes apparent, I have argued, that analytically separating "reproduction" or "planning" from "sexuality" may not be appropriate in understanding post-colonial nationalisms. Especially as development agendas move away from coercive and repressive policies toward programs that attempt to work within existing cultural meanings and nationalist discourses, our analyses need to acknowledge the role of sexuality as it is articulated and contested by post-colonial people and their governments. In Indonesia, sexual spectacles, local resistances and cultural values are entwined in networks of power and passion that challenge us to rethink our own culturally embedded theories.

The Muslim call to prayer echoes out from tinny loudspeakers as dusk covers Jakarta. The stage is now deserted. Lying on the ground, the banners' bright colors seem muted, their messages tracked over and torn by dusty feet. Everyday life continues, as people move through the square on their diverse errands. Meanwhile, the family planning party is already beginning to be reproduced. The next day it will be labelled a *sukses* by the Indonesian press, another example of the pleasures of citizenship and the power of planning. Statistics will have already begun to be tabulated and evaluations formed, so that the party might be reborn in the training manuals Indonesia produces for the third world bureaucrats who come to learn about family planning promotion, and in the reports of development agencies lauding the ingenuity and effectiveness of Indonesia's programs. A few months later, in neighboring Singapore, where population control is linked to intense ethnic anxiety and eugenic aspirations (Heng and Devan 1992), a Jakarta party will be described in wistful tones. In the national newspaper, one event's "accomplishments" are listed in the "neutral," purportedly cross-cultural, language of numbers: "57 women had IUDs inserted in them; eight men signed up for vasectomy; six women signed up for sterilization; 18 infants were immunized; 35 people had their teeth fixed; 20 boys were circumcised; and 312 people had a medical checkup" (Abu Bakar 1993).

Through these narratives, the Indonesian state will have advertised a domestication of sexuality's potentially disruptive force. It will have encouraged the reproduction of its self-proclaimed modernity through the non-reproduction of its citizens, and it will have positioned and represented women so that their bodies might become the site of this "national" development. But to stop at these state stories, to participate in sewing the seams of control so tightly around Indonesian women and men in our analyses, is to be complicit in the reproduction of state power. To ignore the alternative desires and discourses that percolate through Indonesian nationalist space is to ignore Indonesians, and the lives they lead, not as demographic units but as cultural beings.

Acknowledgements

This chapter is based on ethnographic and historical research carried out primarily in Yogyakarta, Central Java, Indonesia. Research was funded by Fulbright IIE; the American Association of Asian Studies Southeast Asia

Council/Luce Foundation; the MacArthur Foundation/Center of International Studies, Princeton University; the Andrew W. Mellon Foundation and the Department of Anthropology, Princeton University. For criticism and comments on drafts of this chapter I thank Hildred Geertz, Vincanne Adams, Daniela Brancaforte, Jeff Himpele, Lauren Leve, Elizabeth Oram, Michele Rivkin-Fish and the participants in the Center for the Critical Analysis of Contemporary Culture, Rutgers University conference on the Cultures and Politics of Reproduction, to whom an earlier draft of a portion of this chapter was presented. I also thank Tamar Mayer for her patience and editorial assistance.

Notes

1. See Foster (1991) for a review of the anthropological literature on nationalism.
2. See Gupta (1992) for an analysis of the instabilities of post-colonial nationalisms *vis-à-vis* Western nationalisms.
3. Examples of this are (unfortunately) too numerous to comprehensively cite. But Chatterjee's (1993) discussion of the Calcutta stage actress Binodini provides an especially poignant description of how, despite her patriotism, a woman was excluded from full membership in the imagined community of India because of her sexuality.
4. For some excellent analyses of the politics of divided nations, see the essays collected in Warren (1993).
5. See Mani (1989), Kelly (1991) and Stoler (1991) for discussions in the colonial context.
6. For example, when U.S. groups on the "Religious Right" exhibit "pornography" in order to garner support for its prohibition, sexual alternatives are introduced into cultural spaces that may previously have been closed to them. And as Heng and Devan (1992) have argued regarding nationalism and the politics of fertility in Singapore, the continual reproduction of nationalism may even require the repeated production of "narratives of crisis" which place "deviant" reproductive and sexual practices at the forefront of national discourse. Heng and Devan suggest that in creating and managing sexual crises the Singapore state's "administration of crisis operate[s] to revitalize ownership of the instruments of power even as it vindicates the necessity of their use" (1992: 343). They also conclude that state power is never totalizing in its ability to control sexuality, because embedded in these attempts at repression is an awareness of sexual pleasure's "non-economy" (1992: 347), its irreducibility to "planning" or "reproduction." Which is not, I would argue, to say that sexuality exists outside of culture, giving off purely biological meanings, or that sexuality escapes relations of power. Perhaps it would be more accurate to say that sexual pleasure operates within a different "economy" than that of classical economic theories presuming a rational, profit-maximizing subject.
7. There are a number of social theorists, including Gramsci and Williams on "hegemony," one could use to make such a point. I prefer a more Bakhtinian reading in order to try to move away from the emphasis on consciousness and agency found in Gramsci (cf. Kaplan and Kelly 1994), and to avoid reifying either "hegemony" or "resistance." According to Bakhtin, the centripetal or centrifugal forces of language can be pushed by active historical agents, but such agency is always partial in its genre constraints and indeed is not necessary for destabilization of "the official" to occur. Despite the existence of power that attempts to stop these

shifts and fix a unitary meaning, there is always the possibility of dialogue through parody or quotation, or in the very nature of an utterance which germinates in a field of allusions to other utterances, past, present and future. One can read Bakhtin as suggesting the possibility of "resistance" in dialogue, not just in a responsive, quid-pro-quo direct revolutionary assault.

8 Foucault's writings on power, especially Foucault (1982), are instructive on this point.

9 Indonesia has not, for instance, constructed itself as Papua New Guinea nor the Netherlands nor China nor Portugal nor the U.S., and it constructs nationalist narratives for both "internal" and "external" audiences (categories that themselves may cross and blur) that assert and reify these different differences. Not only nationalist proclamations of "Indonesian morality" but also the sexual and gendered meanings attributed to the "primitive Other" or the "amoral West" work to define a sense of sexual self for Indonesian citizens. The highly sexualized portrayals in Indonesian national media of "internal others" such as the inhabitants of Irian Jaya or Kalimantan and the persistent discourses on "free sex in the West" are fascinating topics that are, unfortunately, beyond the scope of this chapter. See Sen (1994) for a brief discussion of "the primitive" in Indonesian film.

10 See Martin (1994) for a discussion of new idioms of "flexibility" in U.S. culture and Harvey (1990) for an analysis of new practices of "flexible accumulation" in multinational capitalism.

11 Recognizing these multiple orientations and audiences of nationalisms and the complexities that they engender allows analyses to go beyond Anderson's (1991) notion that nationalism was simply a "modular" concept exported from Europe to the post-colonies where it was translated into local practice, highlighting both the continuing engagements of post-colonial nationalisms with international capitalism and development, and the cultural discourses, including those of sexuality, that are actively implicated in these transnational negotiations. For example, Indonesia advertises itself for a Western corporate audience as "Muslim, but not fanatically so," at the same time as the government attempts to court Muslim elites by, for instance, permitting veiling in public schools, sponsoring a state "Festival of Islam" or publicizing President Suharto's pilgrimages to Mecca. And, like many other Asian countries, it stresses the value to multinational industry of its purportedly educated and "nimble-fingered" female workforce (Ong 1990) at the same time as it promotes an image of the ideal Indonesian woman as a virtuous housewife and mother whose primary concern is with the care of her family rather than work outside the home (Blackwood 1995, Djajadiningrat-Nieuwenhuis 1987).

12 Cited in Gordon (1977) and discussed in Hartmann (1995).

13 For a discussion of Malthus's views on sexuality, the body and population, see Gallagher (1987).

14 See Scheper-Hughes and Lock (1987) for a discussion of the implications of this body/mind split for social theorizing, especially within the field of medical anthropology.

15 As Ginsburg and Rapp argue, "biomedical research is premised on scientific representations of the human body as a universal constant, not accounting for the biological impact of cultural differences and social inequalities. . . . Representations of birth-control devices as universally available and effective are subject to question in the absence of appropriate support services such as access to public sanitation and primary health care, especially for the poor" (1995: 7). Hartmann (1995) reports that mortality rates from I.U.D.s are on average twice as high in the developing world as in the West, and that in poor countries where malnutrition is common, increased menstrual bleeding – a common side-effect of

I.U.D.s and Depo-Provera – often leads to serious anemia. In areas where primary health care is absent or inaccessible, lack of provisions for follow-up care or for removal of devices such as Norplant makes contraception even more dangerously different. Also relevant here are the analyses by Latour (1987, 1988), Traweek (1988), Martin (1994) and Hess (1993) of the cultural embeddedness of science and scientific knowledge and technology.

16 See Duden (1992) for a historical discussion of the conceptual apparatus that turns diverse groups of culturally situated humans into "populations."

17 A number of scholars have analyzed ideologies of "choice" in Western discourses of reproduction (Rivkin-Fish 1994, Strathern 1992, Rapp 1991, Luker 1984). A discussion of how feminists might offer critiques of reproductive politics articulated in terms other than those of rational, individualist choice can be found in Poovey (1992).

18 This split between reproductive and sexual identity is discussed in a different vein in studies such as Lewin's (1995) and Weston's (1991) of lesbians and their relations to Euro-American kinship practices, especially those of motherhood.

19 See Foucault (1978). See also Rubin's (1984) arguments about sexuality's "excess of significance" in Western culture and Stanton's (1992) essay on the centrality of sexuality to Western and psychoanalytic discourses of the self.

20 Here I wish to beg the question of whether in "the West" (a contested, multicultural field in its own right) passion is really so separable from and subservient to planning. But I do think that Bourdieu's (1977) critique of the "objectivist fallacy" in the human sciences, where the tendency to "map" and metatheorize on the part of the analyst of culture is assumed to be shared by the people she studies, is pertinent to this issue. By turning to cultural difference by way of the Indonesian case, I hope to throw our assumptions into sharper relief, but this is not to imply that this could not be accomplished as well by examining the sexual and reproductive politics of "planning" in the West.

21 Herdt (1981), for instance, describes how the Sambia of Papua New Guinea do not construct a sense of identity around an essentialized core of "sexuality." Sahlins (1985) challenges the assumption that sexuality is an essence deeply buried in the self, reversing the Freudian logic holding that discourses of social and political interaction can be understood as discourses standing in for a hidden, underlying stratum of sexuality by describing how in Hawaiian culture sexuality is not the hidden subtext of chants and poetry but the language *behind* which political meanings can be found.

22 See the essays collected in Ginsburg and Rapp (1995), especially Anagnost (1995), Pearce (1995) and Schneider and Schneider (1995). For a classic example of an argument along these lines see Mamdani (1972).

23 Classic arguments regarding Javanese "syncretism" include C. Geertz (1960) and Anderson (1965). Both Geertz and Anderson describe a Javanese history of mixing cultural forms from various eras and places as a result of a Javanese "relativism" or "tolerance" for difference. Pemberton (1994) poses a challenge to these arguments by situating Javanese culture within colonial history, thus highlighting the uneven power dynamics that were at work in the adoption of new cultural practices.

24 In exploring these genealogies, my focus is turned most acutely toward Java, where I conduct my own research, although the discourses and practices that I examine are among those which have sought to create an Indonesian "national culture" subsuming or erasing any local variations. The extent to which, especially under the current government, "Javanese culture" has been identified and displayed as a national model for emulation also makes the historical and contemporary practices of population control in Java of central relevance to an understanding of Indonesian nationalism, although regional differences and

oppositions to a nationalist "Javanism" should not be underestimated. On national "Javanism" see, especially, Pemberton (1994) and Kipp (1993). For ethnographic analyses of the politics of family planning in Indonesian areas other than Java see Tsing (1993) and Robinson (1989).

25 In this chapter I focus most closely on World Bank discourse, which is not to suggest that "Development" is entirely homogenous in its attitudes or effects. As the World Bank's fourth largest borrower, Indonesia's program plays an important role in Bank population discourse, while the Bank's policies are of primary importance in defining the Indonesian state's strategies. See Hartmann (1995) for a discussion of the major organizational players in international funding for population control. Other development organizations funding the Indonesian program include U.S.A.I.D. and the United Nations Fund for Population Activities (U.N.F.P.A.). The role of non-governmental organizations (N.G.O.s) in family planning in Indonesia is limited by state regulations that they work in coordination with B.K.K.B.N. activities.

26 Birth-spacing methods in Java include post-partum and terminal abstinence, lengthy periods of breastfeeding, herbal medicines (*jamu*), spells, and massage to "turn" the uterus (Alexander 1984, McNicholl and Singarimbun 1982, H. Geertz 1961). Hildred Geertz notes that in the 1950s Java people mentioned techniques for effecting abortion as well, but that it seemed very rarely to occur, probably because social sanctions against unmarried mothers were few and the possibilities for fostering children numerous. In contemporary Indonesia abortion is rarely openly available because of strong Muslim opposition, although women's health activists told me that illegal abortions do often occur.

27 The reasons for this population growth remain the subject of sustained debate. See Alexander (1984) and McNicholl and Singarimbun (1982) for overviews of different explanations.

28 Colonial commentaries on "native" masculinity make special and repeated note of the practice of transvestism and "pederasty" in the East Indies. See Anderson (1990: 277 n. 15).

29 See Taylor (1996) for an analysis of the *nyai* theme in several Indonesian novels and films.

30 On the gender politics of Gerwani, the left-wing women's group affiliated with the *Partai Kommunis Indonesia* (P.K.I.), see Wieringa (1992). H. Geertz (1961) and C. Geertz (1960) also contain useful information on gender in the Sukarno era.

31 For discussions of Anderson's characterization of a Javanese idea of "Power" see Brenner (1995), Keeler (1990) and Errington (1989b). I thank Hildred Geertz for discussion on this topic.

32 See also Anderson (1990) for an analysis of the links between sexual control and "Power" in Javanese literature.

33 See H. Geertz (1961) for the classic study of Javanese kinship.

34 This idea seems to be current in Indonesian historical consciousness as well. For instance, Peacock, in his 1987 afterword to his classic work on *ludruk* theatre in the 1960s, performances of which included heavily sexual scenes and transvestite performers, recounts how a contemporary Indonesian "journalist of the New Order" cited his book "as an exposé of bizarre behavior during the Old" (Peacock 1987: 263).

35 Some scholars have challenged whether a reduction in population growth rates in Indonesia can be directly attributed to the family planning program or if they are attributable instead to other social changes (cf. Hull 1980, McNicoll and Singarimbun 1982).

36 See Robinson (1989) for an account of a "safari" in South Sulawesi. Greenhalgh (1994) notes a similar preference by the Chinese government for contraceptive devices whose operation does not rely on the control of the user, who can "forget"

to take pills or who can pull out an I.U.D.

37 See Niehof (1992) for a discussion of the role of midwives, or "traditional birth attendants" (T.B.A.s), as development discourse calls them, in mediating between local and state interests.

38 Despite this shift in rhetoric, however, Indonesia has moved increasingly toward long-term contraceptive methods. In a 1976 survey of contraceptive users in Java and Bali, 56.7 percent were using the pill, 6.8 percent condoms and 0.8 percent Depo-Provera. In a 1987 survey, 31.4 percent were using the pill, 3.5 percent condoms, 21 percent Depo-Provera and 0.8 percent Norplant (I.B.R.D. 1990). A 1995 survey reported 32 percent using the pill, 32 percent Depo-Provera, 22 percent I.U.D.s, 8 percent Norplant and 6 percent sterilization (*Jakarta Post* 1996). These figures are weighted even more heavily toward non-user-controlled methods in islands other than Java and Bali. With the opening of the first Norplant factory outside of Finland, it seems reasonable to expect that this shift will become even more pronounced.

39 Described as occurring in West Sumatra by Blackwood (1995), who notes that these placards also reinforce the state's gender ideology by requiring inhabitants to list "household head" and "wife," despite the matrilineal culture of the area.

40 Smyth (1991) cites figures from the B.K.K.B.N. stating that 90 percent of birth control users are women, and that the rate of condom use in Indonesia is among the lowest of Asian countries. The actual figures for women's responsibility for contraception may well be higher, because the B.K.K.B.N. counts coitus interruptus as a "male" form of birth control.

41 Brenner's argument counters previous analyses that describe ascetic power as primarily the province of men (e.g. Anderson 1990, Hatley 1990, Keeler 1990). Brenner argues that while Javanese men may see themselves as paragons of control, Javanese women may perceive the situation quite differently.

42 In one World Bank report, the authors relate how Indonesia's family planning success was unexpected because of its "large areas of Muslim fundamentalism" (I.B.R.D. 1992: 27). The state's dealings with Muslims are listed as one of six main reasons the Indonesian program has succeeded. The report writes:

> The BKKBN has been unusually successful at working with Muslim religious leaders, in contrast to programs in most other Muslim countries. This has been facilitated by an Indonesian tradition of working out problems through frequent conferences and discussions in which confrontation is avoided and accommodation and consensus are stressed. Muslim leaders were asked for advice and co-opted to a common enterprise. The effectiveness of this approach is evident in the relatively high acceptance rates, even in areas of Java known to be more orthodox in their Muslim identification.
>
> (I.B.R.D. 1992: 142)

43 For an official government history of the negotiations between Muslims and family planning officials, see B.K.K.B.N. (1993).

44 The increasing vehemence with which Indonesian Muslims are asserting that bodies are sexually inflected topographies rather than an assortment of neutral parts has also led to a large increase in the number of women wearing what they call "Islamic clothing" that covers sexually attractive areas. The politics of this shift is extremely complex, and I cannot enter into this here save to point out that "veiling" does not simply constrict women's social or signifying abilities.

45 See also Yang (1988) for a characterization of Chinese family planning as an example of what Foucault calls modern "bio-power."

46 See Williams (1995) and Mayne (1993) for overviews.

References cited

Abu Bakar, M. (1993) "Soft-sell approach includes parties," *The Straits Times* October 9:3.

Abu-Lughod, L. (1993) "Finding a place for Islam: Egyptian television serials and the national interest," *Public Culture* 5, 3: 493–514.

Alexander, P. (1984) "Women, labour and fertility: population growth in nineteenth century Java," *Mankind* 14, 5: 361–372.

Alonso, A. (1988) "The effects of truth: re-presentations of the past and the imagining of community," *Journal of Historical Sociology* 1, 1: 33–57.

Anagnost, A. (1995) "A surfeit of bodies: population and the rationality of the state in post-Mao China," in F. Ginsburg and R. Rapp (eds.) *Conceiving the New World Order: The Global Politics of Reproduction*. Berkeley: University of California Press.

Anderson, B. (1965) *Mythology and the Tolerance of the Javanese*. Ithaca, NY: Cornell University Modern Indonesia Project Monograph Series.

—— (1990) *Language and Power: Exploring Political Cultures in Indonesia*. Ithaca, NY: Cornell University Press.

—— (1991) *Imagined Communities: Reflections on the Origin and Spread of Nationalism*. New York: Verso.

Appadurai, A. (1990) "Disjuncture and difference in the global cultural economy," *Theory, Culture and Society* 7: 295–310.

—— (1993) "Patriotism and its futures," *Public Culture* 5, 3: 411–430.

Appadurai, A. and C. Breckenridge (1988) "Why public culture?" *Public Culture* 1, 1: 5–9.

Bhabha, H. (1984) "Of mimicry and man: the ambivalence of colonial discourse," *October* 28: 125–134.

—— (1990) "Introduction: narrating the nation," in H. Bhabha (ed.) *Nation and Narration*. New York: Routledge, pp. 1–7.

Bhattacharjee, A. (1992) "The habit of ex-nomination: nation, woman, and the Indian immigrant bourgeoisie," *Public Culture* 5, 1: 19–44.

B.K.K.B.N. (Indonesian National Family Planning Coordinating Board) (1993) *The Muslim Ummah and Family Planning Movement in Indonesia*. Jakarta: B.K.K.B.N.

Blackwood, E. (1995) "Senior women, model mothers, and dutiful wives: managing gender contradictions in a Minangkabau village," in A. Ong and M. Peletz (eds.) *Bewitching Women, Pious Men: Gender and Body Politics in Southeast Asia*. Berkeley: University of California Press, pp. 124–158.

Bland, L. and F. Mort (1984) "Look out for the 'good time' girl: dangerous sexualities as a threat to national health," in Formations Editorial Collective (ed.) *Formations of Nation and People*. London: Routledge & Kegan Paul, pp. 131–151.

Bourdieu, P. (1977) *Outline of a Theory of Practice*. New York: Cambridge University Press.

Brenner, S. (1995) "Why women rule the roost: rethinking Javanese ideologies of gender and self-control," in A. Ong and M. Peletz, (eds.) *Bewitching Women, Pious Men: Gender and Body Politics in Southeast Asia*. Berkeley: University of California Press, pp. 19–50.

Chatterjee, P. (1993) *The Nation and its Fragments: Colonial and Post-colonial Histories*. Princeton, N.J.: Princeton University Press.

Cohen, M. (1991) "Success brings new problems," *Far East Economic Review* April 18, 151, 16: 48–49.

—— (1993) "Spreading the word," *Far East Economic Review* April 22, 156, 16: 60.

Corrigan, P. and D. Sayer (1985) *The Great Arch: English State Formation as Cultural Revolution.* Oxford: Basil Blackwell.

Crossette, B. (1994) "A third-world effort on family planning," *New York Times* September 7: 8.

Debord, G. (1994 [1967]) *The Society of the Spectacle*, trans. D. Nicholson-Smith. New York: Zone Books.

Deleuze, G. (1979) "The birth of the social," foreword to J. Donzelot, *The Policing of Families.* New York: Pantheon.

Deleuze, G. and F. Guattari (1983) *Anti-Oedipus: Capitalism and Schizophrenia.* Minneapolis: University of Minnesota Press.

Dirks, N. (1990) "History as a sign of the modern," *Public Culture* 2, 3: 25–32.

Djajadiningrat-Nieuwenhuis, M. (1987) "Ibuism and priyayization: path to power?" in E. Locher-Scholten and A. Niehof (eds.) *Indonesian Women in Focus.* Dordrecht: Foris, pp. 43–51.

Dominguez, V. (1993) "Visual nationalism: on looking at 'national symbols'," *Public Culture* 5, 3: 451–456.

Donzelot, J. (1979) *The Policing of Families*, trans. R. Hurley. New York: Pantheon.

Douglas, M. (1973) *Natural Symbols: Explorations in Cosmology.* London: Barrie & Jenkins.

Duden, B. (1992) "Population," in W. Sachs (ed.) *The Development Dictionary: A Guide to Knowledge as Power.* Atlantic Highlands, N.J.: Zed, pp. 146–157.

Dumont, L. (1986) *Essays on Individualism: Modern Ideology in Anthropological Perspective.* Chicago: University of Chicago Press.

Duncan, C. (1991) "Art museums and the ritual of citizenship," in I. Karp and S. Levine (eds.) *Exhibiting Cultures.* Washington, D.C.: Smithsonian Institution, pp. 88–103.

Enloe, C. (1989) *Bananas, Beaches and Bases: Making Feminist Sense of International Politics.* London: Pandora.

Errington, S. (1989a) "Fragile traditions and contested meanings," *Public Culture* 1, 2: 49–59.

—— (1989b) *Meaning and Power in a Southeast Asian Realm.* Princeton, N.J.: Princeton University Press.

Escobar, A. (1995) *Encountering Development: Making and Unmaking of the Third World.* Princeton, N.J.: Princeton University Press.

Foster, R. (1991) "Making national cultures in the global ecumene," *Annual Reviews of Anthropology* 20: 235–260.

Foucault, M. (1977) *Discipline and Punish: The Birth of the Prison*, trans. A. Sheridan. New York: Vintage.

—— (1978) *The History of Sexuality: An Introduction.* New York: Vintage.

—— (1982) "The subject and power," *Critical Inquiry* 8: 777–793.

Fox, R. (1990) "Introduction," in *Nationalist Ideologies and the Production of National Cultures.* Washington, D.C.: American Ethnological Society Monograph Series no. 2, pp. 1–13.

Gallagher, C. (1987) "The body versus the social body in the works of Thomas Malthus and Henry Mayhew," in C. Gallagher and T. Laqueur (eds.) *The Making of the Modern Body: Sexuality and Society in the Nineteenth Century.* Berkeley: University of California Press, pp. 83–106.

Geertz, C. (1960) *The Religion of Java*. Chicago: University of Chicago Press.

—— (1995) *After the Fact: Two Countries, Four Decades, One Anthropologist*. Cambridge, MA: Harvard University Press.

Geertz, H. (1961) *The Javanese Family*. New York: Free Press of Glencoe.

Ginsburg, F. and R. Rapp (1995) "Introduction: conceiving the new world order," in F. Ginsburg and R. Rapp (eds.) *Conceiving the New World Order: The Global Politics of Reproduction*. Berkeley: University of California Press, pp. 1–17.

Gordon, L. (1977) *Woman's Body, Woman's Right: A Social History of Birth Control in America*. New York: Penguin.

Gouda, F. (1993) "The gendered rhetoric of colonialism and anti-colonialism in Indonesia," *Indonesia* 55: 1–22.

Greenhalgh, S. (1994) "Controlling births and bodies in village China," *American Ethnologist* 21, 1: 3–30.

Gupta, A. (1992) "The song of the non-aligned world: transnational identities and the reinscription of space in late capitalism," *Cultural Anthropology* 7: 63–79.

Gupta, A. and J. Ferguson (1992) "Beyond 'culture': space, identity, and the politics of difference," *Cultural Anthropology* 7: 6–23.

Handelman, D. and L. Shamgar-Handelman (1993) "Aesthetics versus ideology in national symbolism: the creation of the emblem of Israel," *Public Culture* 5, 3: 431–450.

Hartmann, B. (1995) *Reproductive Rights and Wrongs: The Global Politics of Population Control*. Boston, MA: South End Press.

Harvey, D. (1990) *The Condition of Postmodernity*. Cambridge, MA: Blackwell.

Hatley, B. (1990) "Theatrical imagery and gender ideology in Java," in J. Atkinson and S. Errington (eds.) *Power and Difference: Gender in Island Southeast Asia*. Stanford, CA: Stanford University Press, pp. 177–207.

Heng, G. and J. Devan (1992) "State fatherhood: the politics of nationalism, sexuality and race in Singapore," in A. Parker *et al.* (eds.) *Nationalisms and Sexualities*, New York: Routledge, pp. 343–364.

Herdt, G. (1981) *Guardians of the Flutes: Idioms of Masculinity*. Chicago: University of Chicago Press.

Hess, D. (1993) *Science in the New Age: The Paranormal, its Defenders and Debunkers, and American Culture*. Madison: University of Wisconsin Press.

Hobart, M. (1993) "Introduction: the growth of ignorance?" in M. Hobart (ed.) *An Anthropological Critique of Development: The Growth of Ignorance*. New York: Routledge, pp. 1–30.

Hobsbawm, E. (1983) "Introduction: inventing traditions," in E. Hobsbawm and T. Ranger (eds.) *The Invention of Tradition*. New York: Cambridge University Press, pp. 1–14.

Horn, D. (1994) *Social Bodies: Science, Reproduction, and Italian Modernity*. Princeton, N.J.: Princeton University Press.

Hull, T. (1980) "Fertility decline in Indonesia: a review of recent evidence," *Bulletin of Indonesian Economic Studies* 16, 2.

International Bank for Reconstruction and Development (I.B.R.D.: The World Bank) (1990) *Indonesia: Family Planning Perspectives in the 1990s*. Washington, D.C.: I.B.R.D.

—— (1992) *Population and the World Bank: Implications from Eight Case Studies*. Washington, D.C.: I.B.R.D. Operations Evaluation Department.

Jakarta Post (1996) "Family planning participants spurn comfy contraceptives," July 12: 2.

Kaplan, M. (1995) "Blood on the grass and the dogs will speak: ritual-politics and the nation in independent Fiji," in R. Foster (ed.) *Nation Making: Emergent Identities in Post-colonial Melanesia.* Ann Arbor: University of Michigan Press, pp. 95–126.

Kaplan, M. and J. Kelly (1994) "Rethinking resistance: dialogics of 'disaffection' in colonial Fiji," *American Ethnologist* 21, 1: 123–151.

Keeler, W. (1987) *Javanese Shadow Plays, Javanese Selves.* Princeton, N.J.: Princeton University Press.

—— (1990) "Speaking of gender in Java," in J. Atkinson and S. Errington (eds.) *Power and Difference: Gender in Island Southeast Asia.* Stanford: Stanford University Press, pp. 127–152.

Kelly, J. (1991) *A Politics of Virtue: Hinduism, Sexuality, and Countercolonial Discourse in Fiji.* Chicago: University of Chicago Press.

—— (1995) "The privileges of citizenship: nations, states, markets, and narratives," in R. Foster (ed.) *Nation Making: Emergent Identities in Post-colonial Melanesia.* Ann Arbor: University of Michigan Press, pp. 253–274.

Kemper, S. (1993) "The nation consumed: buying and believing in Sri Lanka," *Public Culture* 5, 3: 377–394.

Kipp, R. Smith (1993) *Dissociated Identities: Ethnicity, Religion and Class in Indonesian Society.* Ann Arbor: University of Michigan Press.

Koester, D. (1995) "Gender ideology and nationalism in the culture and politics of Iceland," *American Ethnologist* 22, 3: 572–588.

Latour, B. (1987) *Science in Action.* Cambridge, MA: Harvard University Press.

—— (1988) *The Pasteurization of France.* Cambridge, MA: Harvard University Press.

Lewin, E. (1995) "On the outside looking in: the politics of lesbian motherhood," in F. Ginsburg and R. Rapp (eds.) *Conceiving the New World Order: The Global Politics of Reproduction.* Berkeley: University of California Press, pp. 103–121.

Luker, K. (1984) *Abortion and the Politics of Motherhood.* Berkeley: University of California Press.

McNicoll, G. and M. Singarimbun (1982) *Fertility Decline in Indonesia: Analysis and Interpretation.* Center for Policy Studies Working Papers no. 93, New York: Population Council.

Malthus, T. (1976 [1798]) *An Essay on the Principle of Population*, ed. P. Appleman. New York: Norton.

Mamdani, M. (1972) *The Myth of Population Control: Family, Caste and Class in an Indian Village.* New York: Monthly Review Press.

Mani, L. (1989) "Contentious traditions: the debate on *sati* in colonial India," in K. Sangari and S. Vaid (eds.) *Recasting Women: Essays in Colonial History.* New Delhi: Kali for Women, pp. 88–126.

Mankekar, P. (1993) "Television tales and a woman's rage: a nationalist reading of Draupadi's 'Disrobing'," *Public Culture* 5, 3: 469–492.

Martin, E. (1990) "Toward an anthropology of immunology: the body as nation state," *Medical Anthropology Quarterly* 1, 4: 410–426.

—— (1994) *Flexible Bodies: Tracking Immunity in American Culture from the Days of Polio to the Age of AIDS.* Boston, MA: Beacon Press.

Mayne, J. (1993) *Cinema and Spectatorship.* New York: Routledge.

Morsy, S. (n.d.) "Biotechnology and the international politics of population control: long-term contraception in Egypt," unpublished paper.

Mosse, G. (1985) *Nationalism and Sexuality: Middle-Class Morality and Sexual Norms in Modern Europe*. Madison: University of Wisconsin Press.

Niehof, A. (1992) "Mediating roles of the traditional birth attendant in Indonesia," in S. van Bemmelen, M. Djajadiningrat-Niewenhuis, E. Locher-Scholten and E. Touwen-Bouwsma (eds.) *Women and Mediation in Indonesia*. Leiden: KITLV Press, pp. 167–186.

Ong, A. (1990) "Japanese factories, Malay workers: class and sexual metaphors in West Malaysia," in J. Atkinson and S. Errington, (eds.) *Power and Difference: Gender in Island Southeast Asia*. Stanford: Stanford University Press, pp. 385–422.

Parker, A., M. Russo, D. Sommer and P. Yaeger (eds.) (1992) "Introduction," in A. Parker *et al.* (eds.) *Nationalisms and Sexualities*. New York: Routledge, pp. 1–18.

Peacock, J. (1987) *Rites of Modernization: Symbols and Social Aspects of Indonesian Proletarian Drama*. Chicago: University of Chicago Press.

Pearce, T. O. (1995) "Women's reproductive practices and biomedicine: cultural conflicts and transformations in Nigeria," in F. Ginsburg and R. Rapp (eds.) *Conceiving the New World Order: The Global Politics of Reproduction*. Berkeley: University of California Press, pp. 195–208.

Pemberton, J. (1994) *On the Subject of "Java."* Ithaca, NY: Cornell University Press.

Piotrow, P. T., K. Treiman, J. Rimon III, S. Hee Yun and B. Lozare (1994) *Strategies for Family Planning Promotion*. World Bank Technical Paper no. 223, Washington, D.C.: The World Bank.

Poovey, M. (1992) "The abortion question and the death of man," in J. Butler and J. Scott (eds.) *Feminists Theorize the Political*. New York: Routledge, pp. 239–256.

Rabinow, P. (1989) *French Modern: Norms and Forms of the Social Environment*. Cambridge, MA: MIT Press.

Rapp, R. (1991) "Constructing amniocentesis: maternal and medical discourses," in F. Ginsburg and A. Lowenhaupt Tsing (eds.) *Uncertain Terms: Negotiating Gender in American Culture*. Boston MA: Beacon Press, pp. 28–42.

Rivkin-Fish, M. (1994) "Post-communist transformations and abortion politics: reflections on feminist strategies and 'choice'," *Critical Matrix* 8, 2: 101–125.

Robinson, K. (1989) "Choosing contraception: cultural change and the Indonesian family planning program," in P. Alexander (ed.) *Creating Indonesian Cultures*. Sydney: Oceania Publications.

Rubin, G. (1984) "Thinking sex: notes for a radical theory of the politics of sexuality," in C. Vance (ed.) *Pleasure and Danger: Exploring Female Sexuality*. Boston, MA: Routledge & Kegan Paul, pp. 267–319.

Sahlins, M. (1985) *Islands of History*. Chicago: University of Chicago Press.

Scheper-Hughes, N. and M. Lock (1987) "The mindful body: a prolegomenon to future work in medical anthropology," *Medical Anthropology Quarterly* 1, 1: 6–41.

Schneider, P. and J. Schneider (1995) "Coitus interruptus and family respectability in Catholic Europe: a Sicilian case study," in F. Ginsburg and R. Rapp (eds.) *Conceiving the New World Order: The Global Politics of Reproduction*. Berkeley: University of California Press, pp. 177–194.

Sen, K. (1994) *Indonesian Cinema: Framing the New Order*. Atlantic Highlands, N.J.: Zed.

Siegel, J. (1986) *Solo in the New Order: Language and Hierarchy in an Indonesian City*. Princeton, N.J.: Princeton University Press.

Smyth, I. (1991) "The Indonesian family planning programme: a success story for women?" *Development and Change* 22: 781–805.

Sodhy, L.S., G. A. Metcalf and J. S. Wallach (1980) "Islam and family planning: Indonesia's Mohammadiyah," Chestnut Hill, MA: Pathfinder Fund Pathpapers no. 6.

Stanton, D. (1992) "Introduction: the subject of sexuality," in D. Stanton (ed.) *Discourses of Sexuality*. Ann Arbor: University of Michigan Press, pp. 1–46.

Stoler, A. (1985) *Capitalism and Confrontation in Sumatra's Plantation Belt*. New Haven, CT: Yale University Press.

—— (1991) "Carnal knowledge and imperial power: gender, race and morality in colonial Asia," in M. di Leonardo (ed.) *Gender at the Crossroads of Knowledge: Feminist Anthropology in the Postmodern Era*. Berkeley: University of California Press, pp. 51–101.

Strathern, M. (1992) *Reproducing the Future: Anthropology, Kinship and the New Reproductive Technologies*. New York: Routledge.

Taylor, J. G. (1996) "Nyai Dasima: portrait of a mistress in literature and film," in L. Sears (ed.) *Fantasizing the Feminine in Indonesia*. Durham, N.C.: Duke University Press, pp. 225–248.

Tiwon, S. (1996) "Models and maniacs: articulating the female in Indonesia," in L. Sears (ed.) *Fantasizing the Feminine in Indonesia*. Durham, N.C.: Duke University Press, pp. 47–70.

Traweek, S. (1988) *Beamtimes and Lifetimes: The World of High Energy Physics*. Cambridge, MA: Harvard University Press.

Tsing, A. L. (1993) *In the Realm of the Diamond Queen*. Princeton, N.J.: Princeton University Press.

U.S.A.I.D. (United States Agency for International Development) (1979) *AID's Role in Indonesian Family Planning: A Case Study with General Lessons for Foreign Assistance*. AID Program Evaluation Report no. 2. Washington, D.C.: U.S.A.I.D.

Warren, K. (ed.) (1993) *The Violence Within: Cultural and Political Opposition in Divided Nations*. Boulder, CO: Westview Press.

Weston, K. (1991) *Families We Choose: Lesbians, Gays, Kinship*. New York: Columbia University Press.

White, B. (1973) "Demand for labor and population growth in colonial Java," *Human Ecology* 1: 217–235.

Wieringa, S. (1992) "Ibu or the beast? Gender interests in two Indonesian women's organizations," *Feminist Review* 41: 98–113.

Williams, L. (ed.) (1995) *Viewing Positions: Ways of Seeing Film*. New Brunswick, N.J.: Rutgers University Press.

Wright, P. (1985) *On Living in an Old Country: The National Past in Contemporary Britain*. London: Verso.

Yang, M. (1988) "The modernity of power in the Chinese socialist order," *Cultural Anthropology* 3: 408–427.

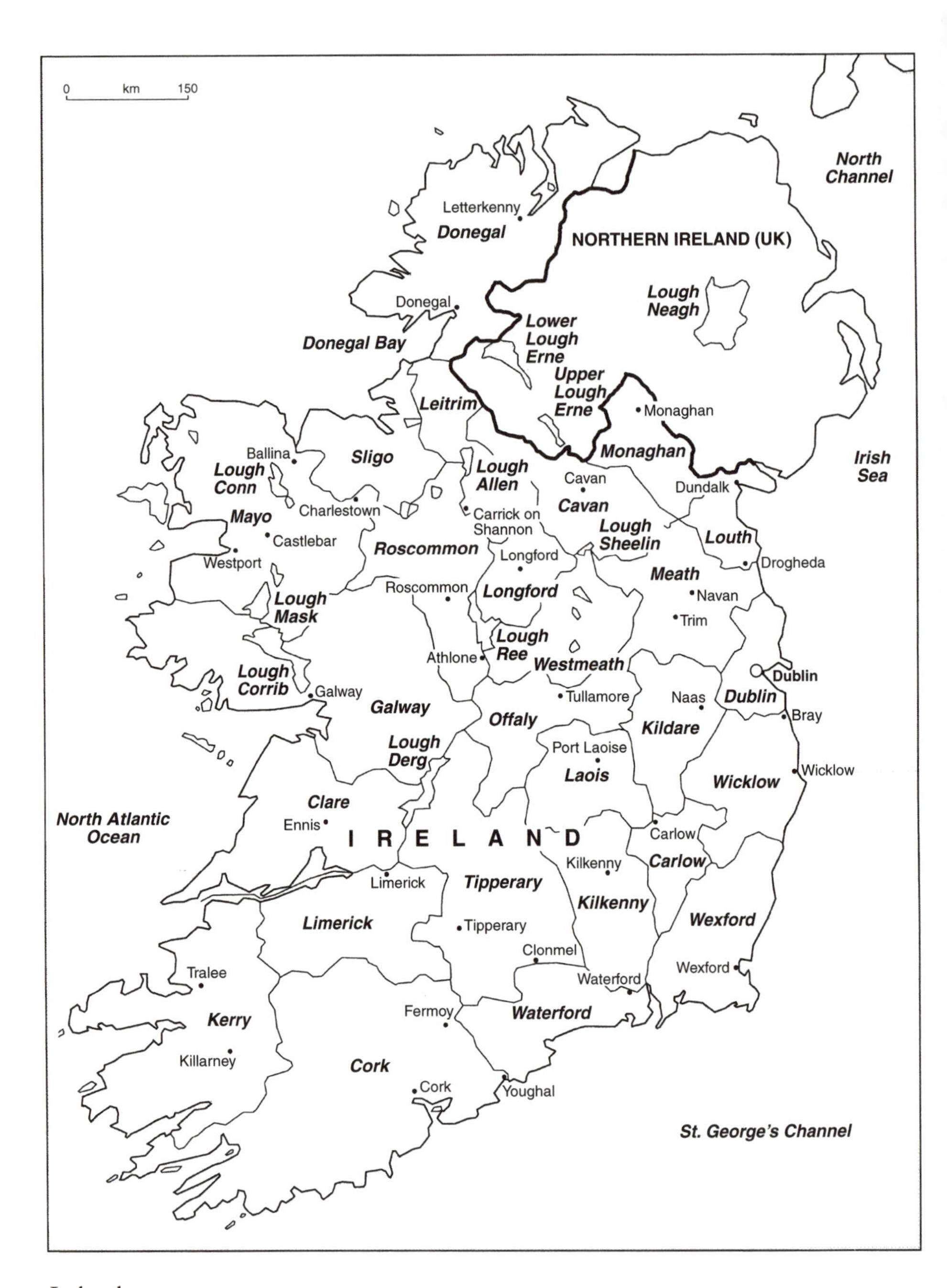

0 km 150
North Channel
Letterkenny
Donegal
NORTHERN IRELAND (UK)
Lough Neagh
Donegal
Lower Lough Erne
Donegal Bay
Upper Lough Erne
Leitrim
Monaghan
Ballina
Sligo
Lough Allen
Monaghan
Irish Sea
Lough Conn
Cavan
Dundalk
Charlestown
Cavan
Mayo
Carrick on Shannon
Lough Sheelin
Castlebar
Louth
Roscommon
Longford
Westport
Drogheda
Meath
Roscommon
Longford
Navan
Lough Mask
Trim
Lough Ree
Athlone
Westmeath
Lough Corrib
Dublin
Galway
Tullamore
Naas
Dublin
Galway
Offaly
Bray
Kildare
Lough Derg
Port Laoise
Laois
Wicklow
Wicklow
Clare
North Atlantic Ocean
Ennis
IRELAND
Carlow
Kilkenny
Carlow
Limerick
Tipperary
Kilkenny
Limerick
Tipperary
Wexford
Clonmel
Tralee
Wexford
Waterford
Fermoy
Waterford
Kerry
Killarney
Cork
Cork
Youghal
St. George's Channel

Ireland

3 Death of a nation

Transnationalism, bodies and abortion in late twentieth-century Ireland

Angela K. Martin

In May 1992, I traveled to the Republic of Ireland for the first time. I was a graduate student in anthropology, living in a small rural community in the West of Ireland and studying local devotional practices associated with the Virgin Mary. I was a newly arrived, inexperienced ethnographer in a country that was in the throes of enormous moral and political change as a result of recent events of which I knew little at best. At the time I did not feel equal to the tremendous task of interpreting the effect of massive national political change on the local community in which I lived. I couldn't see what the Virgin Mary might have to do with any of the national changes taking place in Ireland. Ironically, it was Mary herself – with her curious ability to appear just about anywhere at pivotal and volatile moments in time – who drew me into the realm of politics and who set me on a path toward relevant anthropological interpretation.

One day, only a couple of weeks into my research, I saw a story on the 6 o'clock news. Videotaped footage showed a crowd of two thousand marching through the streets of Dublin. The marchers had a double mission: to protest against the possible legalization of abortion in Ireland and also to urge a vote of "No" against ratification of the Maastricht Treaty in a national referendum to be held the following month (in June 1992). The protesters were clearly angry about something, and scared. They believed that ratification of the Maastricht Treaty (the European Union treaty scheduled to bring a common currency, economic and military integration to Europe by the year 2000) would be the ruin of the Irish "nation."[1] As women began to appear in the crowd carrying large pictures of the Virgin of Guadalupe (known to me only as one of Mexico's most important revolutionary and national symbols), I became quite confused. What was she doing in Ireland in the middle of that throng of protesters? Somehow, women, the Virgin Mary, abortion, the Irish nation and the European Union Treaty were all linked together in a meaningful and potent way for this particular segment of the Irish population. Seeing Guadalupe in this context forever changed the course of my research and drew me into an ongoing investigation of the links among the female body, abortion and the nation in Ireland.

In this chapter I discuss the ways in which abortion has been used as a national boundary issue in contemporary Ireland.[2] In Ireland, the debate over access to abortion has been explicitly linked to attempts to define the moral, political and economic boundaries of the Irish nation within the larger context of the transnationalist European Union (E.U.). The solidification and maintenance of both real and imagined national boundaries involves a disciplining of those bodies onto which the image of the nation has been projected. Through content analysis, I show here that conservative discourses deployed around the abortion and divorce controversies in Ireland between 1992 and 1995 symbolically equate Irish women with the nation, and construct their bodies as Other with respect to the E.U.

In Ireland, many of the gendered and sexualized meanings embodied in the idea of the Irish nation have been institutionalized in state juridical and economic structures. The Irish nation and the state are therefore paradigmatically linked, similarly experienced, and often conflated by Irish people, especially in political contexts where Irish Catholic morals are particularly threatened. Accordingly, women's bodies have been used to determine the moral, political and physical boundaries of the Republic itself, as well as the nation. Such discursive correspondence between the nation-state and gendered bodies materially mediates the ways in which feminine bodies are constructed, disciplined and experienced. To illustrate this last point, I consider the famous 1992 court case (the X Case) in which a 14-year-old girl (Miss X), pregnant as the result of rape, was prohibited from traveling to England for an abortion. Miss X's experience exposes Irish anti-abortion rhetoric as a violent form of nationalist identity politics in which a coherence is established between the materiality of the nation (as it is manifested in state juridical structure) and the materiality of women's bodies.

Women, mimesis, and alterity in nationalist Ireland: historical context

Historically, Irish national identity has been strongly linked to Irish Catholicism. The ideological and material roots of modern Irish nationalism can be found in the Devotional Revolution of the mid-nineteenth century (see Larkin 1984), a process of rapid religious change linked to concurrent economic and political changes that took place in Ireland between 1840 and 1870. This period of change included the deployment of a new series of devotional practices throughout the peasantry, as well as more frequent confession and mass attendance (all of these changes are summarized in Martin 1997). These devotional changes, combined with educational reforms which gave the clergy direct control over the national school curriculum and teachers, significantly increased the Irish Church's ability to regulate marriage, sexual practices and the construction of gender identity in local communities. Irish nationalist clubs and organizations (which were almost exclusively limited to male membership) followed largely on the heels of the Devotional

Revolution and embodied in their doctrine (including their representations of the nation) constructions of masculinity and femininity which mirrored those deployed by the Irish Church in local communities via the discursive regulation of sexuality.

Since the late nineteenth century, nationalist constructions of the Irish nation have thus been paradigmatic of Irish Catholic morality. Subsequently, not only have Catholic and nationalist constructions of masculinity and femininity defined the terms through which the Irish nation would be represented, but also with the establishment of the Irish Free State in 1922 and the writing of the Republic's first Constitution in 1937, these constructions have become embodied in the juridical structure of the state itself. Thus, the first Irish Constitution "enshrined the patriarchal nuclear family as the cornerstone of the new state" (Mullin 1991: 42). The text of the Constitution posited an equivalence between "woman," "the home" and motherhood, and legally restricted married women to the domestic domain by limiting the access of married women to work outside of the home. The welfare of both the Irish state and the nation were explicitly linked to the appropriate placement of women in the home, and with male control over this domain (see Martin 1997).

Irish women have historically also been charged with the labor of representing or embodying the nation. From "Mother Ireland" (Loftus 1990, R. Kearney 1986) to the "land as woman" motif (Lloyd 1993) to the Virgin Mary, Queen of Ireland, women have symbolically represented the purity and tradition of the country. Published works on the importance of feminine representations of the nation in nationalist Ireland are abundant (see, among others, Martin 1997, Nash 1994, 1993, Ap Hywel 1991, Mullin 1991, Cullingford 1989), so I will not repeat their contents here. It is important to note, however, that the modern Irish nation formed at a time when men in Ireland were seeking to establish a nationalist masculine identity in counterposition to Irish colonial feminization under British rule (Goldring 1993, Lloyd 1993, Cairns and Richards 1987), a task in part accomplished by legally limiting Irish women to the domestic sphere (Martin 1997). Indeed, the success of Irish male nationalists' attempts at nation-building, in the public spheres of politics, economy and war or rebellion, was made dependent upon the cooperation of Irish women as mothers and hearth-keepers in the prototypical turn-of-the-century Irish home (Ap Hywel 1991).

It is clear that historically in nationalist Ireland women have almost exclusively borne the labor of representing the nation. It is also important to note that the *labor of representation* discussed here is not merely symbolic or abstract, not merely a necessary consequence of *imagining* the nation (Anderson 1983). The labor of representation also involves very real material consequences for body, self and nation. As noted above, standards of representation are sometimes also codified in law and can have a profound influence on the legal and economic status of those persons charged with representing the nation. We must therefore look in more depth at what it means *materially* for Irish women when they represent the Irish nation and its alterity or difference with respect to Britain, Europe, and/or other Others. In Ireland, such material consequences have historically included the restriction of married women from work outside the home, and from legal access to contraception, abortion, and divorce. More recently they have also included threatened and real restriction of the right of pregnant women to travel out of Ireland.

In *Mimésis and Alterity* (1993), Michael Taussig deals with the importance of mimesis to the representation of difference in Cuna Indian cosmology. The Cuna are a native American group indigenous to Panama and have been much studied by anthropologists. In his examination of the Cuna spirit world Taussig finds an interesting if perplexing "gendered division of mimetic labor" (1993: 177).[3] While Cuna men travel in the world of whites, donning Western clothing and working on the Panama Canal, Cuna women "bedeck themselves as magnificently Other. It is they who provide the shimmering appearance 'of Indianness'" (ibid.). Taussig defines the "mimetic faculty" as "the nature that culture uses to create second nature, the faculty to copy, imitate, make models, explore difference, yield into and become other" (ibid.: xiii). Cuna women, then, mimetically represent and embody the difference of Cuna "Indianness" or Cuna colonial and post-colonial alterity, and exclusively bear the labor of that representation. The gendered division of mimetic labor among the Cuna means that Cuna women represent Cuna alterity not only for the colonial and post-colonial "outside" worlds, but also for the Cuna themselves.

Janice Boddy (1993) applies Taussig's mimetic analysis to the embodiment of cultural aesthetics in women in Northern Sudan. Here "embodiment" refers to the material manifestation of cultural prescriptions, representations or ideals in individuals as a result of a series of discursive cultural practices. "Mimesis" involves the material embodiment of cultural ideals in individuals to the extent that their bodies become a metaphor of (or become paradigmatic of) the cultural ideal so embodied. As Boddy writes, "An embodied aesthetic is productive, in Foucault's sense, of specified bodies, selves, relationships" (1993: 6). In the case of the individual Sudanese woman, for example, surgical removal of her genitalia (infibulation) produces an enclosed female body that is paradigmatically aligned with, and represents through an embodied cultural aesthetic, the interior of the family home, the enclosure within which this home is contained, and the larger, bounded social group (see also Boddy 1989). Infibulation thus becomes "an embodied referent supporting [a woman's] sense of who and what she is" (Boddy 1993: 5).

Boddy borrows her use of the term "aesthetic" from Susan Bordo's (1993) analysis of anorexia nervosa and the internalization or embodiment of cultural norms of femininity expressed through slimness in America. Anorexia nervosa and infibulation both comprise dramatic examples of the material embodiment of cultural standards of femininity. Among the Cuna and Sudanese these standards are also discursively linked to the representation of native difference in colonial and post-colonial settings. Taussig's analysis of the gendered division of mimetic labor among the Cuna, in which women embody "Cuna indianness" as well as standards of femininity, illustrates the extent to which representations of gender and ethnicity are mimetically and materially linked in some contexts.

The mimetic links between women and the nation in contemporary Ireland have generally been structured around the Virgin Mary. The mimetic relationship between women and Mary in Ireland may be difficult to grasp because Mary shapes female corporeality in ways more subtle than Boddy's example of infibulation and "embodied cultural aesthetics" in Sudan perhaps conveys: in Ireland women are not surgically altered to manifest a direct correspondence between symbolic ideal and materiality. Rather, in the Irish case femininity is inscribed and embodied in individuals as a product of the everyday discursive practices that compromise devotion to Mary. Such embodiment affects the ways that Irish women experience their bodies. Most studies of Marian devotion have suggested that Mary's influence on women materializes as a result of encouraged and attempted emulation of an ideal (for example, see Rodriguez 1994, Collier 1986, Orsi 1985, Campbell 1982, Warner 1976). In other words, women are encouraged to represent and manifest the ideal of Mary in their own "essence" – in their behavior, motherhood and their relationships with others. Yet none of these studies illustrated how women embody Mary as a result of cultural practice. The concept of mimesis adds new analytic vigor and brings bodily experience and cultural constructions of feminine corporeality and difference to the fore. Individual women are mimetic of Mary. It is through mimetic performance that Irish women come to embody femininity and, by extension, the Irish nation.

Direct links between the performance of Irish womanhood and Mary's symbolic content can be found in metonymic and metaphoric associations between women and Mary constructed through local familial and devotional practices. Notions of ideal motherhood that inform behavior and the discursive regulation of feminine bodies are obvious links. The performance of identity in home and church spaces and the simultaneous construction and experience of places as gendered (see Martin 1997, 1993b) inform Irish women's mimetic performance of Mary. Mary's appearances in apparition form in various cultural contexts are particularly powerful defining moments and help to clearly construct mimetic links between Mary and earthly women. Apparition performances are "ideal mimetic moments" (see Martin and Kryst 1998) in that they provide a symbolically loaded and dramatic performative context in which female sexuality, ideal motherhood and female bodily boundaries, particularly the boundary between womb and outside world, are foregrounded. In these performative moments, all of the women present, especially the visionary, carry most of the symbolic load or labor of the performance. Women are intensely disciplined to be appropriately "feminine" during these moments, as constructions of femininity are more highly contested than those of masculinity, which are defined by extension. Women are thus mimetically or symbolically linked to Mary via apparition performances.

In Ireland, there are a number of well-known apparition sites, including Knock and Milleray Grotto (see O'Dwyer 1988). There is also a well-known contemporary Irish Marian visionary, Christina Gallagher. She reports having received messages from Mary about the importance of motherhood, the uniqueness (alterity/difference) and importance of the Irish nation, and about Irish political issues such as abortion and divorce (see Petrisko 1995 for an account of the apparitions). Individuals also invoke the Virgin Mary and refer strategically to those areas of her messages where Mary sometimes highlights the uniqueness or specialness of Ireland as the last pure and holy nation in Europe and, perhaps, the world. The timing and context of Mary's appearances, as well as the content of her messages, often work together to foreground the Irish nation's difference or alterity with respect to other nations in Europe. Considering the historical discursive connections between women, the Virgin Mary and the nation in Ireland, it is not surprising that both Mary and Irish women have played a significant role in the negotiation of the Irish nation-state's identity in the contemporary transnationalist context of the E.U.

Recent change: transnationalism and deterritorialization in Ireland

How does it make cultural and political "sense", then, that the Virgin of Guadalupe might be found in the middle of a throng of anti-Maastricht, anti-abortion protesters? To answer this question, we must first consider the implications of the Maastricht Treaty and E.U. membership for the traditional geopolitical boundaries of the Irish state and the moral boundaries of the Irish nation.

The latter half of the twentieth century has seen rapid change in Ireland, but the period since Ireland's entry into the European Economic Community (E.E.C.) in 1973 has been particularly tumultuous. Since the mid-1970s the Irish state has been experiencing legal and economic deterritorialization. 'Deterritorialization' refers to the breakdown of the geopolitical, economic and juridical boundaries of states that occurs under the influence of transnationalism and the globalization of capital (Chavez 1994, Gupta 1992, Appadurai 1991, M. Kearney 1991, Rouse 1991; see also Jameson 1984). As a result of such processes, Irish internal social policies have been challenged in E.U. courts and the Irish national economy has been increasingly regulated by E.U. policy decisions.

Transnational deterritorialization has also affected the Irish nation in that it has undermined the historical precedent set in the early twentieth century in which the economic, juridical and political structure of the state was shaped and determined by the conservative moral composition of the Irish nation.

Since Ireland joined the E.U. the paradigmatic links between Irish state and the Irish nation have weakened. Such weakening in correspondence between the symbolic content of the Irish nation and, particularly, the juridical content of the state, has resulted in the widespread perception in Ireland that the Irish nation itself is dissolving. The breakdown of the Irish nation is further seen as extending from a continuing decrease in the structural and moral differences between Ireland and other E.U. nation-states. As a result, since the mid-1980s national politics in Ireland have been largely dominated by highly gendered and sexualized political issues such as contraception, abortion and divorce – those issues which directly threaten the moral-symbolic content of the Irish nation. And, in fact, Irish national policies relating to these issues are often debated on the grounds of whether or not they continue to signal Ireland's alterity with respect to the rest of trans-national Europe.

As noted, the contested domain has been primarily one of moral issues concerning gender and sexuality, and such contestation has been further linked to the nature of Irish unity (Girvin 1993). Although other E.U. member states have resisted various E.U. policies and further economic integration, Ireland is unique in its resistance to ratification of Maastricht on the almost exclusive grounds of the Treaty's potential influence on internal Irish moral policy. In 1992, ratification of the Maastricht Treaty was strongly opposed by conservatives in Ireland, primarily on the grounds that this treaty would make it easier for E.U. courts to pass and implement judgments on Irish moral issues. These fears were not totally unfounded, as the European Court was already in the process of reviewing Irish laws which criminalized various homosexual acts; the European Court has subsequently found these laws to be in violation of the civil rights of Irish citizens (as European citizens), and has encouraged a corresponding change in Irish juridical structure – a change currently in progress. As Irish legal boundaries have become more porous and vulnerable to penetration from E.U. courts (see MacCurtain 1993: 8), the nation's boundaries, mapped out by moral traditions, have also become more tenuous. What is most interesting for us are the ways in which anxieties over the nation's boundaries have been projected onto the bodies of Irish women and have been materially manifested in constitutional attempts to define the limits of women's bodies. Significantly, in the abortion, Maastricht Treaty and divorce constitutional referenda campaigns, the Virgin Mary has functioned as an important symbol of Ireland's radical alterity with respect to the rest of the E.U., and has also served to foreground the contested nature of women's bodies within these political debates.

Abortion, Mary and Maastricht

Vivid illustration of the ways in which Mary has been used as a symbol of the Irish nation's moral purity and thus difference can be found in the pages of the *Irish Democrat* (*I.D.*), a conservative newspaper founded in the spring of 1992.[4] This newspaper is published in Dublin, but has a readership and contributors in different areas throughout the Republic. In August 1992, after the Maastricht Treaty had been ratified in Ireland, a letter to the editor appeared. At this time the anti-abortion campaign was becoming more heated because of the perceived threat to abortion policy in Ireland from E.U. courts and calls for a public referendum to protect and reinforce the constitutional ban on abortion were being made. The author of the letter suggests that Mary be used as "a rallying point" in the anti-abortion campaign. He voices his concern:

> As we witness more and more the erosion of family values and come increasingly under pressure to go the way of the world, we see all around us people falling foul of the evil regimes and toeing the line of modernism. . . . We have at our disposal the greatest weapon of all, the Mother of God. . . . Now we must let them see that not everyone wants to go the way of the world.
>
> (*I.D.* 1(14): 7, August 16, 1992)

Another letter to the editor (a response to letter above) was submitted by a local Dublin (male) representative to the Dáil (the Irish parliament). Using similar imagery which equates Ireland with moral purity and directly links Marian devotion to that purity, he insists:

> Together, let us storm Heaven with prayer to arrest and reverse the pagan-like onslaught on our Christian and Catholic heritage. . . . Unite in strength; and the Power of Prayer with Our Lady's intercession can break the bonds of Satan which are strangling Ireland.
>
> (*I.D.* 1(16): 6, August 30, 1992)

Both letters convey the impression that the moral boundaries between the Irish nation (as an entity defined by its particular brand of Catholic morality) and the rest of the E.U. are eroding. Described in terms of "erosion of family values" and increasing modernism, Ireland is perceived by traditionalists as in danger of becoming just like the rest of Europe (as a result of membership in the E.U.). The erosion of the nation's moral identity is linked to a disappearing and unique heritage. Both letters simultaneously appeal to Mary to help re-establish and protect the nation's boundaries, and both insist on Mary as a symbol of those boundaries.

In other items published in the *Irish Democrat* around the same time, references are made to messages delivered by Mary when she has appeared in apparition form to various Irish visionaries. These references, in which people often quote directly from messages believed to have been delivered by Mary to her Irish visionaries, serve to strengthen the connection between Mary and the struggle to preserve Irish cultural and moral heritage and, by paradigmatic extension, national boundaries. In an article about Mary's apparition appearance at Milleray Grotto in Ireland in 1985, for example, the author chooses to reproduce these particular parts of Mary's message:

> I love the Irish people . . . I am praying for the people of Ireland . . . Ireland will be saved . . . I want the Irish people to convey my message to the world.
>
> (*I.D.* 1(17): 11, September 6, 1992)

These words emphasize the special relationship between Mary and the Irish moral tradition which is constitutive of the Irish nation.[5] They also suggest that the Irish nation can be saved from the evils threatening other European nations and that Ireland, because of its unique moral tradition, should serve as a moral European leader – a place committed to representing certain values and to carrying those values to the rest of the world.

That Mary is mimetic of Ireland's post-colonial alterity in Europe is especially evident in appeals to apparition personas of Mary. Another September article in the *Irish Democrat* is about an apparition of Mary thought to have appeared at Maastricht, The Netherlands (where the Maastricht Treaty was originally drafted). This article appears in the midst of anxiety among conservative Catholics in Ireland over the successful ratification of the Maastricht Treaty. The writer (gender unknown) notes that:

> The story of Maastricht is one of persistent resistance against attack on the Faith. For centuries the people have had to struggle to keep Catholic belief alive. The existence of the shrine of Our Lady up to the present day, is witness to this. . . . It was always an inspiration to Dutch Catholics when their faith was in danger of being destroyed.
>
> (*I.D.* 1(20): 10, September 27, 1992)

The writer of this article sets up a direct correspondence between devotion to Mary and the preservation of traditional beliefs, especially in the face of religious persecution. Religious persecution in Ireland is historically linked to British colonial oppression there. Such correspondence between devotion to Mary and the preservation of religious heritage is dependent on a cultural logic wherein Mary is mimetic of the Irish traditional values embodied in the nation, meaning she operates as a symbol of Ireland's difference with respect to other European nations.

The same writer also intends that readers identify the historical struggle of Dutch Catholics with the danger to Irish faith and tradition in 1992. Note the content of the prayer included in Our Lady of Maastricht elsewhere in the article:

> We now implore you to come to the help of the Irish people who have joined the Maastricht Treaty, whom we put under your protection to steer their barque through the rough waters which may lie ahead. Pray

> that we may be able to hold on to our rich Irish heritage and traditions and to keep the priority of Christian principles before material gains.
>
> (Ibid.)

There is a great deal of anxiety in this prayer stemming from the general perception that Irish heritage and tradition (or the essence of the Irish nation) are threatened by the Maastricht Treaty. The prayer contains a request that the Virgin Mary help the Irish people retain their unique heritage despite having ratified the treaty. In the context of post-Maastricht conservative anxiety and religious fervor, both of which fueled the November 1992 anti-abortion campaign in Ireland, appeals to Mary acquired added significance, as they directly linked abortion and control over women's bodies to the preservation of the Irish nation.

Mary's appearance in apparition form to make comments on the political issue of abortion hints at manipulation of the mimetic relationship between women and Mary. In November, just before the anti-abortion referendum vote, an article appeared entitled: "WHAT OUR LADY SAYS ABOUT ABORTION" (*I.D.* 1(28): 12, November 22–28, 1992, capitalization in original). The article quotes from Mary's anti-abortion messages, culled from reports from different Marian visionaries around the world. The extended account of Mary's (then) recent appearance to Christina Gallagher, the Irish Marian visionary, illustrates particularly effectively the mimetic links extant among Mary, Irish women as mothers, and the moral tradition of the Irish nation:

> Mrs. Gallagher was upset when Our Lady appeared to her distressed and sobbing, and gave her the following message: "My child, pray with me for Ireland, that my children of Ireland will be spared. Pray with me for all my sons, priests, bishops, cardinals, and daughters, nuns. My child, my heart weeps blood for the abomination – the killing of the unborn." Mrs. Gallagher said Our Lady made it clear that she was pleading with the Irish people to remain steadfast in defense of unborn life and in upholding the law of God.
>
> (Ibid.)

As mentioned previously, Mary's appearances in apparition form are particularly powerful defining moments in that they help to clearly construct mimetic links between Mary and earthly women. In the excerpt from Mary's message quoted above, Mary appears as a suffering mother, distraught over the murder of her unborn children in locales outside of Ireland. Mary appeals to the Irish people that they maintain their difference with respect to the rest of the world by prohibiting abortion in Ireland. In this "ideal mimetic moment," Mary's appearance and message directly link the preservation of Irish alterity or difference with state control over women's reproductive labor and thus the disciplining of women's bodies as Mother-writ-large. Such mapping of tradition (or the Irish nation) onto women's bodies materializes as a result of the direct mimetic correspondence extant among Irish women, Mary and Irish national alterity.

Mimesis and materiality: the X case

We can find a potent example of how the mimetic relationship between women, Mary and the nation helps to materially construct appropriately feminine Irish bodies by traveling back to April 1992 and the X case. In early 1992, a few months before the Maastricht Treaty vote in June, an important rape case that resulted in a legal and ethical challenge to the Constitutional Amendment prohibiting abortion was dominating the popular media and political debate in Ireland (see Smyth 1996). Coined "The X Case" by the press (because the name of the victim was never released), this case involved a 14-year-old girl who had been raped by the father of a schoolmate and had become pregnant as a result. The trouble started for the officials investigating the case only when they learned that "Miss X" was planning to travel to England for an abortion, at which point the Attorney-General of Ireland filed in the courts to prohibit her from traveling. The agents of the state in this case reasoned that the Constitutional Amendment prohibiting Irish women from obtaining an abortion applied to these women wherever they might be.[6] Thus, if the state becomes aware that an Irish woman is traveling elsewhere for an abortion, she may be prohibited from traveling outside of Ireland. The X Case was tied up in the courts for some time until, in April, shortly before it would have been too late in her pregnancy for Miss X to obtain a legal abortion in England, the Irish High Court ruled that she could travel. The ruling came only after Miss X's attorneys convinced the court that she would commit suicide if she was denied permission to travel.

The X Case set the stage in Ireland for 1992's Maastricht Treaty debate and the Abortion Referendum that occurred later in November that same year. For many in Ireland, it seems, the death of Miss X's baby was somehow equivalent to the death of the Irish nation, or was at least a foreboding death knell. For our purposes the events surrounding the X Case provide a poignant example of "mimesis materialized." The loaded mimetic links extant among women, Mary, and the Irish nation comprise a discursive context. In this context it makes cultural sense to negotiate Ireland's place in the E.U. through seemingly unrelated internal political issues such as contraception, abortion and divorce. When the anti-abortion amendment to the Irish Constitution was applied to Miss X and her right to travel was restricted, the state acted to establish a coherence between the materiality of the nation (as embodied in state juridical content) and the materiality of Miss X's body. Indeed, for a few weeks in the eventful spring of 1992, the boundaries of Miss X's body and the boundaries of the Irish nation symbolically became one.

The X Case, and the subsequent High Court ruling, helped define the boundaries and terms for future political debate over changing moral traditions and the morally deteriorating effect of E.U. membership in Ireland. The X Case had a profound impact on the government's handling of the Maastricht Treaty Referendum in June 1992 and helped make necessary the November 1992 Abortion Referendum. Dealing with the issues raised by the X Case influenced the wording of the three separate anti-abortion amendments proposed for ratification and voted upon in November. These amendments dealt separately with the following issues: the right of Irish women to abortion in Ireland; the right of pregnant Irish women to travel; and the question of whether or not an abortion can be had in Ireland in the event that the mother's life is in danger (see Smyth 1996). Finally, the political debate over abortion and the European Union Treaty, as played out across the terrain of Miss X's body, shaped a developing conservative discourse that would later specifically link abortion with liberal calls for legalized divorce in Ireland in the mid-1990s. Irish conservative discourse in both the anti-abortion and anti-divorce referendum campaigns has repeatedly emphasized the mimetic links extant between the unique moral tradition embodied in the Irish nation and the bodies of Irish women.

Recent developments in Irish conservative discourse

In 1995 the main issue dominating the political scene in the Republic of Ireland was a national referendum scheduled for November of that year designed to decide whether or not Irish people were, for the first time in the history of the Republic, to have limited access to divorce. Conservative discourse deployed in the 1995 anti-divorce campaign had much in common with that found in the 1992 anti-abortion campaign, although a number of themes were more clearly and more substantially developed. The passages reviewed above from the *Irish Democrat* in 1992 contain the seeds of the rhetorical terms within which the conservative anti-divorce campaign would later be couched. However, in the conservative discourse of 1995 the links between violent imagery, moralizing, and the threats of modernism or change to the Irish nation's radical alterity are more clear.

Content analysis of a number of conservative pamphlets and newsletters published in 1994 and 1995 before the November 1995 divorce referendum reveals several common themes.[7] In 1994 a pamphlet was published by the Donegal Pro-Life Campaign containing a printed version of an address given by Fr. Denis Faul in early 1994, *The Death of a Nation*. In this document, Fr. Faul discusses the dangers of both abortion and divorce, conflating the two in terms of their impact on Ireland and its unique heritage and way of life. Fr. Faul is able to present divorce and abortion as basically the same by referencing their common threat to something he calls "fertility." For him, high fertility rates, especially in rural Ireland, equal a healthy sense of community and family, and a healthy nation. Both divorce and abortion threaten fertility and reproduction, in that they threaten to sever what he and many others see as necessary links between love, sex and the production of children. Moreover, Fr. Faul explicitly states that the separation of sexual intercourse from reproduction in Ireland represents *the death of the nation.* His stance is a direct extension of Irish Catholic morality – that morality which has historically been embodied in the Irish nation. Contraception, artificial insemination and other "experimentation" on "the reproductive system" are represented by Fr. Faul as part of an overall "culture of death." Interestingly, "the reproductive system" that he refers to is defined by context as exclusively female. Fr. Faul's notion of "fertility" reflects more than a concern with demographic decline, or the death of the Irish nation due to declining birthrate. Rather, "fertility" for him is equally about culturally appropriate reproduction and the continued flowering of Irish Catholic morality.

According to Fr. Faul, the opposite of fertility is disease and deterioration. Indeed, he asserts that decreasing fertility rates indicate that "Ireland has a fatal disease" and that "sense of community is withering." Fr. Faul also asserts that attempts to introduce divorce and abortion into Ireland are linked "with death, with killing persons, with killing personal relationships" in that they threaten culturally appropriate reproduction and thus Catholic fertility. Significantly, in his view, Ireland contracted the disease of "liberalism" (which is also sometimes called "modernism" in these publications) from Europe. Note that it is not the Republic of Ireland that Fr. Faul sees as being at risk, but the Irish nation – a nation being killed by state juridical changes introduced into Ireland via E.U. political imperatives.

Another publication, a newsletter called *Solidarity News*, which began to appear in early 1995 in response to the impending national Divorce Referendum, provides further evidence of the construction of the Irish nation's moral alterity through conservative discourse. This newsletter served as a mouthpiece for the conservative organization Family Solidarity and its leader, Nora Bennis. The articles and commentary included in *Solidarity News* contain arguments against divorce and abortion which are very similar in their nationalistic orientation to those provided by Fr. Faul. In her editorial in the July/August 1995 issue of the newsletter, for example, Bennis argues:

> the family based on father, mother, and children in the indissoluble bond of marriage, is the cornerstone of our society. The values and standards of a nation stem back to the family and the home. If family values are strong, the nation will be strong.

Rather than talk about threats to the Irish way of life (family, community and nation) as disease and deterioration, however, Bennis constantly uses military metaphors to describe and justify the conservative stances against political change in Ireland, using terms like "war," "enemy," "attack," "battle", "rearguard action," "fighting," "combat" and "front" throughout her publication. Irish women and the female body are particularly targeted as strategic to the conservative battle to preserve the Irish nation and its moral alterity with respect to Europe. Abortion, an issue relating specifically to women and reproduction, is defined by Bennis as violence and abuse against women, as rape, and as the murder of women. Yet, her use of military metaphors as a framework for her argument against abortion serves to move the issue of abortion away from the arena of individual conflict over a woman's right to choose. Bennis's use of military metaphors instead displaces the issue of abortion into the arena of national conflict. In the latter arena, individual choice is irrelevant to any national struggle to preserve the Irish nation.

For Bennis and other conservative writers, Europe (and sometimes "radical feminists" influenced by European ideals) is clearly the enemy in the battle to preserve the Irish nation. Another conservative pamphlet published in 1995 by Fr. Peter Byrne called *Don't Blow Out the Candle* further illustrates this point. In it he argues for the preservation of Ireland's unique moral heritage in the face of recent, rapid political and economic change. Fr. Byrne writes vividly about the ills of Europe and the temptation of a European way of life:

> Europe is lost on the mountain of her own wealth. . . . She stumbles in the darkness and like an octopus reaches out her tentacles to crush the life and quench the light of those who oppose her liberal murdering ways. . . . **Ireland is the light to the blind, stumbling tramp that is Europe**.
>
> (Original emphasis)

Frequently using violent terms such as "violation," "maim," "murder" and "destruction," Fr. Byrne asserts that Ireland is under attack from within by groups linked to mainland Europe. Like the other publications reviewed above, he links divorce and abortion to the destruction of the family, the destruction of the nation, and the loss of a unique Irish moral heritage. Sexual activity outside of marriage, sex education in schools, and even health programmes are labeled by Byrne as "the first steps towards abortion and corruption of family values." A prolific writer of widely distributed devotional and political pamphlets, Fr. Byrne's influence on Irish conservative discourse has been far-reaching. He has published at least 21 booklets which can be found in churches, devotional shops and petrol stations throughout Ireland, including titles such as *Woman the Beautiful* and *The High Noon of God*.

The categories of discourse deployed in the political struggle against divorce and abortion in Ireland fall under four main headings: Seduction, Disease, Violence and Alterity. All of these ways of thinking and speaking about divorce and abortion are linked together, used interchangeably, and even embedded within one another. The rhetoric of Seduction, for example, involves the use of sexualized terms, such as "tramp," commentary on sexual activity, sex education and reproduction, and the linking together of notions of disease, perversion and promiscuity in ways that conflate threats to the Irish nation with changes in sexual morality.

The rhetoric of Disease is difficult to separate from that of Violence and Seduction. In it disease is seen as a form of violence that results in the destruction of the family, in the burning down of the proto-typical Irish home and in rape (in the form of abortion). The rhetoric of Violence is most prevalent in Irish conservative discourse, as military metaphors and references to violation and destruction abound throughout conservative publications. The rhetoric of Violence is particularly effective at establishing discursive links between issues like abortion and divorce and threats to the Irish nation. As illustrated above, the use of military metaphors in particular helps to move the issues of abortion and divorce away from an arena of personal choice and into one of national conflict. As a result, issues of individual choice, which could be seen as matters of personal conscience, become matters of national concern. Rather than being threats to individual Catholic moral integrity, abortion and divorce become matters which threaten the moral integrity of the Irish nation. In the rhetoric of Violence, such threats to the moral fabric of the nation by their very nature come from the outside, from that which is not Ireland; from the enemy – that "blind and stumbling tramp" that is Europe.

The Irish nation's alterity, its cultural and moral purity, are further delineated by particular kinds of Irish bodies. The publications reviewed here constantly foreground the necessity of preserving traditional links between family, community and nation. These links, in their historically specific Irish Catholic form, *are* Irish alterity. Considering the historical significance of women and the female body in representations of the Irish nation, and the prevalence of metonymic links among women, the home, the family, the Virgin Mary and the nation in historical and contemporary Irish discourse, it is not surprising that the female body should serve as the material terrain upon and through which Ireland's place in contemporary Europe is negotiated and contested. We come back, yet again, to the material implications of the Irish gendered division of mimetic labor for Irish women.

The Virgin Mary, with her mimetic links to both Irish women and the nation, also remains present in the recent conservative discourse described above. Images of Mary appear either on the cover or inside each of the three publications discussed. On the cover of *Solidarity News*, for example, a large picture of Mary is set beneath a statement of Family Solidarity's objectives.[8] This image depicts the Virgin Mary as crying, holding in her hands a fetus, most likely aborted. Here, as is often the case, women (elsewhere the family, or the nation) are spoken or written about, but replaced visually by the Virgin mother of Christ. Fetuses are also depicted more and more often without any visual reference to a womb or to the mother's body. As Michelle McCaffery has noted, the erasure of real-life female bodies from the picture "ultimately depends on the erosion of female subjectivity, and material female bodies – a concept so skillfully employed by pro-life movements" (1996: 11). As much of the literature reviewed above equates abortion with the death of the Irish nation, it is not too difficult to imagine how the aborted fetus that Mary is cradling in her hands might be read by some people as representative of the Irish nation itself – a nation potentially destroyed by the womb-violating acts of amoral Irish women, such as Miss X.

Conservative discourse in late-twentieth-century Ireland, such as that contained in the *Irish Democrat* and the other publications reviewed above, is not simply the product of a radical, far right fringe element in Irish society. Instead, this discourse has significantly influenced the terms upon which important political and economic changes, such as ratification of the Maastricht Treaty, have been debated on a national scale. Such discourse has had a very real influence on the government's wording of proposed constitutional referenda (see Smyth 1996), on voting patterns, and even on the wording of international treaties (such as clauses attached to the Irish version of the Maastricht Treaty, see p. 82), not to mention on how women experience their place in Irish society and even their own bodies.

Months before the 1995 Divorce Referendum, the media, and those most active in publicly debating the issue, revisited the X Case, asserting that dozens of Miss X's would be running around after the referendum if people (women?) decided to challenge the wording of the amendment in Ireland's courts, or possibly even in the European Court (the latter being the worst case scenario). It became virtually impossible for anyone in Ireland to consider legal access to divorce in 1995 outside of the discursive terms established around the abortion debate in 1992. Both sides in the divorce debate used manipulative campaign scare tactics to highlight women's possible abuse, impoverishment and desertion at the hands of men (and politicians). The most cogent campaign artifact denoting the links between divorce and abortion in conservative discourse was carried by people once again marching en masse in Dublin in 1995 to protest against divorce. Many of these people carried placards stating: "DIVORCE ABORTS MARRIAGE." Considering the discursive links extant among women, the family and the nation of Ireland, many people received the message that divorce would also result in the death of the Irish nation.

Conclusions: discourse, mimesis and nationalist violence

In much of the analysis presented in this chapter, I have tried to show that there are mimetic links among women, the Virgin Mary, and the nation in both historical and contemporary Ireland, and that mimesis has a very real, material impact on the construction of feminine bodies in that country. Conservative Irish discourse, and even liberal and far left political discourses (which are often structured in response to conservative discourse), construct these mimetic links through language and symbolizations which, in turn, influence the juridical structure of the Irish state. When the state and the nation are paradigmatic of each other, as they have historically been in Ireland, the nation is actually manifested itself at the site of the body via the workings of state juridical structures. When women in particular are mimetic of the nation, their bodies are disciplined to correspond to an ideal of femininity embodied in the nation through the application of state laws. Such disciplining could be defined as violent when it violates the civil rights of female citizens or threatens their health. I believe that this violence, when it operates in the interests of the nation, is one among many forms of Irish nationalist violence. Nationalist violence in the Republic of Ireland is therefore not only about policing, terrorism, or war; it is also about forcibly and materially shaping certain bodies in ways that limit the freedom and that threaten the health and overall quality of life of those bodies.

Begoña Aretxaga (1995) and Alan Feldman (1991) are anthropologists who have studied nationalism and violence in Northern Ireland. They both identify mimetic processes as inherent to the production of a culture of terror in the North. Working from Foucault's vision of the body as political institution (see Foucault 1979), Feldman elaborates the endocolonization of society by the state (1991: 86). The term "endocolonization" does not refer to the colonial status of different states or peoples, rather it refers to the colonization of bodies by the state and suggests such colonization is metaphorically equivalent to colonizer/colonized relations of power. Endocolonization thus involves the penetration of bodies by state apparatuses of power. The process of endocolonization rests upon mimesis, as it results in the production of specified bodies that paradigmatically mimic or embody the state. When the state and the nation are paradigmatically aligned, as they have historically been in Ireland, endocolonization also results in the production of specified bodies that are mimetic of or embody the nation. Endocolonization is different, however, from the more general concept of mimesis, as it specifically refers to the discursive regulation of bodies by state institutions. Bodies are thus colonized by the state juridical apparatus and disciplined, via state laws and the carceral system, to where they are manifested in limited configurations.

I believe that parallels in Ireland between the situation of violence in the North and nationalist violence in the Republic can be found by application of Feldman's treatment of endocolonization and torture to the situation of the X Case and the Maastricht Treaty. As Feldman writes:

> The performance of torture does not apply power; rather it manufactures it from the "raw" ingredient of the captive's body. The surface of the body is the stage where the state is made to appear as an effective material force.
>
> (Feldman 1991: 155, emphasis removed)

Thinking of Miss X, a parallel scenario comes to mind. The power of the state was manufactured from the raw ingredients of Miss X's body, and not merely the surface of her body, but her womb as well. Miss X's body was a female body violated, impregnated and incarcerated within the state.[9] True, the state was only clearly active in the last instance, the restriction of Miss X's right to travel, but for our purposes the last instance is the most important and the most profound. Blocking Miss X's travel allowed the state to manifest itself in the physical, corporeal transformation of Miss X's continuing pregnancy. The end product was a pregnant female body which corresponded mimetically to the ideal image of the Irish nation. In other words, nationalist violence, when applied to Miss X's body through state juridical structures, resulted in the mimetic embodiment of the Irish nation in the form of Miss X.

As mentioned above, the death of Miss X's baby also seemed to forebode the death of the Irish nation for many people (as evidenced by the title of Fr. Faul's pamphlet). Thus, the X Case radically affected the Maastricht Treaty referendum debate which followed. As the transnational European Union Treaty, Maastricht was to further dissolve state boundaries and Irish national identity. Not surprisingly, in Ireland Maastricht was debated almost exclusively on the basis of the perceived impact it would have on the "Right to Life Amendment" to the Irish Constitution. Each E.U. country considering ratification of Maastricht had the right to attach special "Protocols" or protection clauses to the text of the Treaty before approving it. In Ireland, the *only* Protocol attached to Maastricht was one protecting the Eighth Amendment. As Albert Reynolds, then An Taoiseach (Prime Minister), wrote in the government pamphlet introducing the terms of the Treaty to the people:

> the Maastricht Treaty is *not* about introducing abortion into Ireland. To emphasise the fact that it would not interfere with our Constitutional provisions in that area, the Government negotiated last November [in 1991] a special Protocol – a protection clause – confirming that the Treaty would have no effect on the application in Ireland of the right to life article in our Constitution.
>
> (*A Short Guide to the Maastricht Treaty* 1992: 2, original emphasis)

What this statement indicates is that through a trick of mimesis, the state attempted to assuage people's fears about the dissolving boundaries of their Republic by assuring them that, regardless of that dissolution, the boundaries of female bodies in Ireland would remain intact, and, thus, so would the boundaries of the Irish nation.

In transnationalist Ireland, the endocolonization of Miss X's body by the state and the mimetic violence that this scenario implies, comprise efforts to preserve and constantly recreate a unique Irish nation somehow immune to the processes of deterritorialization which all E.U. countries are experiencing in the late twentieth century. It is important to note that most Irish women do not experience the mimetic relationship between themselves, the Virgin Mary and the nation in such a dramatic and public way. But this does not mean that the gendered division of mimetic labor in Irish society has any less of a profound or enduring influence on the way "typical" Irish women experience their bodies, or on the everyday discursive regulation and construction of feminine corporealities. One of the implications of the analysis presented in this chapter is that as states become more deterritorialized under the influence of transnationalism (and as nationalist identity becomes more threatened), the endocolonization of certain bodies becomes more intense, especially in those states where a gendered division of mimetic labor has historically existed in representations of the nation. In the Irish case, deterritorialization of the state has led to a violent, coercive, and intrusive re-territorialization of feminine bodies by the Irish nation.

Acknowledgments

I thank the following for their invaluable comments on several different incarnations of this chapter: Mary Anglin, Christine Gailey, Timi Mayer and Heidi Nast.

Notes

1 Although the Maastricht Treaty is a treaty between numerous *states* rather than *nations*, many people in Ireland see it as directly affecting the status of the Irish nation, not just the structure of the state. Typically in the literature on nations and nationalisms "the state" and "the nation" are treated as separate, distinguishable entities. For example, Gellner defines the state as an

> institution or set of institutions specifically concerned with the enforcement of order (whatever else they may also be concerned with). The state exists where specialized order-enforcing agencies, such as police forces and courts, have separated out the rest of social life. They *are* the state.
>
> (Gellner 1983: 4, original emphasis)

But the nation is theorized by Gellner as separate from state institutions or structure, as many nations have "emerged without the blessings of their own state" (Gellner 1983: 6). Benedict Anderson further reinforced the state–nation dichotomy when he defined the nation as an "imagined political entity" (Anderson 1983: 6) having little to do with material structures or institutions.

I do not acknowledge the state and nation in Ireland as separate, distinct entities. In Ireland the state, as a series of disciplinary institutions and practices, is discursively linked to the nation. Since many of the gendered and sexualized meanings and practices through which the nation is manifested are themselves institutionalized in state laws and the Irish Constitution, principally as a result of nationalist political imperatives (see Martin 1997), the nation and the state are paradigmatic of each other. The "Nation" is both a real and imagined entity. It is a collection of representations structured by, and structuring, a series of discursive practices that materially influence the way the citizens of the nation experience themselves, each other, and other "Others."

2 This chapter expands upon analyses previously presented in two different conference papers: "Gender and nationalist violence in the Republic of Ireland," a paper presented at the American Anthropological Association Annual Meeting in Washington, D.C., November 15–19, 1995, in the invited session Surviving Gendered Violence; and "The Virgin Mary and the gendered division of mimetic labor in contemporary Ireland," a paper presented at the American Association of Geographers' Annual Meeting, March 14–18, 1995, in the invited session Postcolonialism, Gender and Identity.

The research is based on twenty months of master's thesis and dissertation fieldwork in western Ireland between 1992 and 1996, the longest trip lasting fourteen months in 1995–96. The research was supported by a National Science Foundation Ethnographic Research Training Grant awarded through the University of Kentucky Anthropology Department in 1993, a Small Grant from the Wenner Gren Foundation for Anthropological Research awarded in 1995, and a National Science Foundation Dissertation Improvement Grant awarded in 1996.

3 Taussig uses the wooden figurines that are a central part of Cuna healing rituals, figures carved to resemble Westerners but imbued with Cuna spirits, as a point of departure in his study of mimesis and colonial alterity.

4 The items from the *Irish Democrat* considered in this analysis fall chronologically between the ratification of the Maastricht Treaty in Ireland (June 18, 1992) and the subsequent abortion referendum in the country (end of November, 1992), which was designed to strengthen and clarify the "right to life" amendment (section 40.3.3) of the Irish Constitution.

5 They also echo the words of many of the people I interviewed about Mary in the West of Ireland (see Martin 1993a). People often called Mary the Mother of Ireland or the Mother of Us All, and linked her historic importance in Ireland directly to the rosary, which they believed to be the primary devotional practice responsible for the preservation of Irish Catholicism during the years of the anti-Catholic Penal Laws in the seventeenth and eighteenth centuries. In actuality, the rosary's role in this preservation is a fiction, as the rosary was not introduced into Ireland until the nineteenth century (see O'Dwyer 1988).

6 The Eighth Amendment to the Irish Constitution, known as the "right to life amendment," prohibited abortion in the Republic, unless the mother's life was in danger. It was ratified by public referendum in 1986.

7 Although I review only three publications here, pamphlets and newsletters such as these can be found in numerous churches, bookshops and even at the cashier counters of petrol stations throughout Ireland.

8 The following is the statement of Family Solidarity's objectives which appears above the described picture of Mary:

> Solidarity is a grassroots movement of individuals, prayer groups, pro-life, pro-family groups whose core is the restoration of Christian family values in Ireland. Solidarity is unashamedly rooted in Christ and powered by prayer. Members can either commit to prayer only or prayer and action.

9 There is no conflict here between resistance to violation of the Irish nation extant within conservative Irish discourse and the violation of Miss X's body by state apparatuses of power. The latter is constructed by conservatives as necessary to the preservation of the Irish nation. Irish female bodies can be violated as long as the context is appropriate.

References cited

Anderson, B. (1983) *Imagined Communities: Reflections on the Origin and Spread of Nationalism*, London: Verso.

Ap Hywel, E. (1991) "Elise and the great queens of Ireland: 'femininity' as constructed by Sinn Fein and the Abbey Theatre, 1901–1907," in T. O'Brien Johnson and D. Cairns (eds.) *Gender in Irish Writing*, Milton Keynes: Open University Press, pp. 23–39.

Appadurai, A. (1991) "Global ethnoscapes: notes and queries for a transnational anthropology," in R. G. Fox (ed.) *Recapturing Anthropology, Working in the Present*, Santa Fe, N.M.: School for American Research Press, pp. 91–210.

Aretxaga, B. (1995) "Dirty protest, symbolic overdetermination and gender in Northern Ireland ethnic violence," *Ethos* 23, 2: 123–140.

Boddy, J. (1989) *Wombs and Alien Spirits: Women, Men and the Zar Cult in Northern Sudan*, Madison, WI: University of Wisconsin Press.

—— (1993) "Aesthetics, politics, and women's health in Northern Sudan and beyond," a paper presented at the Annual Meeting of the American Anthropological Association.

Bordo, S. (1993) *Unbearable Weight: Feminism, Western Culture, and the Body*, Berkeley: University of California Press.

Cairns, D. and S. Richards (1987) "'WOMAN' in the discourse of celticism: a reading of 'The Shadow of the Glen'," *Canadian Journal of Irish Studies* 13, 1: 43–60.

Campbell, E. (1982) "The Virgin of Guadalupe and female self-image: a Mexican case history," in J. L. Preston (ed.) *Mother Worship*, Chapel Hill, N.C.: University of North Carolina Press, pp. 5–24.

Chavez, L. (1994) "The power of the imagined community: the settlement of undocumented Mexicans and Central Americans in the United States," *American Anthropologist* 96, 1: 52–73.

Collier, J. (1986) "From Mary to modern woman: the material basis of marianismo and its transformation in a Spanish village," *American Ethnologist* 13, 1: 100–107.

Cullingford, E. B. (1989) "'Thinking of her . . . as . . . Ireland': Yeats, Pearse and Heaney", *Textual Practice* 4, 1: 1–21.

Feldman, A. (1991) *Formations of Violence: The Narrative of the Body and Political Terror in Northern Ireland*, Chicago: University of Chicago Press.

Foucault, M. (1979) *Discipline and Punish: The Birth of the Prison*, New York: Vintage.

Gellner, E. (1983) *Nations and Nationalism*, Ithaca, N.Y.: Cornell University Press.

Girvin, B. (1993) "Social change and political culture in the Republic of Ireland," *Parliamentary Affairs* 46, 3: 380–398.

Goldring, M. (1993) *Pleasant the Scholar's Life: Irish Intellectuals and the Construction of the Nation State*, London: Serif.

Gupta, A. (1992) "The song of the nonaligned world: transnational identities and the reinscription of space in late capitalism," *Cultural Anthropology* 7, 1: 63–79.

Irish Government (1992) *A Short Guide to the Maastricht Treaty*, Dublin.

Jameson, F. (1984) "Postmodernism, or the cultural logic of late capitalism," *New Left Review* 146: 53–93.

Kearney, M. (1991) "Borders and boundaries of state and self at the end of empire," *Journal of Historical Sociology* 4, 1: 52–74.

Kearney, R. (1986) "Myth and motherland," in Field Day Theatre Company (ed.) *Ireland's Field Day*, Notre Dame: University of Notre Dame Press, pp. 51–80.

Larkin, E. (1984) *Historical Dimensions of Irish Catholicism*, Washington, D.C.: Catholic University Press of America.

Lloyd, D. (1993) *Anomalous States, Irish Writing and the Post-Colonial Moment*, Durham, N.C.: Duke University Press.

Loftus, B. (1990) *Mirrors: William III & Mother Ireland*, Dundrum, Co. Down: Picture Press.

McCaffery, M. (1996) "Foetal subjectivity and the rise of paternal rights," *Irish Journal of Feminist Studies* 1, 1: 11.

MacCurtain, M. (1993) "The role of religion in Ireland: the historical dimension," *Social Compass* 40, 1: 7–13.

Martin, A. (1993a) "The Virgin Mary: gender, religion and politics in contemporary Ireland," unpublished master's thesis, University of Kentucky.

—— (1993b) "Gender and religious symbolism: the Virgin Mary and ritual space in the Republic of Ireland," *Southern Anthropologist* 20, 3: 23–33.

—— (1997) "The practice of identity and an Irish sense of place," *Gender, Place and Culture* 4, 1: 89–119.

Martin, A. and S. Kryst (1998) "Encountering Mary: ritualization and place contagion in postmodernity," in H. Nast and S. Pile (eds.) *Places Through the Body*, New York: Routledge.

Mullin, M. (1991) "Representations of history, Irish feminism, and the politics of difference," *Feminist Studies* 17, 1: 29–50.

Nash, C. (1993) "Remapping and renaming: new cartographies of identity, gender and landscape in Ireland," *Feminist Review* 44, summer: 39–57.

—— (1994) "Remapping the body/land: new cartographies of identity, gender and landscape in Ireland," in A. Blunt and G. Rose (eds.) *Writing Women and Space, Colonial and Postcolonial Geographies*, New York: Guilford Press, pp. 227–250.

O'Dwyer, P. (1988) *Mary. A History of Devotion in Ireland*, Dublin: Four Courts Press.

Orsi, R. (1985) *The Madonna of 115th Street, Faith and Community in Italian Harlem, 1880–1950*, New Haven, CT and London: Yale University Press.

Petrisko, T. (1995) *The Sorrow, the Sacrifice and the Triumph: The Apparitions, Visions and Prophecies of Christina Gallagher*, privately published in Ireland.

Rodriguez, J. (1994) *Our Lady of Guadalupe: Faith and Empowerment Among Mexican-American Women*, Austin: University of Texas Press.

Rouse, R. (1991) "Mexican migration and the social space of postmodernism," *Diaspora: A Journal of Transnational Studies* 1, spring: 8–23.

Smyth, A. (1996) "'And nobody was any the wiser,' Irish abortion rights and the European Union," in R. A. Elman (ed.) *Sexual Politics and the European Union: The New Feminist Challenge*, Oxford: Berghahn, pp. 104–130.

Taussig, M. (1993) *Mimesis and Alterity: A Particular History of the Senses*, New York: Routledge.

Warner, M. (1976) *Alone of All Her Sex: The Myth and the Cult of the Virgin Mary*, New York: Knopf.

Former Yugoslavia

4 Sexing the nation/desexing the body

Politics of national identity in the former Yugoslavia

Julie Mostov

Gender and nation are social constructions which intimately participate in the formation of one another: nations are gendered, and the topography of the nation is mapped in gendered terms (feminized soil, landscapes and boundaries, and masculine movement over these spaces). National mythologies draw on traditional gender roles, and nationalist discourse is filled with images of the nation as mother, wife and maiden. Efforts at nation-building which seek to "recover" the unique character and purity of the nation and celebrate its ancient roots and historical continuity generally describe the nation as timeless and changeless, as a "natural" set of bonds binding people to one another (Mostov 1995b). Thus, "the nation" naturalizes constructions of masculinity and femininity: women physically reproduce the nation, and men protect and avenge it. At the same time, this notion of nation collectivizes and neutralizes the sexuality of female (and, to some extent, male) members of the nation.

This chapter is an exploration of the ways in which a particular politics of national identity seeks to eroticize the nation in sensual and spiritual imagery while, at the same time, repressing the sexuality of actual women and men. I situate this exploration of nation, gender and sexuality in the former Yugoslavia and in the context of a politics of national identity which I call ethnocracy (Mostov 1996b). Ethnocracy is a particular type of rule in which power is concentrated in the hands of leaders successful in promoting themselves as uniquely qualified to define and defend the (ethno)national interests, and in which the ruled are collective bodies defined by common culture, history, religion, myths and presumed descent. The strategies and practices of ethnonational leaders can be seen as reconfigurations of power relationships, aiming to transform the social and political landscapes. A crucial aspect of these reconfigurations is the reduction of political subjects and political space (Mostov 1994).

In the case of the former Yugoslavia (and much of post-communist Central and Eastern Europe), the collective entity class is replaced by the ethnonation, and a new hierarchy of national guardians emerges. These would-be rulers redraw or create territorial and symbolic national boundaries, and set the penalties for transgressing these borders. They exaggerate the differences between those on either side of the boundaries and celebrate the common identity among those within, demonizing the ethnic or national Other and denying individual difference among their "own." Eroticism and sexuality play important roles in this politics of national identity, as I hope to show in this chapter. For ethnocratic strategies in the former Yugoslavia fuse eroticism of the nation with sexually repressive gender roles and patriarchal culture. I argue that these strategies seek, on one hand, to sensualize the broad idea of the nation and, on the other hand, to desensualize or "desex" its particular living parts.

In the first section of the chapter, I outline the links among gender, nation and sexuality. I draw on my own work in the former Yugoslavia and that of feminist theorists. This is not an empirical study, but, instead, an attempt to develop an idea about the politics of national identity which has roots in one context but resonates in others.

In the second section, I suggest that in this politics of national identity one can see the linking of two processes: one, which sensualizes the nation and national imagery; and another which desensualizes individual members of the nation. On the one hand, the metaphors and images of national glory and heroic struggle create a kind of eroticism, which is highly symbolic, abstract and seductive. On the other hand, traditional national spokesmen (clergy and national elites) espouse a politics of reproduction and demographic control, backed by de-eroticized images of chaste and faithful wives, patriotic mothers and pure, wholesome heroes. These messages coincide with the use of explicit sexual images or metaphors to enhance military recruitment strategies and negative propaganda about the enemy Other.

In the third and fourth sections, then, I suggest how the strategy of sexing the nation as a whole while desexing individual members of it serves would-be ethnocrats well by supporting social and political coercion and undermining individual agency.

Gender and nation

Gender and, in particular, proper gender roles become symbolic boundaries of the nation in the nationalist discourse of would-be ethnocrats in the former Yugoslavia.[1] Women's bodies actually become boundaries of the nation (Mostov 1995a), for not only are they symbols of the fecundity of the nation and vessels for the nation's reproduction, but also they serve as territorial markers. Mothers, wives and daughters designate the space of the nation and are, at the same time, the property of the nation. As markers and as property, mothers, daughters and wives require in turn the defense and protection of patriotic sons.

The gendering of boundaries and spaces (landscapes, farmlands and battlefields) in the former Yugoslavia makes possible the use of the sexual imagery of courtship, seduction and violation. The nation is adored and adorned, made strong and bountiful or raped and defiled, its limbs torn apart, its womb invaded. Feminine spaces are caressed and nurtured or occupied and trampled by masculine actors. Feminine spaces remain open to invasion – and this image of vulnerability is particularly inviting to ethnocrats or those engaged in crafting nationalist rhetoric and expanding national boundaries or in waging war on behalf of the nation. The vulnerability and seductiveness of women/borders (space/nation) require the vigilance of protectors or border guards. Thus, just as the territory of the nation must be protected by male soldiers and national leaders, women's bodies must be protected by fathers, husbands and the (national) state.

The need to protect women inevitably comes to include, as well, the need to monitor women's actions. As mothers, women are reproducers of the nation. In this role, women are heroines and symbols of virtue, fertility, strength and continuity. Conversely, women who refuse to have children or who have children with members of other nations become potential enemies of the nation, traitors to it, collaborators in its death. Women also remain vulnerable to invasion and defilement: as symbols of the nation and potential mothers, they could become objects of the ethnic/national Other's desire and vessels for his offspring. The enemy status of the Other women lies in their potential as reproducers, to multiply the number of outsiders, to conspire to dilute and destroy "our" nation with their numerous offspring. It follows, then, that while "our" women are to be revered as mothers, all women's bodies must be controlled.

Variations of struggles for power by new or would-be guardians of the nation are, according to this model, played out over the feminine body: over the feminine space of the nation – battlefields, farmlands and homes – and over actual female bodies. Moreover, claims to territory and sovereignty based on the numerical strength of the dominant (majority) nation render demographic and reproductive policies high priorities of the state. Women of the majority nation must be encouraged to reproduce, and to do so only with members of their own national group. Women of the minority (from the point of view of the majority) nation(s) should be discouraged from reproducing. These policy priorities are reinforced on the symbolic level by metaphoric figurations of the nation as mother (homeland/Motherland) and as female body. These variations are reinforced by sexual imagery shaped by stereotypical gender roles. The feminine is passive, the masculine is active. The Motherland provides a receptive and vulnerable image in contrast to the active image of the Fatherland, which is the force behind government and military action – invasion, conquest and defense.

This gendered national imagery recognizes women as a symbolic collective. "The nation as mother" produces an image of the allegorical mother whose offspring belong to the entire country's guardians, heroes and martyrs. Individual mothers are celebrated primarily as instances of this collectivized image. Their pain, suffering and sacrifices are recognized only as part of the nation's sacrifice; their individual plights are relevant only to this extent. Women as reproducers are recognized either as members of the majority or minority nations, as members either of the collective "our" women or of "their" women. And ultimately, the rape and violation of individual women becomes significant in nationalist discourse and the politics of national identity primarily as a violation of the nation and an act against the collective men of the enemy nation (Mostov 1995a, Kesić 1994, Zajović 1994). Only framed as the plight of "our" women does individual women's suffering threaten or offend the nation.

Sexual fantasies also follow this collectivizing of "our" women and of "their" women: the enemy male is figured as threatening to invade the national space and abduct "our" women, to steal our identity, to dilute "our" culture. Each side fantasizes about invading the space of the other, robbing the alien society and installing its own culture (Mostov 1995a: 517, Salecl 1992). The Other's men are seen collectively as sexual aggressors, and "our" women become the object of male temptation. "Their" women are forbidden prizes, and as such, a potential site for warfare.

Seen this "ethnocratized" way, the sexuality of individual women presents a potential threat to the nation, as an "entry" point for invasion. Moreover, individual women become potential suspects in border transgressions. At the same time, it is the collective – "our" women – that represents the national tragedy, that gains the sympathy of the people and in whose name the nation suffers.

Eroticizing the nation

The nationalist imagery supporting the power struggles of various competing national leaders in Serbia and Croatia in their preparations for war, appeals to the manhood of potential recruits by incorporating traditional gender roles for "real" men and erotic messages about a feminized nation. The sensualizing of the nation serves as a seductive displacement of erotic energy and, thus, as both an effective recruitment strategy and mechanism of sexual repression. The latter is manifested in messages that discourage unregulated sexual activity and limit the proper behavior of "our" women to motherhood and defense of national culture and values. Sexual repression facilitates the demographic policies of ethnocrats and reiterates gender divisions which, in turn, support the hierarchical and patriarchal structure of ethnocracy.[2]

Sensualization of the nation, articulated first in literary essays and historical studies by increasingly emboldened "national elites" in Serbia in the mid-1980s, and later in Croatia, involves evocation of heroic figures from the distant past, stories of suffering and sacrifice, and visions of glory by a defiant nation. In verse, song and prose the respective national literary elites spin tales of betrayal and tragedy, unique gifts and messianic roles (Karabeg 1997, Gojković 1996, Mostov 1996b, Radić 1996, Mostov 1995b, Žanić 1995, Zirojević 1995, Tašić 1994). Soon would-be ethnocrats pick up these motifs and embellish them with "facts" supplied by national historians about atrocities committed against "their" nation and promises of triumph, unity and freedom from foreign oppression. These national spokesmen move easily between the everyday world of politics and the otherworldly spheres of heroes, sacred spaces, martyrs and traitors – naturalizing the patriarchal values expressed in their romantic vision of the nation. Their messages are carried to the public in the late 1980s and early 1990s over the state television and radio, in popular dailies and at public meetings in Serbia and Croatia.[3] The national space (Motherland) must be protected by new heroes, willing to join in the nation's age-old battle against the forces of evil.

This romantic image of the national guardian revives the masculine roles of traditional patriarchal society. The manliness or virility associated with the guardians and heroes of the nation is tied to the battlefield and sexual abstinence.[4] The male actors leave the women behind and display their virility through bravery. It is in the company of other men that they demonstrate their manliness and potency. The erotic nature of these warrior/guardians is captured in their moments of sacrifice and death.

In his work on contemporary war propaganda in the former Yugoslavia, ethnographer Ivan Čolović identifies four images of masculinity associated with national warriors in Serbia: fearless warriors; warriors with "maiden's souls"; sons of the Motherland; and, finally, dead warriors.[5] I draw upon these four images here to illustrate the redirection of sexual passion and sensuality on the symbolic plane toward service to the nation and the de-eroticism of the physical body.

A dominant characteristic of the national warrior expressed in contemporary war propaganda in Serbia and Croatia is his virility. He is brave, ready to expose himself to the dangers of warfare and tests of his manhood. Initiation into the world of real men comes with readiness to demonstrate this bravery to comrades on the battlefield. "Let those who are men enough, come with us," shouts an officer of the J.N.A. (Yugoslav/Serbian Army), in front of television cameras urging his soldiers to cross a minefield. A Croatian military magazine echoes this theme: "Tiger, join the army if you are a man!" (Čolović 1996a: I). Those who shy from the dangers of combat are "sissies" or "mamas' boys."[6] Masculinity is expressed in the "man's" world of the battlefield. While these expressions can be seen as typical of military rhetoric, they enter into the strategic language of ethnocracy as expressions selected for public consumption on television or in popular magazines and newspapers.

At the same time, the warrior with the body and the courage of a man may be depicted as a warrior with "hero's blood" and a "maiden's soul" (Čolović 1996a). This image draws on the traditional, patriarchal Montenegrin or Balkan model of "heroic shyness" – a strong and courageous man, shy with strangers, particularly women, and monk-like in sexual conduct – and provides a mechanism for a romantic vision of warfare without either women or sexual activity. Manhood, according to this traditional model, is tied to heroic cults of battle, ancestors and community, and has nothing to do with passion or love for women. The warrior's maiden soul is also a metaphor for sexual chastity and purity, the sublimation of sensuality, the redirection of erotic passion toward the higher purpose of the nation, and freedom from "base" sexual urges.

Another image of asexuality among national warriors is that of the boy, still "untainted by sexual experience," captured in the television clips of new recruits or photos of young soldiers with captions describing the sorrow in their eyes for the homeland and their "boyish faces." These boys, however, are not "mamas' boys" or "sissies" but masculine, pure sons of the nation, "sons of the Motherland." Serb soldiers fighting in the Krajina under a unit called "Wolves from Vlašić" were described in a local military newspaper with the following lines of verse: "Who are these boys who carry the sadness of their homeland in their eyes?/ Warriors with boyish faces" (*Krajiški vojnik* no. 1 Krajiškog korpusa Vojske Republike Srpske, February 1995).

Even those warriors who have lovers or wives become chaste again as they head for the battlefield, leaving their women, heterosexual eroticism and carnal love behind. Yet, their sexual desire is not always sublimated in battle, but often, instead, redirected to their instruments of warfare and put into the energy with which they engage in battle. In photos and over the airwaves young male soldiers are described as embracing rifles instead of women. A local report on Croatian troops in Bosnia included these lines: "We are struck by a boy . . . 17-year-old Juro Ivaković, who instead of his first girlfriend hugs an 84 (automatic weapon), and instead of his first cigarette, lights up a Chetnick (Serbian) tank" ("U posjeti jednoj HVO brigadi," *Zmaj od Bosne* November 3, 1994).

Finally, the image of the warrior is often a memory kept alive in the rhetoric of ethnocrats (and national elites), for the warrior's stature is achieved through extraordinary acts of bravery and sacrifice. The sacred son of the nation is in this case the dead hero. The slain, sexually innocent or sexually restrained warrior becomes the property of the nation, his sacrifice a celebration of the national spirit, his death a page in the national history, and his grave a boundary of the nation-state. He sacrifices his life on the battlefield, spilling his "virgin" blood in the soil, returning to the body of the nation – regenerating her and marking her with his bodily remains. His soul then becomes part of the heavenly community of male heroes. Sexual abstinence and, finally, death provide the ultimate erotic experience.

These warrior images provide the core notions of masculinity for nationalist rhetoric even during peacetime. Heroic sacrifices and sacred spaces marked by fallen warriors provide ethnocrats with the stuff of national mythology and models of virility, sexual conduct and patriarchal values. They are readily available to distinguish "our" healthy, noble men from the diseased, bestial and effeminate others or from weak and cowardly traitors. Moreover, they provide a mechanism for transcending earthly sensuality, and redirecting sexual passion to the higher purpose of the nation.

The erotic appeal of the national imagery is directed primarily to men. Women can participate in this national sexual fantasy by identifying with men, with male pleasure (or male participation in the heroic). They may participate in it indirectly by supporting war efforts and national policies, guarding their own chastity during the struggle, and giving themselves to the heroes (husbands and fathers) when they return from war. And they may also imagine themselves in the role of national heroes on the battlefield.[7] Given the social roles proscribed for women in the politics of national identity, these fantasies become a plane of alienated pleasure for women. It is pleasure in that which denies a woman's sexual pleasure, an idea which instrumentalizes her body and threatens her life with violence.[8]

Namely, for women, this national imagery of warriors and heroes eroticizes submission to the collective (Blagojević 1995, Papić 1995, Feldman *et al.* 1993). Paired with the epic heroes, we have brave mothers who sacrifice their sons and husbands for the nation and who tend the wounds of the fallen warriors, and faithful wives who keep the hearth burning and bear the future generation of heroes. Mourning mothers, daughters, sisters and wives, widows dressed in black, victims of rape and torture, refugees packed into trucks with crying children, all are symbols of the national tragedy and reasons for national revenge.

While the language of nation-building used by various competing ethnocrats in the former Yugoslavia hails a glorious past, rich and tragic history, and spiritual values, the popular press also paints pictures of sexually active soldiers enjoying the adoration and favors of young women. The latter are portrayed in the local media for the sake of recruiting otherwise unenthusiastic warriors.[9] Their sexuality becomes a source of metaphors for battle. The sexually charged aggressive behavior of "our" males can be directed toward the enemy (the ethnonational Other, e.g. Croatians or Muslims) – penetrating lines of defense, invading and occupying their space. In the text that accompanies their pictures we read how "our" soldiers defend the Motherland (the feminine body of the nation, and mothers, sisters and wives) against sexually aggressive, bestial violators of the nation/body, invaders of the national space.[10]

As a complement to the young, bold and swaggering soldiers enjoying camaraderie in the field that appear in popular propaganda, beautiful, young women are also portrayed as adoring and admiring "maidens" or playful girlfriends (Čolović 1994a, Dragičević-Šešić 1994). Some of them are even women warriors, dressed in fatigues, but displaying sure signs of their femininity. For the most part these images are part of recruitment propaganda directed toward men: an illustrated article on the Bosnian government Army in *Ljiljan,* for example, carries the headline "Beautiful women liberate the most beautiful country," and notes what a pleasure it is "to share the company of blond, black, and red-haired, impeccably neat girls with discrete make-up, who clutch automatic rifles in their tender girls' hands." A 1993 cover of a Croatian Serb military magazine *The Army of the Krajina* captured this image with a picture of sexy female volunteers, dressed in revealing fatigues, clutching their rifles with red-polished nails (Čolović 1996a: II).

Although these messages appear to contradict the chaste images of national heroes and sexless wives/mothers they are not necessarily problematic from the standpoint of the ethnocracy. For one of the characteristics of a successful ethnocrat is the ability to make contradictory messages appear complementary. The heroic, chaste warrior images appeal broadly on the level of idealized nationalist rhetoric and values and provide a larger framework for authoritarian rule and aggressive nationalism. The various images of attractive young men and women are linked to particular moments and to local, popular culture and appeal directly to recruits who may be untouched by the lofty rhetoric of the nation. Moreover "healthy" heterosexual sexual interest fits the rural image of national purity constructed in the politics of national identity. The girlfriends of patriotic soldier or paramilitary forces are portrayed as healthy outdoor girls, not "feminists" or androgynous city girls. The wholesome sexuality of girls from Pale (Bosnian Serb military headquarters) represents the vitality and purity of the nation in contrast to the decadent, undefined sexuality of urban Belgrade.[11]

Still, sexual exploits figure minimally in the portrayal of young soldiers as the brave and bold protectors of the nation. Such exploits have no place in the ideas of masculinity and femininity promoted as desirable national models and serve a special and limited purpose tightly linked to military recruitment. The reality of sexual aggression and sexual violence is not part of this ideal picture. The Other's men are perpetrators of violence against "our" women. It is this violence against "our" women as members of the collective nation that elicits the ethnocrats' concern. Actual women's bodies are important here as part of the collective body. The nationalist discourse denies the specificity of female experience by giving larger meanings to the signifier of rape; that is Bosnia (Croatia, Serbia) is violated by the Serbian (Muslim, Albanian) rapist (Mostov 1995b, Kesić 1994, Zajović 1994). While rape may fit into battle plans as an extension of warfare – designed to discourage any return to old homesteads or to humiliate and break the will of the Other – it does not fit into the nationalist heroic rhetoric or erotic imagery. Mass rapes such as those reported by women in Bosnia are not about sexual pleasure. They are about the invasion of the Other's boundaries (the occupation of his symbolic space, property and territory) and the violation of his "manhood."[12] Rape, in this context, figures as a sexual violation of boundaries and, by extension, violation of the autonomy and sovereignty of nation.

Images of purity and sacrifice, of chastity and innocence provide national models that are more in tune with the practice of transferring the sensual to the symbolic plane of the nation, imagining the nation as an object of pure love, and desexing the physical body.

Controlling the body/controlling choice

While competitors in struggles to define and defend national interests draw on myths of cultural uniqueness, heroic sacrifice, and national values to gain power, the successful ethnocrat fortifies his advantageous relationships of power by calling for a (re)commitment to traditional values. Nostalgia for strict moral codes and promotion of "traditional" restraint in sexual practices are not just a matter of preserving family values, romanticized village life, and traditional gender roles but, equally important, a way of preserving political control.

The traditional culture of the South Slavs provides a foundation for ethnocratic manipulation of negative views of female sexuality. The standard of sexual conduct was especially rigid for women and reinforced the lines of patriarchal authority (Bringa 1995, Levin 1989). Church and social regulations assumed that even the most chaste woman was a potential danger to a man's salvation. Virtuous women were portrayed in religious and public discourse as completely nonsexual or devoted mothers. Monastic celibacy was seen as a superior lifestyle for men as well. As in the warrior imagery, sexual abstinence was linked to manliness and heroic sacrifice (Levin 1989: 59–60). Sexual desire and love were viewed as antithetical feelings – even within marriage (Levin 1989: 61–65).

Medieval church regulations and teachings survive loosely in national mythology in the former Yugoslavia and provide a source (as traditional ways do for all societies) from which authoritarian leaders can draw in establishing the requirements of social order. A would-be ethnocrat in the region today can reinforce demographic policies, social control and authoritarian models of behavior by evoking traditional values and at the same time recall the unique quality of these values with respect to the decadent ways of others (the West, former Communists, cosmopolitans, antiwar activists, feminists, Muslims, Croatians and Serbs). "Let's return to *our* traditional values." This, he can argue, will create order and stability out of the chaos of political disintegration and social turmoil. Control of sexual conduct, thus, becomes a road to national recovery.

Community practices and religious and social norms tied to ethnonational customs provide one mechanism for supervision of women's bodies. Women are assigned special roles in the physical and cultural reproduction of the community, but the very importance of these roles to the continuity of the community and the assumed vulnerability of women to seduction and violation makes these roles at once affirming and constricting. Tone Bringa captures this predicament when she writes about life in a Muslim village in Central Bosnia (since destroyed by the war): "although men are seen as the ultimate moral guardians, it is women who literally embody this morality. Women's bodies and movement in space are symbolically defined and protected" (Bringa 1995: 86).

According to Bringa, the "morality" at the center of Bosnian Muslim national identity (in the cultural and religious sense) is determined by what is referred to as the "social environment." Reproduction of this environment, particularly the household and its division of labor, requires "channeling and therefore controlling sexuality" and socializing new members (Bringa 1995:

87). In village life, brides enter the home as strangers. As their loyalty to their new home is not taken for granted, it is reinforced by elaborate moral injunctions. Yet, the continuity of the household also becomes the responsibility of these same new members. The vulnerability of the household as a unit is "literally embodied in the woman who was at once the outside to and the main reproducer of the unit" (Bringa 1995: 91).

In my own analysis of gender and the politics of national identity, I read the household in the above passages as nation. That is, I argue that this ambiguous status – in which women are at once crucial to the continuance of the household (nation), but always suspect (potentially disloyal) – figures as an important aspect of the ambiguous relationship among gender, nation and sexuality. This ambiguity appears to be a common fate of women in the politics of national identity.[13] In the discourse of ethnocracy, women serve as symbols of national virtue and purity, even as they remain vulnerable to contamination. Women embody the homeland, but remain potential strangers in it. The precariousness of a woman's place in the very home/nation that constitutes at the same time her designated space underlines both the danger to women of exclusion and the pressures on them to conform. Their marginality is always with them – national culture and values give them a place in society, but always remind them of the potential risk of their falling from the margins within.

This marginality can be effectively used in the politics of national identity. Ethnocrats rely upon the historically grounded fears of their populations and, therefore, work to identify and nurture threats to the nation. Threats about the decreasing numbers of the nation, the loss of distinct culture, language and religious freedom provide the core of mobilizing discourse. These threats are effectively expressed through feminine metaphors for the nation that follow from women's precarious situation as sexual beings. Women are mothers, daughters and wives – symbols of purity, nurturers and transmitters of national values, and reproducers of the nation's warriors and rulers, but also victims – vulnerable to seduction, open to physical invasion and contamination, and symbols of territorial vulnerability and national defilement.[14] Finally, according to the discourse of ethnocracy, women can be traitors and actively participate in the weakening of nation.

At the outbreak of war with Yugoslavia, for example, President Tudjman of Croatia blamed the tragedy of the Croatian nation on "women, pornography, and abortion," calling women who have abortions "mortal enemies of the nation" (Salecl 1992: 59). Likewise, the Serbian Patriarch Pavle a few years later warned that women's selfishness was bringing a "plague" of low birthrates down upon Serbia (Mostov 1995a). In Bosnia the Reis-ul-ulema, Mustafa Čerić, issued a *fetva* (edict) calling on Muslim women there to have five children and condemning mixed marriages as a betrayal of one's faith and culture (Garmož 1994: 7). The so-called "white plague" of low birthrates is a repeated theme in the state-controlled newspapers throughout the former Yugoslavia; the solution to this "biological death" of the Serbian

(Croatian/Slovenian/Muslim) nation is for women to have more children. Note the following selection of headlines from the Serbian popular press: "Third child saves the nation" (Smiljić in *Novost* May 31, 1993); "How to defeat the white plague" (Iković in *Borba* June 30, 1993); "Medals to Serb women for bearing children" (*Borba* May 11, 1993); "Give birth, give birth, and just give birth!" (Pacivić in *TV Novosti* April 1990); "In 50 years there won't be any Serbs" (*Demokratija* July 17, 1997).

Women in the former Yugoslavia are at once put "on a pedestal" and celebrated as biological reproducers of the nation[15] – and publicly scolded for abdicating their reproductive responsibilities. In newspaper and magazine articles such as the ones under the above headlines, women are reminded of their moral duty to bear children for "their" nation and scolded for being selfish and hedonistic.[16] In turn, the Other's women become enemies as reproducers, multiplying the number of outsiders, conspiring to dilute and destroy the nation with their numerous offspring. (Both Croats and Serbs warn of the rapid increase in birthrates of local Muslims and Albanians (Milić 1993: 112–113).)[17]

National demographic policies replace or assume the function of community elders in reiterating women's centrality to the survival of the nation and the dangers of allowing women to stray from their assigned role in preserving and promoting the family/nation. The National Program for Demographic Revival, created under the Croatian Ministry for Reconstruction and Renewal, is committed to promoting motherhood and the family in the interest of spiritual and demographic revival of the nation. In 1992, the Croatian government, following President Tudjman's initiative for a "spiritual renewal of the nation," established the Ministry for Renewal with a special Department for Demographic Renewal. Don Ante Baković, one of the most militant, nationalist voices against abortion, contraception and feminist groups in Croatia, was the head of the department. Among other things, the department's program proposed a strategy for the development of an "ethnically clean" birthrate in Croatia. Confronted by international pressure, the liberal press and women's groups, this department was removed from the ministry. It was subsequently reformed as an N.G.O., the Croatian Population Movement. In the meantime the Program was renamed the National Program for Demographic Development within the Ministry for Reconstruction and Renewal (1996). The current program warns of the demographic threat to the Croatian nation and the need to take action to stop the "national hemorrhage." Strategies to increase the birthrate must be backed up, according to the Program, with strong pro-family propaganda and incentives. Otherwise, the Program warns, Croats could become a minority in their own country (see report from B.a.B.e. Women's Human Rights Group in Zagreb, Croatia, February 7, 1996; Network of East–West Women: February 8, 1996).

Those promoting these birthrate revival policies remind us that the highest calling for women in the nation is that of motherhood. This theme is reiterated in many of the collected papers of the Serbian Academy of Sciences and Art (SANU)'s two conferences on population policies.[18] A number of the papers cite hedonism, possessive individualism and lack of moral responsibility and patriotism as reasons for the low birthrate. Suggested policies and programs designed to protect the continuity and growth of the nation (covered by provocative headlines in the popular media) serve the double role of promoting "desirable" numbers of majority and minority populations and policing both impermissible boundary transgressions (intermarriage) and sexual activity.

Sex as agency

Sexuality, I argue, particularly women's sexual activity, presents a threat to established social order. Sexuality is an expression of individuality and potentially involves numerous challenges to authority and social roles, as well as the chance for creativity, emotional bonding and pleasure. At the same time, sexuality offers opportunities for policing, internalization of moral and social injunctions, fears, prejudices and inhibitions.[19] Constraints on men's and women's sexuality are constraints on personal agency – and ethnocrats or would-be national leaders are keen to reduce the political agency of citizens and expressions of individual "self-determination." This attitude toward political agency fits well in the former Yugoslavia with the rhetoric of return to national values. This is not to suggest that ethnocrats there seriously expect people to refrain from engaging in sexual activity, but, rather, that they discourage sexual activity outside of proscribed norms – sex with the members of the wrong ethnic group or in conflict with demographic goals; sex not directed toward procreation; "abnormal" relations or activities – and register it as a reason for social exclusion.[20]

Messages aimed at desensualizing procreative activity share the airwaves and magazine pages in the former Yugoslavia with sexually provocative depictions of young women and suggestive sexual activity. On the one hand, the popular media owned or promoted by the state (in, for example, Belgrade or Zagreb) reminds women of their duty and calling as virtuous mothers and as reproducers of the nation; on the other hand, it showcases popular singers in tight-fitting, short, sexy outfits and barely clad photo models.[21] Vesna Kesić notes the even more striking juxtaposition in Croatia of the celebration of motherhood and family values with defense of massage parlors and similar entertainment enterprises in the interest of Croatia's economy (Kesić 1994). A mixture of messages – emphasis on reproduction and clean family living along with the trappings of sexuality (suggestive clothing and music) – creates a tension which can be redirected as violence toward ethnic others and internal traitors or internalized as guilt for national failures.[22]

At the same time, stories that combine motherhood, duty to the nation, romance and sex appeal – such as those about the marriage and family life of top country singer Cece Ražnatović (wife of the infamous Arkan)[23] – create what one student of this phenomenon calls a "neofolk" attachment to "national values" (Dragičević-Šešić 1994). Sexuality is packaged in the Serbian country/folk music (sometimes called turbo folk) and in the Croatian rock-based equivalent, both of which include a good dose of nationalist rhetoric, romanticized lifestyles of the new rich (and new criminal underground), and soft pornography. Two pieces in an independent magazine (*Vreme*) in Serbia describe the lifestyles of the children of the new ruling class in Belgrade as characterized by disco clubs, drugs, fast cars, guns, expensive clothes, revealing outfits for women and flashy jewelry. Sex, according to the article, "is considered less a source of pleasure than a means of demonstrating power." "It's a male world in which women are only a status symbol" and are "happy with their subordinate role" (Ristanić and Nikolić 1997). According to these articles, the traditional male–female division of roles is a key feature of this lifestyle, as is the cult of male physical strength: "the woman is to be the dazzling beauty of television videos, while the man is to be a he-man." Male strength is demonstrated through traditional symbols of virility: muscles, closely cut hair, and an unseen but implied arsenal of weapons. The children of the new "rich and powerful" claim that after getting their fill of the fast life they plan to settle down to patriarchal family life as described in the country/folk songs enjoyed by the somewhat older generations (Dragičević-Šešić 1997). This kitsch packaging of national values, like the literary and political messages about heroic suffering and sacrifice in the politics of ethnocracy, diminishes the sensuality of sexual activity and, likewise, discourages individual autonomy or agency.

Ethnonational discourse and practice in the former Yugoslavia thus seeks to transfer sensuality to the experience of national belonging. This eroticizing of social relations within the embrace of the nation serves an important depoliticizing function in ethnocratic state-building strategies.

"The nation" in the language of ethnocrats is portrayed as a natural community; identification with and loyalty to the nation does not involve choice but, instead, acceptance of the obligations of belonging and the mission of the nation, as articulated by its guardians. Accordingly, ethnocratic state-building strategies seek to empty the public space of political subjects, to reduce the categories of political subjectivity, and to limit access to institutions of social power. Government attempts to prevent independent media and limit access to state-controlled television stations are obvious examples of this and part of the everyday politics of the former Yugoslav communities (Petković 1997, Djilas 1997).[24]

The ethnonational model of belonging (or exclusion) is based on acceptance of "natural" bonds and roles (as in "natural" gender roles in sexual reproduction) defined by tradition and interpreted by national leaders. Activities such as consensual sexual relations not necessarily linked to procreation challenge this traditional model of social relations. They suggest social ties based on reciprocity and individual agency. These are the kinds of ties associated with republican or democratic government in which there is an implicit compact among members of the community. The bonds created through this compact are based on mutual recognition among the participants as competent choosers and bearers of rights and obligations. An authoritarian society, such as an ethnocracy, cannot countenance ties based on personal agency or "choice." Bonds must be those of collective "belonging" (Mostov 1996a, 1995b).

In order to discourage sexual agency and its parallels in the political sphere (some form of reciprocity and sharing in the exercise of social choice), the bonding offered by nation must be enticing. From our earlier descriptions of male images in the national iconography and politics of national identity, it is possible to see the seductive force of national mythology for men.[25] In the politics of national identity the nation becomes a lover and mother to men – demanding of loyalty and sacrifice, comforting, yet vulnerable, needing protection and requiring revenge. The nation provides a framework for a male–male world, for eroticism without women, sanctioned by heterosexuality. "Women are translated into a trope of ideal femininity, a fantasmatic female [the Nation/Motherland] that secures male–male arrangements and an all male history."[26] Off the battlefields, males can develop this sense of belonging as sports fans or members of team clubs, through military reserves (weekend warriors), and public institutions.[27]

Because it means imagining herself as male or imagining the pleasure of male guardians/warriors, a woman's attachment to this national mythology is a denial of her sexuality and alienation of pleasure, or the sublimation of pleasure in the acts of reproducing and nurturing the nation's sons, tending to its wounded, remaining faithful to its protectors. Ethnocrats (like community elders) cannot (do not) take women's loyalty for granted, or assume that the bonds of belonging formed so indirectly – through husbands and sons and sacrifice on the home front – will ensure commitment to the nation. Thus, ethnocrats, I argue, need to impose moral injunctions on women, which include restrictions on women's sexual autonomy and which raise the threat of exclusion from the community.

In the aftermath of war and under economic conditions marked by inflation, unemployment, empty pension funds and a breakdown of social services, women in the countries of the former Yugoslavia have reason to be disenchanted with nationalist rhetoric. As refugees, targets of ethnopolitical politics and symbols of defeat, women may find it harder to be swept up by new promises of national glory, or seduced by feelings of belonging. Under such conditions they are more likely to be moved by the need for security and fear of being excluded from community resources and protection.[28] Ethnonational strategies, however, encourage women to see submission to collective goals, not only as a solution to their economic situation, but also as something positive: acceptance of their natural role, an important contribution to the recovery of traditional values, and the purest form of emotional (spiritual) satisfaction. The politics of national identity bolsters this rhetoric with restrictive demographic and reproductive policies and social and religious codes of conduct. Injunctions such as "our" women should behave this way and "those" women who question "our" way are traitors or sexual misfits provide back-up to the romantic and erotic appeal of the nation.[29]

Conclusion: desexing the body and sexing the nation

Ethnocrats draw on the resources of national writers, poets and historians and the mass media to eroticize the nation, and then they reap the benefits of this mechanism of control. The effectiveness of their attempts to rule as guardians of the national interest depends in part upon the degree to which people internalize this displaced pleasure and, of course, the social, historical and political conditions which provide the more or less fertile ground for distillation of national myths and messages. Yugoslav ethnocrats' national program offers ecstasy through service and sacrifice, bonding in battle, and the chance of "belonging" to the whole. It promises social identity and gender identity, as well as political protection and economic security. On one level, there is the thrill of joining in battle, facing danger and death with ancestors to defend the national space. On another level, there is the need to establish a secure identity and economic survival. The two levels of motivation work together for the ethnocrat in mobilizing the population, However, as the battlefield and the myths of national struggles lose their seductiveness and the promise of economic security fades, the ethnocrat needs to use more oppressive and open mechanisms of social control, from control over the media to the use of police and paramilitary groups in public spaces to secure order and maintain power.[30] This is where repressive reproductive policies or rhetoric about motherhood and duty to the nation, voiced by government officials, religious leaders and "national intellectuals," play an important supportive role. Women are to internalize the desirable national image of mother and wife, of desexualized members of the community. They represent the image of the pure nation. Yet, at the same time, women are the objects of scorn for transgressions (intermarriage or abortion) and the weakness of sexually active men. The burden of maintaining the nation during "peacetime" falls disproportionately – and in contradictory ways – upon women.

Gender-based divisions of labor and norms of conduct figure centrally in romanticized mobilizing strategies of ethnocrats which draw on myths and heroic pasts and historic fears of domination by others, and in strategies for maintaining boundaries of social interaction and lines of authority. The sensualizing of the nation requires repeated rituals (for example, celebrations of the Battle of Kosovo), threats of renewed warfare or biological extinction, and actual battles. It is effective in masking the complementary desexualizing of individuals which takes place both on the symbolic level and on the actual level in reproductive politics and messages that reinforce traditional social hierarchies and authoritarian family and political relations. The seductive power of national discourse draws women as well as men into national battles and momentarily blurs the desexualizing character of ethnocracy. While the eroticized images of the nation fade or lose their appeal, the norms become accepted guidelines and practices for "our" way of life, convenient methods of social control, and dangerous mechanisms for maintaining gender and ethnic inequalities in a politics of national identity.

Notes

1 I am referring, specifically, to Serbia, Croatia, Bosnia and Herzegovina, and Montenegro. Most of my examples come from Serbia, as I have had greater access to the print and electronic media, popular and academic literature, and political actors in Serbia. Despite cultural (largely religious) and historical differences among the former republics of Yugoslavia named here, the similarities in the mechanisms and messages of the politics of national identity and the links between gender and nation in this politics are quite significant. There are no fundamental poetic or thematic differences between Serbian and Croatian epics from which many of the respective Serbian and Croatian national myths are drawn (Žanić 1994: 9). Major heroic figures may be different (e.g. Serbian Saint Sava and Croatian King Zvonimir), but the use of these figures in contemporary nationalist rhetoric in Serbia and Croatia is quite similar, and motifs, such as the "mountain peaks of Romanija" (contested region in Bosnia and Hercegovina), occupy a similar place in the respective Serb and Croatian (as well as Bosnian Muslim) romantic visions of their national histories (legends, folklore and verse), contemporary political discourse and nationalist ideology (Čolović 1994b: 130–131, Žanić 1994).

2 Ethnocracy is like a "superfamily" which reduces political subjectivity to collective acceptance of the dictates of a state father.

3 I draw on my own reading/viewing of electronic and print media, speeches by politicians and works by "national intellectuals," as well as observations of daily life during my lengthy stays and travel in the region from 1987 to 1997. The masculine imagery of heroism used by "national" intellectuals and the state-controlled press in glorifying war and inciting ethnic/national conflict in Serbia and Croatia has been well documented in Serbian, Croatian and English sources (Djordjević 1996, Gojković 1996, Nenadović 1996, Lalić 1995, Žanić 1995, Čolović 1994a, 1994b, Thompson 1994, Zajović 1994, Feldman *et al.* 1993, Popov 1993).

4 Sexual activity finds its way onto the stage of warfare and into nationalist rhetoric. But it is not included in the language of heroism or character of the hero.

5 Čolović notes that the value of these characteristics often depends on whether they refer to "our" warriors or "theirs." He uses images of warriors described in newspaper and magazine articles in Croatia, Serbia and Bosnia 1990–95, other popular texts (including songs), and historical and ethnographic works (Čolović 1996a). While Čolović focuses on Serbia, he includes examples of these images from the media in Croatia and Bosnia. Authors doing this research in Croatia have identified similar images (Karabeg 1997, Feldman *et al.* 1993). The masculine types that Čolović presents also coincide with those described by Mosse in his classic text *Nationalism and Sexuality* (Mosse 1985) and are not, I suggest, all unique to the former Yugoslav Republics.

6 Note, for example, a news report from Vukovar, which includes a criticism of politicians and critics in Belgrade: "In the first lines, the officers are there together with their fellow soldiers, showing by their example how real patriots fight, those who are loyal to their country. This must be made known because of those who stand at a distance and lecture others, seeking safety under the skirts of their mothers or wives, or their parties, insulting the people who for many months, without relief, are fighting for our freedom" (Lalić 1995: 97).

7 Women have traditionally joined men on the battlefield disguised as men. There are many stories and myths of women going off to war, hiding the secret of their sex. In doing so they join the ranks of the desexed or "virgin" warriors. Actual participation of women in paramilitary and military units in the wars in Croatia and Bosnia has been quite different, including work in kitchens, hospitals and administration, as well as "unromantic" shooting from the trenches. In war propaganda these women participate out of duty, in the absence of enough "real men" (Mostov 1996a, Nikolić-Ristanović 1997a, Habjanović-Durović 1993, Zajović 1992).

8 C. L. Innes recounts how Irish poet Eavan Boland writes "of her alienation from the rhetoric and imagery of much of Irish nationalist poetry and its simultaneous appeal to her" (Innes 1994: 13).

9 The comic strip adventures of Captain Dragan, leader of a Serbian paramilitary unit and his ninja-like warriors, were extremely popular during the summer leading up to the war in Croatia. Captain Dragan and his brave band of soldiers were often drawn sitting around the campfire with their beautiful and shapely girlfriends (Mostov 1995a, Čolović 1994a). In Croatia, soldiers with Ray Ban-like sunglasses, earrings and black headbands in the current rock style made it into the press to help recruit young men who might not buy the "heroic" images, and rock music was regularly used to accompany war propaganda songs (Feldman *et al.* 1993).

10 The following are just a few examples of the terms used for the invading "other": by Croatians, we have Serbian "apostles of evil" and "vampires"; and after 1994, Muslim "criminal hordes" and "mujahadin" with hellish plans for an Islamic fundamentalist B-H; and in Serbia: Croatian "cut-throats" and *ustase* "evil doers" and Bosnian "jihad warriors" and "Alija's wanton hordes" (Thompson 1994: 192, 188, 102).

11 In the rhetoric of would-be national leaders in the former Yugoslavia, rural–urban dichotomies play an important role today in constructing acceptable gender roles, sexual mores and social relations. Traditional rural life is described as chaste, healthy, godly and safe. Urban life is portrayed as wanton, diseased, ungodly and dangerous: the city is a place of loose behavior, lost morals, boundary crossings and transgressions. The city is home to criminals and outsiders, as well as effeminate men and reckless, seductive (perhaps, masculine) women (Bogdanović 1994, Vujović 1996).

12 This understanding, common to the patriarchal traditions of all sides in the war in Croatia and Bosnia, accepts the linkage of manhood and national character with the ability to protect one's woman/nation, as expressed by Bosnian Muslim religious leader, Dr. Mustafa Čerić: "If only more Bosnian Muslims had been fundamentalists, then their mothers and daughters would not have been raped and they would not have been slaughtered" (Arsenić 1993: 8).

13 In her excellent discussion of Chinese writer Xiao Hong, Lydia Liu writes of Hong's dilemma in facing two enemies, Japanese imperialism and patriarchy. This is eloquently expressed in her response to her lover's yearning for home: "as far as I am concerned it always comes down to the same thing: either riding a donkey and journeying to an alien place, or staying put in other people's homes. I am never keen on the idea of homeland. Whenever people talk about home I cannot help but be moved, although I know perfectly well that I had become 'homeless' even before the Japanese set their feet on that land (Manchuria)" (Liu 1994: 47).

14 Vesna Kesić cites examples of articles from the Zagreb popular daily *Globus* with titles like "Croatia is a fallen woman" and "The Motherland becomes an untrustworthy, amoral, disgraced woman," as well as the following passage: Croatia has "experienced a moral falling, the kind that only a woman can experience – for

there are no easy men. Only easy women give themselves without a struggle and accept this as their inescapable fate. . . . In contrast, men fight" (Kesić 1994: 11).

15 In the post-Communist transition, nationalist rhetoric gives women a central role in the cultivation, nurturance and transference of national values in exchange for lost economic, social and political rights. Women regain their "natural" mission and revered place in the home as wives and mothers and in the nation as modest and chaste symbols of its religious/spiritual values. In Serbia in 1993 the Patriarch's fund for promoting the continuity of the nation gave out "Majka Jugović" awards, commemorating the Battle of Kosovo in 1389, to women with four or five children (Rajić 1993). Majka Jugović is the stoic Mother of Serbian epic poetry, who sacrifices her nine sons on the battlefield in a tragic, but heroic defeat against the enemy Ottoman empire.

16 Professor Radomir Lukić, arguing that the question of birthrate is the most important national question, suggests taxing those who have fewer than three children. Giving birth is a woman's duty to the community. "After all, it is easier for women to give birth than for men to go to war and die" (*Politika* April 14, 1990: 18–19). In a key note address to the Society which goes by the name of the nine sons of Mother Jugović, Serbian folk artist and nationalist "spokesperson," Milić od Mačve, repeated his sure fire plan for 2.5 million young Serbian women to bear 7.5 million babies in five years by having one child after another, with a "respectable rest of a few months" in between (Milić 1997).

17 According to authors, in such articles as "Give birth, give birth and just give birth" and "A-Bomb over Kosmet," Albanians in Kosovo have seized upon demographic domination as their main weapon against Serbia (*TV Novost* April 1990: 18–19; *Politika* July 1, 1994: 16).

18 These papers are published in two large volumes: *Problem politike obnjavljanja stanovnistva u Srbiji* (1989) and *Osnovni populacione politike: ciljevi, institucije, mere* (1997). In the second volume a few articles by women, which cite economic and social conditions for low birthrates rather than lack of national feeling or hedonism, are included at the end (Blagojević 1997).

19 This has become a Foucauldian "truism."

20 For example, an article in the Zagreb daily *Globus* labeled as "witches" five professional women who dared to be critical of the policies of the ruling party. The author of the article pointed out their "failures" to marry Croatians, to remain married, or to have children. Their sexual deviations were proof of their political deviancy (*Globus* December 11, 1992: 33–34).

21 I refer particularly to the singers of "newly composed" folk music which is akin to some American country music.

22 For example, Bosnian Serb nationalists blame the diluted or contaminated culture of westernized Belgrade for "feminizing" Serbs there, weakening their national spirit and resolve, and religious Muslims in Bosnia point to the secular nature of Bosnian culture as an explanation for defeats there.

23 The independent magazine *Vreme* carried this quote from the popular media by Cece about her 8-month-old baby and "daddy" Zeljko Ražanatović (infamous "baby-faced" paramilitary leader in the wars in Croatia and Bosnia; suspected criminal, police collaborator and war criminal; and sometimes Serbian nationalist politician): the baby "loves it when Željko and I sing, but he listens seriously to his daddy when he tells him about Kosmet [kosovo, Jim], Serbia's holy land. All in all, the baby is sweet and looks like me" (*Vreme* August 9, 1997: 29). The *Financial Times* correspondent Laura Silber covered their wedding, which was treated by some like a national fairy tale. According to Silber, Cece the recording artist, "with moves that would make Elvis blush," vows to bear Zeljko five children (to add to his current seven) (Silber 1995: II).

24 See note 3 on state control of the media in Serbia, Croatia and Bosnia during the wars. Most of the anti-war and human rights groups in these three countries write regularly about limited access to the media in their publications, for example, the bi-monthly bulletin, *Voice*, of the Centre for Antiwar Action, or the monthly magazine, *Pravo na sliku & reč/The right to pictures & words* of the Independent Media Union in Belgrade, and the anti-war magazine, *Arkzin*, in Zagreb.

25 The assumption is that this applies only to heterosexual men – "real" men who know the order of things and engage in sexual relations for the purpose of procreation. Homosexuality undermines the natural hierarchy of male domination over women and threatens the continuity of the nation. The enemy males are sometimes characterized as effeminate and war resisters, peace activists and non-nationalists (supporters of civic alternatives) are lumped together as "gays."

26 These quotes "written under the sign . . . heterosexuality" refer to the Church or "ancient mother," which functions as, I suggest, the nation (or Motherland) does in the politics of national identity on the territory of the former Yugoslavia (Goldberg 1992: 63). Written under the sign of an ancient mother, those male arrangements are secured beneath a spirituality that is nominally female and which serves a normative heterosexuality.

27 Everyone (including party leaders, government officials, the police, the media and thousands of fans) held his breath on two of the biggest nights of the summer of 1997, when rival Belgrade and Zagreb soccer teams ("grave diggers" and "bad blue boys") met at their respective home stadiums. Listening to the sports commentators, it seemed like victory would improve the condition of the nation (from pensions arrears and strikes to international pressure from The Hague). A few commentators couldn't resist the temptation to return to the front: "There they were on their knees . . . the Yugoslav top player fell last night like Belgrade did in 1941, silently, without a shot" (Bunjac 1997: 15). Still, there were congratulations all around, when, despite losses by each of the teams at the other's stadium, there were no significant conflicts. Rather the male comaraderie of sport transcended the battlefield. The start of war in Croatia was marked by violence between Serb and Croatian soccer fans (Čolović 1996b); perhaps the stadium is the venue for public reconciliation.

28 This is clear in the voices of refugee women, some of whom were swept up in the earlier nationalist euphoria that accompanied the breakdown of the former Yugoslavia (Mertus *et al.* 1997, Nikolić-Ristanović 1997a, 1997b).

29 Staša Zajović records some of the comments made to members of Women in Black by people walking by during their regular weekly public vigils. "The most common qualification directed at those of us who protest is: idle women don't you have something else to do but stand there? They express anger because women are involved in politics, because it is allegedly a public sphere intended for men – 'Go home and wash pots.' Women who are publicly active in politics are 'whores' but also child killers, 'barren women,' 'nymphomaniacs and lesbians.' . . . 'You are not Serbs, you are Yugoslavs' because 'if you were Serbs, you would give birth to Serbian heroes for Serbian revenge'" (Zajović 1997: 20–21).

30 See note 28. Over 100 local radio stations have been closed in Serbia and Croatia during 1997–98 by the ruling ethnocratics, access to state television remains extremely limited and the messages are equally controlled by the state monopoly (Djilas 1997, Petković 1997). Violence and police intimidation are used by the police and government in Serbia to keep students and other protesters from moving through Belgrade (Kandić and Stojanović 1997, Kandić 1997) and the regime refuses to put any constraints on the obstruction of legal processes and violent attacks on private citizens by extremist parties (Čurčurević-Lukić 1997, Jevtić 1997, Jovanović 1997, Malešević 1997, Milunović 1997).

References cited

Arsenić, R. (1993) "Kraj 'prirodnog savezništva'," *Politika* (January 10): 8.
Blagojević, M. (1995) "Women and war: the paradox of self/sacrifice or the anatomy of passivity," in M. Blagojević, D. Duhaček and J. Lukić (eds.) *What Can We Do for Ourselves?* Belgrade: Centre for Women's Studies, Research and Communication.
—— (1997) "Neradjanje: pasivni otpor žena," in *Osnovni populacione politike: ciljevi, institucije, mere*, Demografski zbornik, Knjiga IV, Belgrade: SANU.
Bogdanović, B. (1994) *Grad i smrt*, Belgrade: Beogradski krug.
Bringa, T. (1995) *Being Muslim the Bosnian Way: Identity and Community in a Central Bosnian Village*, Princeton, N.J.: Princeton University Press.
Bunjac, M. (1997) "Noc nacionalnog pomirenja," *Naša Borba* (August 1): 15.
Čolović, I. (1994a) *Bordel ratnika: folklor, politika i rat* (2nd edition), Belgrade: Biblioteka XX veka.
—— (1994b) *Pucanje od zdravlja*, Belgrade: Beogradski krug.
—— (1996a) "Društvo mrtvih ratnika," *Republika* 145–146 (August): I–IV.
—— (1996b) "Fudbal, huligani i rat," in N. Popov (ed.) *Srpska strana rata*, Belgrade: Republika.
Čurčurević-Lukić, S. (1997) "Strah nad gradom," *NIN* (August 8): 29–30.
Djilas, M. (1997) "Sve gušći medijski mrak," *Naša Borba* (August 9–10): XVI.
Djordjević, M. (1996) "Književnost populističkog talasa," in N. Popov (ed.) *Srpska strana rata*, Belgrade: Republika.
Dragičević-Šešić, M. (1994) *Neofolk kultura: publika i njene zvezde*, Sremski Karlovci: Izdavačka knjižarnica Zorana Stojanovića.
—— (1997) "Kičasti imidž," *Vreme* (August 2): 16–17.
Feldman, L. C., I. Prica and R. Senjković (eds.) (1993) *Fear, Death and Resistance – An Ethnography of War: Croatia 1991–1992*, Zagreb: Matrix Croatia, X-Press.
Garmož, Z. (1994) "Već sam izdao fetvu da svaka muslimanka rodi petoro djece," *Globus* (February 18): 7.
Gojković, D. (1996) "Trauma bez katarze," in N. Popov (ed.) *Srpska strana rata*, Belgrade: Republika.
Goldberg, J. (1992) "Bradford's 'Ancient Members' and 'A Case of Buggery . . . Amongst Them'," in A. Parker, M. Russo, D. Sommer and P. Yaeger (eds.) *Nationalisms and Sexualities*, New York: Routledge, pp. 60–76.
Habjanović-Durović, L. (1993) "Ja sam vojnik-četnik," *Duga* (January 16–29): 31.
Iković, N. (1993) "Kako pobediti belu kugu," *Naša Borba* (June 30): 10.
Innes, C. (1994) "Virgin territories and motherlands: colonial and nationalist representations of Africa and Ireland," *Feminist Review* 47 (summer): 1–14.
Jevtić, Z. (1997) "I dalje sve neizvesno," *Naša Borba* (August 8): 13.
Jovanović, J. (1997) "Tehnika zavodjenja terora," *Danas* (July 26–27): 5.
Kandić, N. (ed.) (1997) *Spotlight On: Law Enforcement Abuses in Serbia and Montenegro*, Belgrade: Humanitarian Law Center.
Kandić, N. and L. Stojanović (eds.) (1997) *Spotlight On: Political Use of Police Violence during the 1996–1997 Protests in Serbia*, Belgrade: Humanitarian Law Center.
Karabeg, O. (1997) "Srbi kao Jevreji – Hrvati kao Židovi," *Naša Borba* (August 9–10): 7.
Kesić, V. (1994) "Od štovanja do silovanja," *Kruh i Ruže* (Zagreb), 1 (spring): 1–14.
Lalić, L. (1995) *Three TV Years in Serbia*, Belgrade: Independent Media Union.

Levin, E. (1989) *Sex and Society in the World of Orthodox Slavs, 900–1700*, Ithaca, N.Y.: Cornell University Press.
Liu, L. (1994) "The female body and nationalist discourse: *The Field of Life and Death* revisited," in I. Grewal and C. Kaplan (eds.) *Scattered Hegemonies: Postmodernity and Transnational Feminist Practices*, Minneapolis: University of Minnesota Press.
Malešević, K. (1997) "Strah od drugog i drugačijeg," *Republika* 169–170 (August): 15–16.
Mertus, J., J. Tesanović *et al.* (eds.) (1997) *The Suitcase: Refugee Voices from Bosnia and Croatia*, Berkeley: University of California Press.
Milić, A. (1993) "Women and nationalism in the former Yugoslavia," in N. Funk and M. Mueller (eds.) *Gender Politics and Post-Communism*, New York: Routledge.
Milić od Mačve (1997) "Kako do petnaest i po miliona Srba," *NIN* (July 11): 31–32.
Milunović, M. (1997) "Sprečen povratak Barbalića u stan," *Naša Borba* (August 1): 1, 3.
Mosse, G. (1985) *Nationalism and Sexuality: Respectability and Abnormal Sexuality in Modern Europe*, New York: Howard Fertig.
Mostov, J. (1994) "Democracy and the politics of national identity," *Studies in East European Thought* 46 (June): 9–31.
—— (1995a) "'Our women'/'their women': symbolic boundaries, territorial markers and violence in the Balkans," *Peace and Change* 20, 4: 515–529.
—— (1995b) "The use and abuse of history in Eastern Europe: a challenge for the 90s," *East European Constitutional Review* 4, 4: 69–73.
—— (1996a) "Endangered citizenship," in M. Kraus and R. Liebowitz (eds.) *Russia and Eastern Europe After Communism*, New York: Westview Press.
—— (1996b) "La formation de l'ethnocratie," *TransEuropéennes: revue culturelle internationale* 8: 35–44.
Nenadović, A. (1996) "*Politika* u nacionalističkoj oluji," in N. Popov (ed.) *Srpska strana rata*, Belgrade: Republika.
Nikolić-Ristanović, V. (1997a) "Promene u svakodnevnom životu žene i ekonomija preživljavnja: model žrtvovanja," in V. Nikolić-Ristanović *et al.* (eds.) *Zene Krajine i rat, exodus i izbeg listvo*, Belgrade: Institut za Kriminilostvo i Sociološke Israzibanje.
—— (1997b) "Žrtve nasilja i activne učesnice rata," in V. Nikolić-Ristanović *et al.* (eds.) *Zene Krajine i rat, exodus i izbeg listvo*, Belgrade: Institut za Kriminilostvo i Sociološke Istraživanje.
Pacivić, Z. (1990) "Radjati, Radjati i samo Radjati!" *TV Novosti* (April): 18–19.
Papić, Z. (1995) "From state socialism to state nationalism: the case of Serbia in gender perspective," in M. Blagojević, D. Duhaček and J. Lukić (eds.) *What Can We Do for Ourselves?*, Belgrade: Centre for Women's Studies, Research and Communication.
Petković, S. (1997) "Vlast opstaje samo izludjivanjem naroda," *Naša Borba* (August 4): 8.
Popov, N. (1993) "Srpski populizam: od marginalne do dominantne pojave," *Vreme* (separate) 135 (May 24).
Radić, R. (1996) "Crkva i srpsko pitanje," in N. Popov (ed.) *Srpska strana rata*, Belgrade: Republika.
Rajić, L. (1993) "Koliko naroda, toliko neba," *Vreme* (May 31).
Ristanić, B. and Z. Nikolić (1997) "Srpska zlatna mladež," *Vreme* (August 2): 14–17.
Salecl, R. (1992) "Nationalism, anti-Semitism, and anti-feminism in Eastern Europe," *New German Critique* 57 (fall).

SANU (1989) *Problem politike obnjavljanja stanovništva u Srbiji*, Demografski zbornik, Knjiga I, Belgrade: SANU.
—— (1997) *Osnovni populacione politike: ciljevi, institucije, mere*, Demografski zbornik, Knjiga IV, Belgrade: SANU.
Silber, L. (1995) "The folk star and the tiger," *Financial Times* (August 18–19): II.
Smiljić, N. (1993) "Treće dete – spas naroda," *Novost* (May 31).
Tašić, P. (1994) *Kako je ubijena Druga Jugoslavija*, Skopje: AI.
Thompson, M. (1994) *Forging War: The Media in Serbia, Croatia, and Bosnia-Hercegovina*, New York: Article 19 International Centre Against Censorship.
Vujović, S. (1996) "Nelagoda od grada," in N. Popov (ed.) *Srpska strana rata*, Belgrade: Republika.
Zajović, S. (1992) "The war and women in Serbia: patriarchy, language, and national myth," *Peace News* (March).
—— (1994) (ed.) *Women for Peace*, Belgrade: Women in Black.
—— (1997) "Sexism, nationalism and militarism always go together," in S. Zajović (ed.) *Women for Peace*, Belgrade: Women in Black.
Žanić, I. (1994) "Navrh Gore Romanije . . . ," *Erasmus* 7 (June): 8–20.
—— (1995) "The curse of King Zvonimir and political discourse in embattled Croatia," *East European Politics and Society* 9 (winter): 90–122.
Zirojević, O. (1995) "Kosovo u istorijskom pamćenju: mit legende, činjenice," *Republika* (March): 9–24.

Articles

(1991) "Materinska mobilizacija," *Vreme* (February 25): 36–38.
(1993) "Bela kuga," *Naša Borba* (October 11).
(1993) "Medalje Srpkinama za radjanje," *Naša Borba* (May 11).
(1993) "Beba ili hleba, odlučite sam," *Naša Borba* (March 27–28): 11.
(1994) "A-bomba na Kosmetu," *Politika* (July 1): 16.
(1994) "Moze li se zaustaviti demografsku explosiju na Kosmet?" *Politika* (September 27).
(1996) "Bitka za treće dete," *Politika* (October 13): 13.
(1997) "Za 50 godina Srba neće biti" (interview with Dr. Marko Mladenović), *Demokratija* (July 17): 7.

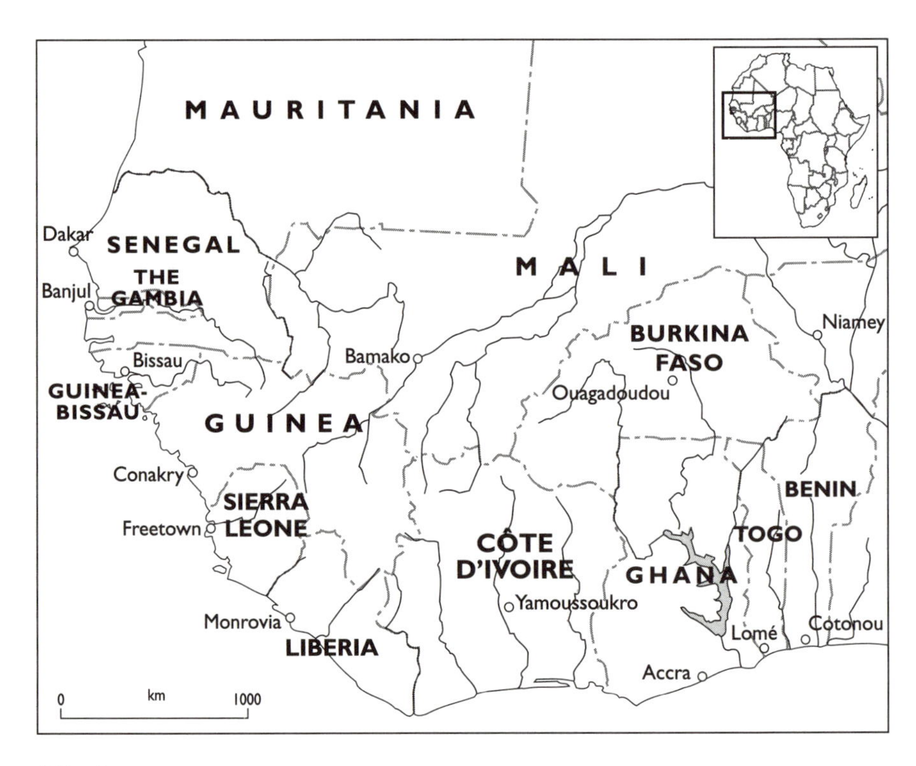
MAURITANIA
SENEGAL
Dakar
THE GAMBIA
Banjul
Bissau
GUINEA-BISSAU
GUINEA
Conakry
SIERRA LEONE
Freetown
Monrovia
LIBERIA
MALI
Bamako
BURKINA FASO
Ouagadoudou
Niamey
CÔTE D'IVOIRE
Yamoussoukro
GHANA
Accra
TOGO
Lomé
BENIN
Cotonou
0
km
1000

Liberia

5 Uneasy images

Contested representations of gender, modernity and nationalism in pre-war Liberia

Mary H. Moran

In most of the scholarly literature on Africa, nationalism is contrasted, either explicitly or implicitly, with ethnicity or "tribalism." There is a curiously unexamined evolutionism to this view, as if ethnicity were a historically earlier, more "traditional" and certainly more deeply felt form of group identity. Nationalism then becomes the equivalent of "modern," and its failure to supersede and replace other attachments becomes in turn the source of much concern. Even though many supposedly "primordial" tribal identities have been shown to be of recent origin and to have strategic political uses (see especially Ranger 1983: 248–249), the binary opposition "nationalism/ethnicity" still drives much analysis of the state in Africa. This chapter examines the often unspoken association between nationalism and modernity or, to use the favored terminology in the West African nation of Liberia, "civilization."

In Liberia, the binary "civilized/native" is often elided with nationalism/ethnicity; "civilized" Liberians are believed to owe their primary allegiance to the nation-state while "natives" privilege the ties of "tribal" or ethnic affiliation. But in Liberia, the term "civilized" also incorporates other meanings and values – making, at best, for an uneasy and contradictory association with nationalism. Officially and unofficially, having two categories of citizens raises questions of loyalty, identity and legal rights. For example, the Liberian legal system has, for more than a century, allowed some minor crimes to be tried by native chiefs when the parties involved are considered "natives" while the same crimes, committed by "civilized" persons, are tried in statutory courts with a system of procedures and penalties based on Anglo-American law. Some individuals have been known to try their luck first in a statutory court, and then redefine themselves as "native" and seek a more favorable outcome elsewhere. Obviously, these categorical identities are more flexible and situational in practice than in the often rigid definitions of Western analysts.

Similarly, men and women, both "native" and "civilized," seem to stand in different positions as citizens in Liberia. The state, as a bureaucratic structure, and the nation, as an imagined collectivity, require and impose differently gendered subjectivities. In keeping with the theme of this volume, I argue that it is with the addition of a gendered analysis that the contradictions of nationalist discourse and its relationship to "civilization" in Liberia are most thoroughly exposed and the ambivalences surrounding both national identity and "civilization" are made visible. The difficulty and ambiguity of assimilating "civilized," modern, *female* citizens into existing discourses of the nation which have been noted in a variety of times and places are crystallized in my analysis of images of women from a major Monrovian newspaper in the early 1980s.

Methodology and sources

For fifteen months in 1982–83 I lived in Liberia, conducting research on gender and prestige hierarchies in a small regional center about two days overland from the capital, Monrovia. It was in this small community that I became fascinated with the Liberian constructions of "civilized/native" and their articulation with cultural constructions of gender (see Moran 1990). While I shall discuss the cultural meanings and behaviors associated with "civilized/native" below, it is important to note that my fieldwork took place in a small face-to-face community. Although not ethnically homogeneous, the area was identified by local inhabitants as the home of the Glebo people, one of a number of groups speaking dialects of the Grebo language. My initial analysis of the "civilized/native" dichotomy took place within the specific, local context of Glebo beliefs and practices (Moran 1992, 1990, 1988).

At the same time, I was aware that these terms transcended purely local usage and were part of a broader discourse of class, status and identity in Liberia. An extensive literature exists linking these particular terms and meanings to Liberia's history of colonial settlement by free Black Americans in the early nineteenth century. The rural and urban implications of these conceptions for understanding upward mobility, capitalist penetration and the mystification of class and ethnic relationships in Liberia have been thoroughly analyzed (Brown 1982, Tonkin 1981, 1980, 1979, Frankel 1964). All these authors noted that to be "civilized" both in rural Liberia and among recent migrants to Monrovia was to be in a highly valued position, one that implied both moral and material advantage over "natives." My work on this concept was the first to address the differential manner in which men and women achieved and maintained "civilized" status. This chapter extends a gendered analysis to the operation of "civilization" on the national level.

While conducting traditional ethnographic research, I also collected national newspaper accounts relating to Liberian women. I subscribed by mail to the *Daily Observer*, the major independent paper in Monrovia. Over the course of the same fifteen months, I was able to accumulate an extensive clipping file, including additional publications that I collected on my occasional trips to Monrovia. Newspapers and other textual materials have provided anthropologists with a rich source of evidence for the analysis of gender and nationalism in contemporary Africa. Classic studies by Little (1980, 1973) and Schuster (1979) emphasized the scapegoating of urban women in African newspapers as sexually uncontrolled, spiritually dangerous, and, most importantly, *unpatriotic* parasites or "folk devils" (Schuster 1979: 140–153). More recent work by Bastian (1995, 1993) on the popular press in Nigeria provides a wonderfully nuanced account of the ambivalence surrounding modernity and urban life and its representation in highly gendered narratives about the bodies of men and women. In addition to documenting shifting views of gender, however, Akhil Gupta has argued that

> In the study of translocal phenomena such as "the state," newspapers contribute to the raw material necessary for a "thick" description. . . . Obviously, perceiving them as having a privileged relation to the truth of social life is naive; they have much to offer us, however, when seen as a major discursive form through which daily life is narrativized and collectivities imagined.
>
> (Gupta 1995: 385)

No more than do our own, African newspapers do not simply "reflect" ongoing social life. Rather, they are one way in which that life is constructed, given meaning, contested, and changed. The imagining of collectivities, whether these are based on gender, ethnicity or common membership in the nation, is never "finished"; it is rather a constant state of negotiation and struggle. The analysis of newspapers can help us capture specific moments in this process, but we must be careful to avoid investing them (or indeed, any text) with too much stability over time.

In this chapter, I examine a series of cartoon images of "civilized" women in Monrovia during the early 1980s. I was intrigued to find in the images of urban, "civilized" women in the Monrovian papers many of the same negative, scapegoating motifs described by analysts of other African countries. I found this curious because the understanding of "civilization" as a positive and desirable status for women was so overwhelming at the site of my rural fieldwork. Specifically, urban "civilized" women are often pictured as greedy, sexually uncontrolled, parasites who divert men from their responsibilities and impede the progress of national development. I wondered why "civilization" could mean such different things for urban and rural women. In both contexts, women seemed to aspire to this status, but in the urban press at least, their public image was far more negative. I argue that it is precisely because "civilization" is more easily elided with the modern nation-state in the context of the national capital that images of urban "civilized" women take on a negative, dangerous character.

Further, I argue that the cartoons represent not only a critique of urban women, but also of the military government during a period when press censorship made open opposition impossible. It is important to note that these images date from the early years of Liberia's first military government;

a time when the head of state, Samuel K. Doe, was a young master sergeant with a tenth grade education. I suggest that the negative image of urban "civilized" women masked a critique of Doe and his pretensions to "civilized" status. In this instance, a discourse of female citizenship becomes a mask for questioning the legitimations of the state. In 1985, Doe became President of the Republic of Liberia in what was widely believed to be a fraudulent election. In 1989, armed opposition to his repressive and autocratic rule began, leading ultimately to a protracted civil war in which about 200,000 Liberians lost their lives. This chapter focuses on the early 1980s, the historical moment before the collapse of the state, when many possibilities seemed open and the definition of the nation and its citizens appeared unusually fluid. The cartoons reflect this sense of uncertainty in that they are neither consistent nor univocal. Rather, they seem to contain contradictory messages which can be read in multiple ways.

Before looking at the specific case of Liberia, we must examine the relationship between gender and nationalism in general. Gender and sexuality have been implicated in the construction of competing nationalist and subnational identities by a number of scholars across the disciplines, including those in this volume (for other examples, see Anthias and Yuval-Davis 1989, Parker *et al.* 1992, Enloe 1995, 1993, 1988). It is clear from this research that the construction of citizens of either gender is an ongoing and often contested project for any nation-state. For my purposes here, I would like to make three points about this relationship. First, both gender and nationalism represent categorical identities that are easily naturalized and essentialized. As Anderson (1991) has noted, in the modern world everyone must "have" a nationality just as they "have" a gender. Such naturalization increases the emotional investment individuals bring to defending or challenging these identities, and increases the political stakes for everyone. Second, both gender and nationalism are tied to the highly charged issue of reproduction, from the level of individual human bodies all the way to the continuation of "the people" however defined. This association between biological and social reproduction can in turn increase the level of tension and potential conflict between women and men over such issues as sexual autonomy, mobility and economic independence.[1] Enloe has noted that "it is precisely because sexuality, reproduction and child rearing acquire such strategic importance with the rise of nationalism that many nationalist men become newly aware of their need to exert control over their community's women" (1995: 22). Third, both gender and nationalism, while locally defined and enacted, are situated within global processes of commodification and identity production. Local constructions are in at least implicit dialogue with globalized media images of masculinity, femininity and "national culture" as they are represented in popular forms like movies, music, and television shows, increasingly available even in the most remote locations. Whether this engagement takes the form of resistance ("*Our* women are not like those loose American women in the films") or emulation ("In order for the nation to be modern and progressive, our women must abandon their backward ways"), it is usually signaled by specific items of clothing or material culture. As these items circulate globally, they become markers of identification with or resistance to the "modern" nation-state. All three of these dynamic points of intersection between gender and nationalism – the tendency toward naturalization; the association with reproduction; and the situatedness within globally circulating systems of commodities – are visible in the cartoon images discussed below.

Nationalism in the Liberian context

In Liberia, nationalism is not the fully naturalized belief in common descent or peoplehood associated with a particular ethnic group (with or without ambitions to become a state) but instead what Anderson (1991) has called "official nationalism." The concept as Anderson defines it refers to the ideological extension of a naturalized identity to all occupants of politically defined territory, or to "stretching the short, tight, skin of the nation over the gigantic body of the empire" (1991: 86). In other words, official nationalism is a product of deliberate policies of absorption and incorporation; we might also think of it as compulsory membership in the entity defined by the geographical boundaries of the state. This distinction between "naturalized" and "official" nationalism raises the question of whether or not it is possible to distinguish analytically the "state" from the "nation" in all contexts. When state institutions are put to the task of producing collective identities for their subject populations, nationalism is clearly no longer an "imagined community" in the purely voluntary sense. Although he is primarily interested in the official nationalisms of pre-modern Europe, Anderson mentions that these state policies were "refracted into non-European cultures and histories . . . picked up and imitated by indigenous ruling groups in those few zones (among them Japan and Siam) which escaped direct subjugation" (1991: 110).

He might have added, "among them Liberia." This small country on the West African coast is unique in sub-Saharan Africa in that it was never formally dominated by a European colonial power. Founded in the early nineteenth century as a "benevolent experiment" in the resettlement of free people of African descent from the United States, Liberia became the first independent black republic in Africa in 1847. The process of "nation-building," generally understood as commencing in Africa in the post-World War II period, thus began in Liberia a full century earlier. In his masterful multipart article, "A tribal reaction to nationalism," Warren d'Azevedo (1970–71, 1970, 1969a, 1969b) uses the term "nationalism" to describe "almost one hundred and fifty years of slow insistent absorption of a heterogeneous population into a national entity" (1969a: 4; see also Martin 1969). The Liberian settlers, who never constituted more than 2–3 percent of the total population, used a combination of military conquest, trading partnerships, strategic marriages and adoptions to create links with rural indigenous elites, and ideological constructions of their own version of manifest destiny to maintain their political and economic dominance until their government was toppled by a military coup in 1980. Liberian nationalism, therefore, has always been "official" in the sense of being a projection of the state; no sense of nationhood predates the arrival of the settler minority. This official nationalism was closely tied to such identifying markers as literacy, fluency in English (the national language), employment in the wage sector, at least nominal membership in a Christian church, residence in urban areas, especially Monrovia, and the accumulation of Western products; in other words, with "civilization" as defined by the American settlers. Under their cultural hegemony, Liberia was imagined as an outpost of Christianity, democracy, and Euro-American capitalism on the "Dark Continent." In fact, even as other independent countries emerged around them in the 1950s and 1960s, Liberia's leaders continued to define themselves as more American than African (as indicated by the use of the term "Americo-Liberian," "settler" or "pioneer" for self-identification) (see Dunn and Tarr 1988, Liebenow 1987).

It is probably safe to say that until quite recently, for the majority of the population other forms of personal and group identity were more central than a sense of being Liberian. Although the lack of a color bar between the colonizers and the colonized allowed many individuals to pass into the elite through marriage, adoption and patronage, most maintained a geographically defined affiliation. Rather than "tribe" or ethnic group, however, these local identities were built around either small clusters of towns with their accompanying farmlands (in the south and east) or loosely structured, often multi-ethnic and multilingual chiefdoms (in the north and west). Unlike the isolated ethnic villagers of Africanist stereotypes, many indigenous Liberians in the eighteenth, nineteenth and early twentieth centuries had extensive knowledge of other peoples, including traders from the northern savannahs and Europeans along the coast. They were aware of a variety of ways of being "civilized" and acquired many "civilized" traits, including proficiency in European languages and access to Western commodities, without the mediation of the settlers (see Tonkin 1979). The official nationalism of the settlers began to take on salience only in the post-World War II period, when a booming economy based on the export of iron ore and rubber encouraged permanent labor migration and the spread of schools and other state institutions in the interior.

Official nationalism in Liberia was thrown into question with the 1980 coup. The new military leaders, young men from a variety of indigenous backgrounds and with modest levels of formal education, were faced with redefining the nation-state they had just acquired by force in a way that would both justify and celebrate their own actions. Ordinary Liberians, of both indigenous and settler origins, suddenly found new possibilities open to them in both government and the private sector, as many in the former ruling class became political refugees abroad. The influx of moderately educated people of indigenous descent, who could never have hoped to rise so far under settler dominance, into government ministries and other state institutions created for many new questions of national identity (see Moran 1996). Should the new government construct itself in class, ethnic and/or universalizing terms? Should one ethnolinguistic group, out of more than sixteen recognized "tribes," be elevated to represent the totality, and, if so, which one? Without the settler class in control, was Liberia still a "civilized" nation? Was it now a more authentic African one? These issues were of evident concern to urban, and, to a lesser extent, rural Liberians during my fieldwork in 1982–83.

"Civilization" in Liberia: the local context

Liberia's peculiar history of American settlement and early independence resulted in a conception of "civilization" with clear roots in the nineteenth century. Although originally describing the cultural differences between African-American settlers and indigenous people, the term "civilized" came to be used locally to include educated, well-employed and/or Westernized sophisticates of all backgrounds. Neither an ethnic category nor a class fraction, "civilized" Liberians may best be understood as sharing "status honor" in the Weberian sense (Moran 1990). Each of Liberia's indigenous ethnic groups developed its own sector of "civilized natives" whose connections to "non-civilized" kin and friends (and sometimes spouses) take a variety of forms. My initial work, for example, was with the "civilized Glebo" of Cape Palmas, who trace their own history to the coming of Episcopalian missionaries who they claim preceded the black American settlers to the region (Moran n.d.).[2] By the mid-nineteenth century, separate "civilized towns" of educated, Christian Glebo had sprung up, by their very presence challenging the settlers' claim of cultural superiority.

In all of its specific, local manifestations, "civilization" includes standards of dress, personal hygiene and home decoration, as well as commitment to "civilizing" institutions such as churches, schools and the state bureaucracy. Virtually all analysts (and, as Tonkin notes, nearly every writer on Liberia in the twentieth century has felt compelled to discuss this concept) have emphasized the positive, moral aspect of the term (see Moran 1992, Tonkin 1981, Frankel 1964). In the case of "civilized natives" like the Glebo, this moral loading has constituted an implicit critique of Liberia's long history of political dominance by the settler upper class. By claiming that they could be both "civilized" and "native," educated Christians of indigenous background presented a challenge to the acceptable route of upward mobility; absorption into the settler group through adoption of a fictive genealogy which denied "native" origins.

The American settlers, like European colonialists, based their claims to territory and their right to administer "natives" on the objective fact of their own "civilization." Unlike Europeans, however, they could not ground these claims in an explicit ideology of racial superiority. The settlers had no choice but to work within a discourse of civilization as acquired rather than innate, but this presented problems of how to regulate access to powerful institutions. "Civilization" as an ideological construct never completely masked the realities of the Liberian class structure, in which a small group of settler families controlled the majority of both state bureaucratic and private financial institutions (see Brown 1982). It was possible, however, for indigenous people who chose to maintain their "tribal" identities to simultaneously invoke the moral, universalizing aspects of being civilized. "Civilized Glebo" who suffered discrimination and segregation in employment, residence and even church congregations at the hands of the settlers (see Martin 1968) still held themselves to be the model of "true civilization" for both their "native" kin and the local settler community. Although deeply committed to their own version of this complex of values and lifestyle, they clearly recognized that it was not enough to grant them access to state power on either the local or national level.

From this point on, I shall use the terms civilized and civilization – without quotation marks – to refer to these particularly Liberian constructions. It is precisely the moral, positive aspect of these terms as they are deployed locally which may be called into question when the person designated as civilized is female. For the implications of civilization, depending as they do on access to an income from the wage sector, have historically been fraught with contradictions for women.

Among Liberia's indigenous populations, cultural constructions of femininity cast "native" women as breadwinners. Under a "female farming" regime of shifting dry rice cultivation, women work on land that belongs patrilineally to their husbands or fathers. Although an individual woman may achieve significant levels of independence, the ideal dictates that she direct her economic energies toward the support of her children, not toward personal consumption. Financial independence for native women is an aspect of their obligations to kin and household, not a route to competition with men. Civilized women, in contrast, are defined by the fact that they do *not* participate in farm labor. Held to be "not strong" enough for strenuous work, civilized women were ideally to be dependent housewives, fully occupied with the care of home, children, and other household members such as servants and foster children. Through their daily domestic practice, including the laborious care of such status markers as clean, well-pressed Western-style clothing (especially children's school uniforms), these women both produce and reproduce the status honor of the entire household as well as the next generation of civilized people (Moran 1990, 1992). The accumulation and display of Western commodities, not to mention the typical dependence on government payrolls, symbolizes the integration of civilized Glebo into national Liberian life, even when such people maintain residence in "native" towns to which they have patrilineal affiliation.

Civilized status is not formally assigned in Liberia; like Weber's view of status honor, it is the consensus of the community which determines who is or is not civilized. While rural men cannot be involuntarily stripped of civilized status once it is attained, women are in a more vulnerable position. A woman loses her status as civilized by engaging in the "wrong" type of work, such as subsistence farming or public marketing, precisely the productive and cash-generating activities associated with "native" women. This shift in status is signaled by a change in clothing from the Western-style dress, worn only by civilized women, to the wrap-around cloths or *lappas* (which may be worn by a woman in any social category, but is especially associated with "native" women). So strong is this connection between clothing style and communally recognized status that gossip about women who "used to be civilized" but, out of necessity, "tied *lappa* and made market," was common in the strained economic times of the early 1980s. For example, during my fieldwork my foster mother was planning to visit a young kinswoman who had been raised in her house and had gone to live with her husband in Côte d'Ivoire. Before she could leave, she received word that the young woman's husband had lost his job. Stranded in a foreign country, the woman had begun selling in the public market place to support them both; as a consequence, she was no longer seen as civilized. My foster mother decided to postpone her trip, because to see the young woman wearing *lappas* would "embarrass them both too much." She hoped that the loss of status would be a temporary one, but knew that some women spent many years as "natives" who "used to be civilized." The shift in dress symbolized the differential experience of civilized status for women and for men.

The constraints that civilized women seemed willing to endure in return for prestige were enforced by the sanctions usually found in small, face-to-face communities: gossip, exclusion, loss of reputation and respect. The prestige of civilized status was further enhanced by such local institutions as the Christian churches and their affiliated women's clubs and organizations – open only to civilized women. In the rural context, therefore, the dilemma of how a woman could be civilized without being somehow "too modern" (and therefore sexually suspect) was neatly solved: she could be civilized as long as she was economically dependent upon a man, her modernity thus controlled and circumscribed. When this economic control was loosened, as with financially independent market women, the status was simply withdrawn. The few well-educated women who were employed as teachers or clerks provided rare examples of civilized women who could afford to be without husbands. Competing with men for scarce jobs and operating in the public domain of professional work, these women frequently had to defend their respectability. And it was, not surprisingly, these women who were under the most pressure to live morally exemplary lives and who were frequently targets for gossip and speculation. Most women who wished to be considered civilized, however, had neither the education nor the personal connections for highly coveted white-collar employment. Unless they could formalize a relationship with a well-employed man, women were reduced to patching together a precarious subsistence by relying on kin, friends and the fathers of their children. Entering the market, as in the example above, was a desperate last resort. One woman tearfully told me that she would "tie *lappa* and make market, never wear dresses again," if this was the only way to keep her children in school; in other words, to give her children at least a chance at civilized life.

In the more anonymous urban context, with its different and varied opportunity structures, the constraints of the locally constituted community lose their power and the outward symbols of civilization (especially clothing) are increasingly commodified. Here, the gender dimension of civilized status becomes a highly visible field of contestation and struggle in such media as the newspapers, which come to replace or at least augment word-of-mouth gossip. Obviously, people not only read newspaper features but talk about them as well. Bastian (1993) has noted the textual interpenetration of oral and print narratives in the eastern Nigerian city where she worked, "Stories that appeared in the tabloids one week might very well make their way back to me as choice Onitsha gossip the next – with names and situations altered to suit the local taste" (1993: 131). Gupta's insight, that newspapers become a form through which "daily life is narrativized and collectivities imagined" (1995: 385), directs us to take very seriously the images that appear in their pages. When, as we shall see, narratives about civilization contain embedded narratives of "the nation" – and when these narratives are plainly gendered – they are all the more crucial.

The cartoons: civilization in the urban context

In addition to the usual feature stories and photographs, Monrovian newspapers in the 1980s carried line drawings that seemed to speak directly to the lives of struggling urban residents. Widely accessible to people of even minimal literacy, the dialogue was written in the "Liberian English" of working-class urban neighborhoods rather than the more standard English of the newspapers' other articles and editorials. These cartoons were not "comics" in the sense of being intentionally humorous or installments of an ongoing narrative. Neither were they exactly analogous to "editorial" or "political" cartoons, since they date from the period of military rule when overt political commentary was illegal, and press harassment and restrictions on journalists were increasing. Rather than directly portraying or satirizing government officials, they depict and comment upon the personal and domestic problems common to urban dwellers: schoolgirl pregnancy, marital infidelity and economic competition.

In 1982–83 a series of cartoons by the artist "Black Baby" appeared portraying urban women as duplicitous and manipulative. Although they depicted the lives of working-class people, the cartoons were also of interest to Monrovia's intellectual and academic elite, eliciting extensive comment and discussion among my colleagues at the University of Liberia, particularly the women. I was present when a young female university administrator confronted the newspaper's publisher, complaining that the cartoons were sexist. The publisher defended his artist by saying: "But Black Baby's not making them up, he hears them on the bus!" The publisher and the cartoonist believed, apparently, that they were representing the authentic voices of working-class Monrovians (since wealthier people ride in taxis or private cars, not on buses) and providing a forum for working-class concerns. The fact that they were read by such a wide range of urban residents makes it possible to view the cartoons as a form of cross-class discourse.

Schuster (1979: 142) has noted that African journalists occupy a rather ambivalent class and political position. Usually young men with above average levels of education, journalists, especially those at independent or opposition newspapers, are generally paid less than similarly educated colleagues in government service. Although they are in a position to sharply observe the wealthy and powerful, journalists' daily experience of urban life more closely resembles that of the working class: an unending struggle for decent housing, reliable transportation and access to other basic services. Male journalists can find themselves competing, both professionally and personally, with similarly educated women who, by deploying their sexuality, may seem to have an advantage in gaining access to jobs and the patronage of powerful figures (Schuster 1979: 146). Moreover, these desirable women are likely to be uninterested in romantic relationships with crusading journalists of modest means. Schuster suggests that the vitriolic newspaper campaigns against urban women, especially educated, modern women observed in so many African countries, can be traced to this ambivalence on the part of male journalists. While I have no direct evidence of the experiences that motivated Black Baby to produce his cartoons, these themes of urban competition and female duplicity were also visible in the reporting of news, editorials, and in the letters from readers chosen for publication in the *Daily Observer* and other Monrovian newspapers.

I have selected eight examples to illustrate the most common themes (see Plates 5.1 to 5.8). What is immediately interesting is the transformation of the category of civilized, which in the rural context had a clearly moral, positive aspect, to something much more ambiguous and even sinister. As they are drawn in these cartoons, the women are marked with the outward signs not only of civilization but also of wealth and modernity: Western clothing, hairstyles and jewelry. They are metropolitan sophisticates rather than the upstanding, educated but still clearly indigenous rural elites delimited by the term in rural towns. Except when cast as the suffering wives of men lured astray, urban women are usually portrayed as predators and aggressors. In contrast with the material represented in the earlier literature (Little 1980, 1973, Schuster 1979), however, these cartoons are not monolithically misogynist; while their critique largely targets women for neglecting family and job responsibilities, wasting money, and not thinking of the future it also sometimes includes men. The cartoons thus offer several different visions of civilized womanhood and female citizenship.

Plates 5.1, 5.2 and 5.3 resemble the images common in other studies of urban African women which characterize women as sexually and economically predatory. Yet these three cartoons also comment on the depressed and uncertain nature of the urban economy, as with the young man who can afford only $50.00 toward his girlfriend's rent (Plate 5.1), the miniskirted woman who notes that "money business hard these days" (Plate 5.2) and the recently unemployed man who finds his young girlfriend losing interest in him (Plate 5.3). In Plate 5.4, we see the confrontation between a man and his wife, whose *lappa* suit marks her as probably "native"; the traditional woman who puts her priorities appropriately on home and family. The man addresses her as "Madam," the polite title for an adult "native" woman of relatively high status; prominent market women are usually addressed in this manner. The husband, who has apparently been missing for some time, leaving his family "dying" for lack of financial support, excuses his own behavior by pointing out that "those street girls are not easy," a reference to the essentialized understanding of men as physically unable to resist displays of female sexuality. All four of these cartoons attribute to women an unfair advantage in the shrinking economy by virtue of their sexuality and their ability to lure men away from home and paycheck.

More pointedly, in Schuster's (1979) account of newspapers from Lusaka in the early 1970s, such women were portrayed as betraying the country by drawing resources and men's attention away from the common goal of national economic development. Here we see the tendency toward naturalization of both gender and nationalism as categorical identities referred to above. This naturalizing discourse casts women as the enemies of national development, less because of their actions than because of their very essence. The equally essentialized patriotic Liberian, concerned primarily with the overall good of the country, can therefore never be a woman. Furthermore, this discourse extends a moral judgment on all who would use material resources for their own selfish and "unproductive" purposes. In a period of increasing restrictions on the free press and a country-wide ban on "political activities," such comments may have been veiled references to the military government then controlling access to government jobs, patronage and wealth. Cloaking this political commentary as a critique of civilized womanhood therefore serves multiple purposes for the cartoonist and his readers.

But the images contained in the cartoons are neither stable nor consistent from day to day or week to week. The contested and fluid nature of these constructs is obvious in the next two plates. Plate 5.5 actually acknowledges and comments on the scapegoating of women, noting that men "never blamed themselves one day, but always blaming us." And in Plate 5.6, a fatherly patriarch holds out the prospect of redemption and extolls the potential of educated, civilized women's contributions to national life: "The time has come when women must no longer be armchair citizens, sitting back and criticizing or gossiping. Try to be a good woman so your man and society will be proud of you." Here, an explicit reference is made to the

Plate 5.1 *Daily Observer*, May 18, 1983

Plate 5.2 *Daily Observer*, November 2, 1982

Plate 5.3 Daily Observer, February 28, 1983

Plate 5.4 Daily Observer, March 22, 1983

local image of the civilized woman, the upstanding pillar of society, whose position depends on, reflects and represents not only the status of her man but also of the nation as a whole. Recall that for the civilized Glebo of Cape Palmas, it is the domestic labor of women which reproduces the conditions necessary for civilized life: an ordered home, clean, pressed clothing, well-trained children (Moran 1992). Being only an "armchair citizen" is like being an "armchair housewife," a contradiction in terms which threatens not only the family ("your man") but "society" as well. Ironically, the women in this drawing wear almost smirking expressions; although they mouth a respectful thanks and promise to be "change[d] women," there is an air of defiance in their stance.[3]

The claims of patrilineal kin groups, represented by either the husband or the father, over sexual access to and the fertility of young women is a theme throughout the cartoon collection which illustrates my second point about the intersection between gender and nationalism, taking the form of an almost obsessive concern with reproduction. There is a sense of nostalgia expressed here for a traditional past in which women's sexuality was controlled rather than deployed, and women "knew their place." Ethnographic sources on indigenous political competition suggest that sexuality, either in the form of arranged marriages or other kinds of liaisons, has long been a factor in the local level strategies used by both women and men, rather than a recent or specifically urban innovation (see Bledsoe 1980, Hoffer 1974). Nevertheless, heightened concern with the conditions of reproduction seems especially likely in a time of uncertainty about the future. The cartoons comment extensively on the dangerous potential of unwanted pregnancy to disrupt a young woman's education, her future prospects for civilized status and her potential contribution to the nation.

The educational system in Liberia, like elsewhere in Africa, is stratified into government-funded public schools, for which students must often pay tuition, parochial or "mission" schools, and a largely unregulated sector of private vocational institutions. Many of these private programs, which offer classes at night, promise to provide students with job-related skills like typing or accounting, but lack the equipment and trained faculty to do so. Girls enrolled in government or parochial schools who become pregnant must leave as soon as their condition is discovered and are not allowed to return after giving birth. The only education options open to these young women are the private night schools, which, although costly, do not confer recognized degrees.

Since parents send children to school when they are deemed "big" rather than at a set chronological age and the educational career of any child may be delayed or disrupted by fluctuations in household income, it is not uncommon to find 18 and 19 year olds in the fifth or sixth grade. Young women who are forced by motherhood to leave school at this point are usually ideologically committed to a civilized identity but lack the skills and education to maintain it themselves. It was exactly this subset of civilized Glebo women with whom I worked most closely in 1982–83. These were

Plate 5.5 *Daily Observer*, September 21, 1982

Plate 5.6 *Daily Observer*, March 10, 1983

the women who were so fearful of having to "tie *lappa* and make market" in order to support themselves and their children. From the point of view of the parents, not only had the investment in a girl's tuition, uniforms, shoes, books and supplies been wasted, but also she was now likely to become a financial dependant rather than a contributor to the household. Unless she could extract a reliable stipend from the father of her child, she had few options for remaining civilized. Knowing the likelihood of this disruption, parents were frequently reluctant to invest in the education of daughters.

The cartoons present two models of schoolgirl behavior: a positive one to be emulated, and a negative one to serve as a warning. In Plate 5.7 the young woman's appropriation of her own sexuality results not only in her father's anger but also, significantly, in the loss of her own status as a student ("Is this the kind of graduation you think I want for you to have?"). In Plate 5.8, the female student rejects a young man's approach, but grounds her refusal in the name of her parents' investment in her education ("Do you think I am coming to allow any boy to spoil me for this '83'. My parents spend a lot on me"). The message to young women is quite clear: virtuous civilized status is possible only at the cost of economic and sexual independence. Like the evil temptresses pictured in Plates 5.1, 5.2, and 5.3, pregnant schoolgirls misappropriate and waste the resources of both their families and the nation at large. In effect, they threaten the orderly reproduction of the nation by exerting control over a resource (their own sexuality) not rightfully theirs.

But what of those women who, in the rush to modernity, insist on flouting the rules of respectability? The possibility of simply stripping women of civilized status, so effective in the small community, appears to be useless in the urban context of increasing commodification of human bodies and social relations. How can a woman be spoken of as one who "used to be civilized" when she refuses to participate in her own demotion by giving up dresses for the market women's *lappas*? And why should she, when, as the cartoon in Plate 5.2 notes, she can always "freak those men's minds at least to get something from them"? The "something" women can get from men in return for sex is economic support, not only enough to survive but also to prosper. Again in Plate 5.2, the woman in the cartoon attributes this power to her ability to "dress that kind of way," in other words, to the deployment of commodified sexuality within a Western aesthetic of beauty. The cartoonist has lavished considerable effort on the details of this woman's costume, jewelry and hairstyle. In effect replacing the local gossip networks, the popular press condemns such misappropriation of power by opening to question assumptions about civilization as a desirable status. A poem, published in 1983 in the same newspaper as the cartoons and reprinted several times by popular demand, expresses this combination of ambivalence, attraction and repulsion felt toward the civilized woman.[4] Describing a woman wearing a short, tight red dress with matching shoes, makeup and expensive jewelry (in effect, the woman in Plate 5.2), the poem comments:

Plate 5.7 *Daily Observer*, September 22, 1982

Plate 5.8 *Daily Observer*, January 18, 1983

She's civilized, so she says
But her body she turns into a commodity
Whose price is determined by negotiation
Shattering all traces of dignity.

The last stanza of the poem concludes:

And she's very proud of her role
Despising others she considers old-fashioned
But if civilization is measured by this yardstick
I want none of it

(J. Dio Hne, "The spoilt child," *Daily Observer*, February 18, 1983)

In this image of the civilized woman as a prostitute, the consequences of loosened sexual control are equated with commodification and the proliferation of images inspired by the Western media. Clothing, which once served to enforce the standards of female behavior by limiting Western dress to economically dependent and sexually controlled women, can no longer contain the forces of rampant commodification. The "miscarriage of society, the destroyer of people's dreams," as the poet describes this independent woman, can represent only the most amoral and alienating effects of civilization, not the "idealism, enthusiasm, and passion which it [civilization] really evokes for many" outside of the urban context (Tonkin 1981: 323). The title of the poem, "The spoilt child," reinforces the images of selfishness, waste and promotion of individual gain over communal responsibility given visible form by the cartoons. How, these cartoons ultimately ask, can a commodified female body produce authentic human beings to carry forth the national agenda?

Conclusion: gender, civilization and nationalism in Liberia

The cartoons discussed above show clearly how naturalization, commodification and an overdetermined emphasis on biological and social reproduction contribute to the construction of both gender and national citizenship in Liberia. In the early 1980s the contradictions between the pre-coup official nationalism, grounded in nineteenth-century ideas of civilization, and the role of the state as a means of resource allocation were becoming increasingly exposed. While Liberian newspapers personified such contradictions in cartoon images of the civilized woman, ironically it was a man, Samuel K. Doe, who enacted them in real life. One way of explaining the ongoing salience of the civilized–native dichotomy in Liberia is by highlighting its ability to occlude relations of political domination. Noting the tension in pre-coup Liberian political discourse between the myth of pure settler descent and the need for political collaboration with the indigenous elite, David Brown (1982) argues that the cultural elaboration of civilized status is a form of mystification. According to this view, allowing some "natives" to become civilized and giving them a stake in maintaining this status was actually a way of obscuring the fact that one "ethnic" group, the settlers, constituted a privileged class (Brown 1982: 299–301).

Based on this analysis, Brown predicted that the military coup of 1980, in which soldiers of indigenous background overthrew the settler oligarchy, would require "a radical redefinition of the legitimations of the state" (Brown 1982: 302). Brown was aware, however, that because the coup occurred so suddenly and unexpectedly, "the circumstances in which the Americo-Liberian elite was overthrown in 1980 prevented the mobilization of competing ideologies . . . so that the implications of the conceptual ambiguities in the ideology of the ruling class were never followed through" (ibid.). In other words, while the official nationalism of the pre-coup period held out the promise of inclusion to all who achieved civilized status, in reality it blocked the upward mobility of civilized "natives" like the Cape Palmas Glebo. The inherent contradictions embedded in the framing, validation, and enforcement of claims to civilized status were, by 1982, becoming publicly exposed in a manner that was hard to ignore. Indeed, the contest over the meaning of civilization intensified in Liberia in tandem with increasing awareness of the instability of state institutions and with the new deployment of "tribal" ethnicity as a political tool.[5]

What the young soldiers took by force in April 1980 was not the civilized nation of Liberia but the *state*: that collection of apparatuses, institutions and technologies of power which, in the post-colonial African context, is the major route to resource acquisition and the accumulation of wealth and so the primary prize in any political struggle. Samuel Doe, who emerged from the small group of coup plotters to become chairman of the ruling council and later, through a rigged election, President of Liberia, had only a tangential and ambiguous claim to civilized status. A member of the most obscure and remote ethnic group in the country who possessed a tenth-grade education, Doe at first tried to appropriate civilized status but only opened himself to ridicule. After receiving an honorary degree from the University of Korea, he insisted on being addressed as "Dr. Doe" and later "earned" a degree from the University of Liberia by importing faculty to the Executive Mansion for private courses. Only 26 years old at the time of the coup, Doe quickly gave up the camouflage uniform of a non-commissioned soldier for three-piece suits and wireframed glasses in an effort to look older and more statesmanlike. Yet, while Doe attempted to solidify his claim to civilization, his wife, well known as a former Monrovian market woman, was constrained by a different set of rules. As we have seen, civilized women may temporarily fall to "native" status, but the process does not work in reverse. No unschooled market woman would dare to try and pass herself off as civilized, even when suddenly elevated to the role of First Lady. Mrs. Doe was therefore careful never to appear in public in Western dress. Even so, by 1982, people were already commenting that Doe had become "fat" since taking office – a clear reference to the belief that he was "eating the money" or using state resources for personal gain. One Liberian commentator remarked in 1986 that Doe "desperately yearns to be considered a 'civilized' man. . . . [but] He cannot grasp its finer points and nuances nor the intricacies of its symbols" (Liberty 1986: 45).

The experience of wrestling with and attempting to redefine the nuances and symbols of civilization permeated Liberian life in the 1980s, from those overheard on the bus by Black Baby to the highest office in the land. I argue that what was going on in the pages of Monrovian newspapers, therefore, was not only a contest about male control of media images or even the depressingly familiar scapegoating of urban women. Equally important, what was at stake was how the civilized nation of Liberia would constitute itself and its citizens. The contradiction between the negative representation of civilized womanhood in the newspaper cartoons and the positive values accorded to the civilization by the Glebo of Cape Palmas (and by other small communities throughout the country) is a displacement of the contradiction between civilization and the official nationalism of pre-coup Liberia. Just as the journalists and cartoonists who controlled the production and dissemination of media images seemed unable to project a unified, hegemonic representation of "civilized" womanhood, the government of Samuel Doe was unable either to dispense with or to productively modify the concept of civilization, even so as to include the top leadership. Like the cartoon civilized women, or the "spoilt child," Doe was directing resources away from their proper roles: the reproduction of households, kin groups and the nation. His pose of civilization was no more convincing than that of the poet's streetwalker, but like her he had the raw material power to ignore those who laughed at his pretensions. Yet, he was also trapped by the official nationalism that linked the entity "Liberia" to civilization. Doe could not say of civilization, "I want none of it," without laying bare his own instrumental manipulation of the state for personal gain.

By 1986, Samuel Doe was finding that his claims to civilized status could not protect him from other ambitious young soldiers like himself. Faced with pressure from counter-coups and organized civilian opposition, Doe fell back on ethnicity, promoting members of his own group, the Krahn, to high-ranking government positions and surrounding himself with ethnically homogeneous military units. Eventually he began taking revenge not only on his rivals but also on unrelated members of rivals' ethnic groups, making utterly transparent the exclusive ethnic status of those who controlled the state apparatus and, therefore, of the opportunities for accumulating wealth. In effect, Doe was moving away from the official nationalism of civilized Liberia toward its supposed opposite, the "native." In the elision of dichotomies, civilized became to the nation what "native" was to "ethnicity." Doe began to deliberately manufacture and promote essentialized "tribal" rivalries in a desperate effort to maintain power. Predictably, his actions led to the emergence of a number of ethnically defined resistance groups with no political ideology beyond opposition to Doe himself, led by men whose sole objective was to replace Doe as head of state. Although even the most cursory analysis shows that such politicized ethnicity in Liberia is a product of the period since 1983, the Western media continue to describe such tragedies as the result of "ancient tribal hatreds." Not surprisingly, since, as we have noted here, the essentializing of gender and other identities tends to occur together, the increasing militarization of Liberian political life has shaped the ideological construction of gender, with new forms of violent masculinity taking privileged positions (Moran 1997, Enloe 1995).

Samuel Doe lost his struggle to become civilized and, eventually, his life in 1990, as rebel factions closed in on Monrovia and a multinational West African force intervened to prevent any one group from claiming victory. The war dragged on for seven years after Doe's death, leaving over 200,000 people dead, at least 1 million displaced, and the descent of Liberia into stateless, ungovernable violence and chaos. Only in 1996 was a peace agreement signed by all the warring factions, and elections were held in July 1997. Less encouraging is the fact that the winner of the election for president, Charles Taylor, is the man responsible for starting the war in the first place.

The newspaper images which both amused and annoyed Monrovia residents in the early 1980s represent only a moment in this crisis of redefining the Liberian nation. The cartoon representation of civilized women, whose relationship to both national identity and civilization was historically more tenuous and ambiguous than that of men, provides a striking metaphor for this moment in Liberian national history. As the country emerges from the long nightmare years of the 1990s, the question of how male and female citizenship will come to be defined remains very much open.

Acknowledgments

Research in Liberia in 1982–83 was supported by a National Science Foundation Graduate Fellowship and a Hannum-Warner Travel Grant from my undergraduate institution, Mount Holyoke College. Early versions of this chapter were given at meetings of the American Anthropological Association and American Ethnological Society. I am grateful to Tamar Mayer for her comments and patience as an editor and to Anne Pitcher for a close, critical and very helpful reading. Jordan Kerber, as always, provided crucial editing and computer assistance.

Notes

1 Of course, there are contexts, particularly in struggles for "national liberation," in which men and women collaborate in resistance against a common enemy.

Women may willingly embrace essentialist and/or restrictive gender constructions under such circumstances. Alternatively, Enloe (1995) suggests that moments of national redefinition and even militarization may open new avenues for women's autonomy and influence in national life. Thanks to Anne Pitcher for reminding me of this point.

2 All available historical evidence, however, points to the arrival of the settlers before that of the Episcopalian missionaries. I have argued in an unpublished paper that there are strategic reasons for the Glebo chroniclers to insist on their version of these events (Moran n.d.).

3 Again, I am indebted to Anne Pitcher for this observation.

4 Permission to reprint the cartoons and to quote from the poem "The spoilt child" was generously granted by Mr. Kenneth Best, publisher of the *Daily Observer*.

5 For more on the deployment of ethnicity during this period, see Moran (1996).

References cited

Anderson, B. (1991) *Imagined Communities*. London: Verso.

Anthias, F. and N. Yuval-Davis (eds.) (1989) *Woman–Nation–State*. London: Macmillan.

Bastian, M. (1993) "Bloodhounds who have no friends: witchcraft and locality in the Nigerian popular press," in Jean Comaroff and John Comaroff (eds.) *Modernity and its Malcontents*. Chicago: University of Chicago Press, pp. 129–166.

—— (1995) "Mami Wata and the lure of Mrs. Money: spirits and dangerous consumption in the Nigerian popular press," paper presented at annual meeting of the African Studies Association, Orlando, FL.

Bledsoe, C. (1980) *Women and Marriage in Kpelle Society*. Stanford, CA: Stanford University Press.

Brown, D. (1982) "On the category 'civilised' in Liberia and elsewhere," *Journal of Modern African Studies* 20, 2: 287–303.

d'Azevedo, W. (1969a) "A tribal reaction to nationalism (Part 1)," *Liberian Studies Journal* 1, 2: 1–22.

—— (1969b) "A tribal reaction to nationalism (Part 2)," *Liberian Studies Journal* 2, 1: 43–63.

—— (1970) "A tribal reaction to nationalism (Part 3)," *Liberian Studies Journal* 2, 2: 99–115.

—— (1970–71) "A tribal reaction to nationalism (Part 4)," *Liberian Studies Journal* 3, 1: 1–19.

Dunn, D. E. and S. B. Tarr (1988) *Liberia: A National Polity in Transition*. Metuchen, N.J.: Scarecrow Press.

Enloe, C. (1988) *Does Khaki Become You? The Militarization of Women's Lives*. London: Pandora/HarperCollins.

—— (1993) *The Morning After: Sexual Politics at the End of the Cold War*. Berkeley: University of California Press.

—— (1995) "Feminism, nationalism, and militarism: wariness without paralysis," in C. Sutton (ed.) *Feminism, Nationalism, and Militarism*. Arlington, VA: American Anthropological Association/Association for Feminist Anthropology, pp. 13–32.

Frankel, M. (1964) *Tribe and Class in Monrovia*. London: Oxford University Press.

Gupta, A. (1995) "Blurred boundaries: the discourse of corruption, the culture of politics, and the imagined state," *American Ethnologist* 22, 2: 375–402.

Hoffer, C. (1974) "Madam Yoko: a leader of the Kpa Mende confederacy," in M. Rosaldo and L. Lamphere (eds.) *Woman, Culture, and Society*. Stanford, CA: Stanford University Press, pp. 173–187.

Liberty, C. E. Z. (1986) "Report from Musardu (letter to an American friend): reflections on the Liberian crisis," *Liberian Studies Journal* 11, 1: 42–81.

Liebenow, J. G. (1987) *Liberia: The Quest for Democracy*. Bloomington: Indiana University Press.

Little, K. (1973) *African Women in Towns*. Cambridge: Cambridge University Press.

—— (1980) *The Sociology of Urban African Women's Image in African Literature*. Totowa, N.J.: Rowman & Littlefield.

Martin, J. (1968) "The dual legacy: government authority and mission influence among the Glebo of eastern Liberia, 1834–1910," unpublished Ph.D. dissertation, Boston University

——.(1969) "How to build a nation: Liberian ideas about national integration in the later nineteenth century," *Liberian Studies Journal* 2, 2: 1–19.

Moran, M. (1988) "Women and 'civilization': the intersection of gender and prestige in southeastern Liberia," *Canadian Journal of African Studies* 22, 3: 491–501.

—— (1990) *Civilized Women: Gender and Prestige in Southeastern Liberia*. Ithaca, N.Y.: Cornell University Press.

—— (1992) "Civilized servants: child fosterage and training for status among the Glebo of Liberia," in K. Hansen (ed.) *African Encounters with Domesticity*. New Brunswick, N.J.: Rutgers University Press, pp. 98–115.

—— (1996) "Carrying the queen: identity and nationalism in a Liberian queen rally," in C. B. Cohen, R. Wilk and B. Stoeltje (eds.) *Beauty Queens on the Global Stage: Gender, Contests, and Power*. New York: Routledge, pp. 147–160.

—— (1997) "Warriors or soldiers? Masculinity and ritual transvestism in the Liberian civil war," in L. Lamphere, H. Ragone and P. Zavella (eds.) *Situated Lives: Gender and Culture in Everyday Life*. New York: Routledge, pp. 440–450.

—— (n.d.) "What happened when? Strategies of chronology in Glebo historical discourse," unpublished paper.

Parker, A., M. Russo, D. Sommer and P. Yaeger (eds.) (1992) *Nationalisms and Sexualities*. New York: Routledge.

Ranger, T. (1983) "The invention of tradition in colonial Africa," in E. Hobsbawm and T. Ranger (eds.) *The Invention of Tradition*. Cambridge: Cambridge University Press, pp. 211–262.

Schuster, I. M. G. (1979) *New Women of Lusaka*. Palo Alto, CA: Mayfield.

Tonkin, E. (1979) "Sasstown's transformation: the Jlao Kru, 1888–1918," *Liberian Studies Journal* 8, 1: 1–34.

—— (1980) "Jealousy names, civilised names: anthroponomy of the Jlao Kru of Liberia," *Man* 15 (n.s.): 653–664.

—— (1981) "Model and ideology: dimensions of being civilised in Liberia," in L. Holy and M. Stuchlik (eds.) *The Structure of Folk Models*. London: Academic Press, pp. 305–330.

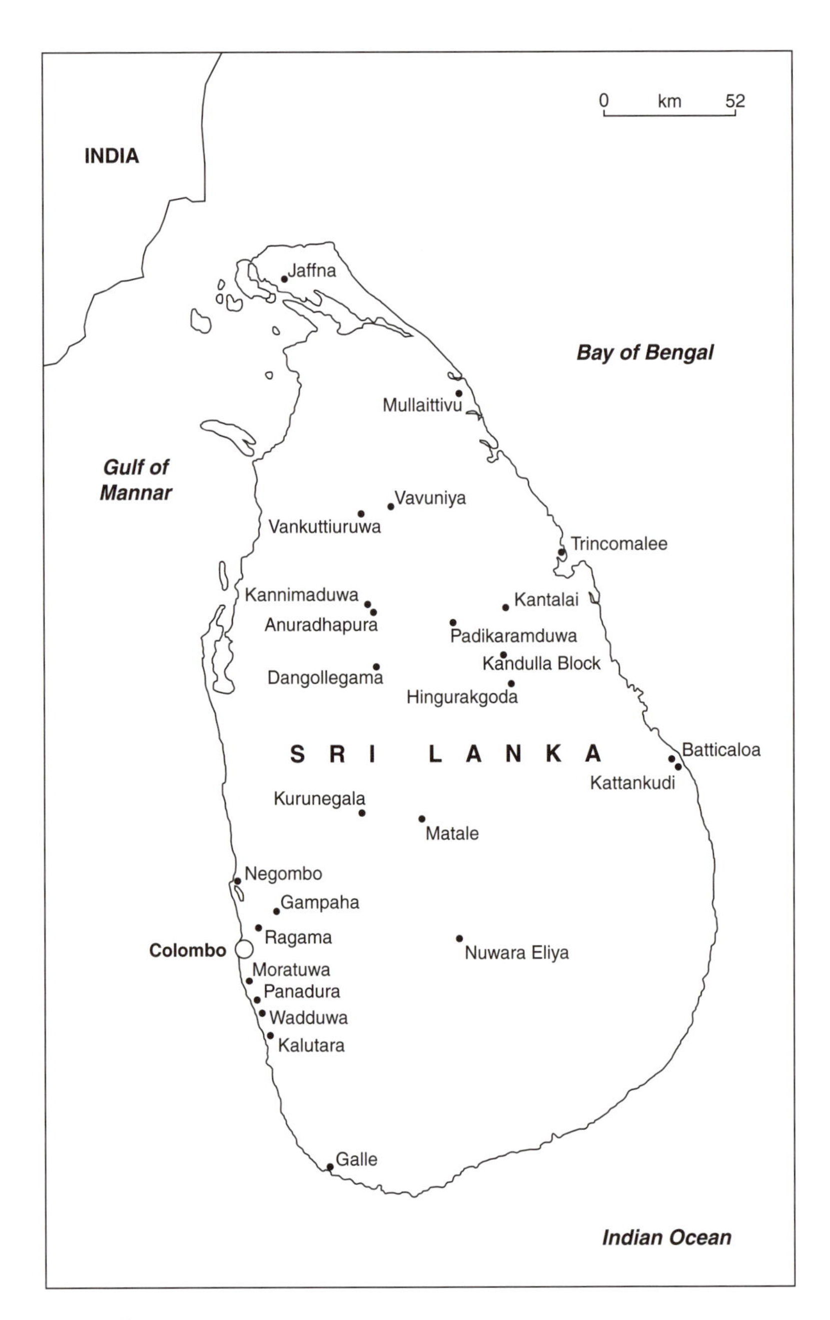
0
km
52
INDIA
Jaffna
Bay of Bengal
Mullaittivu
Gulf of
Mannar
Vavuniya
Vankuttiuruwa
Trincomalee
Kannimaduwa
Kantalai
Anuradhapura
Padikaramduwa
Kandulla Block
Dangollegama
Hingurakgoda
S R I L A N K A
Batticaloa
Kattankudi
Kurunegala
Matale
Negombo
Gampaha
Ragama
Colombo
Nuwara Eliya
Moratuwa
Panadura
Wadduwa
Kalutara
Galle
Indian Ocean

Sri Lanka

6 "Am I a woman in these matters?"[1]

Notes on Sinhala nationalism and gender in Sri Lanka

Jeanne Marecek

> When one speaks . . . of a nation . . . as a homogenous political entity, . . . [one assumes that] people homogeneously inhabit any given piece of territory.
>
> (Luxemburg 1976: 135)

Sri Lanka, known as Ceylon until 1972, is a small island in the Indian Ocean, 40 miles off the southeast coast of India. It is a country in which four of the world's major religions (Buddhism, Hinduism, Islam and Christianity) are represented and in which three languages (Sinhala, Tamil and English) are recognized as official languages. Sri Lankans identify themselves as members of one or another so-called "ethnic" groups – Sinhala, Tamil, Muslim and Burgher being the most prominent. The uneasy and sometimes violent relations among these groups are referred to by locals as "ethnic tensions"; indeed, even the fifteen-year civil war fought by Tamil separatists against the Sinhala-dominated state is usually referred to as the "ethnic conflict."

It is ethnic groups within Sri Lanka and the relations and tensions among them that are the focus of present-day discourses of national identity in Sri Lanka. Although I use the terms "ethnic group" and "ethnic conflict" in deference to local custom, "ethnic" in this case is actually a convenient misnomer, as these groupings are not uniformly constituted by ethnicity. Sri Lanka's Muslims for instance, trace their ethnic origins to a number of different regions of the world, while "Muslim" identity is based on religious affiliation. Nor are the categories of "ethnic" identity as stable, bounded or internally homogenous as the term "ethnic group" might imply. Among Tamils in Sri Lanka, for instance, place of residence and migration history serve as significant markers of difference.[2] Among Sinhala-identified people, religion (Buddhist or Christian) has demarcated an acrimonious divide at various times in the past. Indeed, within every ethnic group, cross-cutting factors such as language, religion, caste, region and urban versus rural residence make for multiple identifications.

Recounting the violent intergroup conflicts of the recent past, Valentine Daniel points out that in Sri Lanka, "race, religion, and language have formed an unholy alliance that is charged by its claimants, adherents, and speakers, respectively, with the mission of dividing the nation's citizens" (1996: 17). Scholars of history, anthropology, linguistics, sociology and Pali and Buddhist studies have brought considerable intellectual energy to assessing the linguistic, religious and racialized claims of popular nationalisms in Sri Lanka, amassing evidence to support or rebut those claims. In the main, this scholarship has regarded such issues as the Aryan ancestry of the Sinhala people, the territorial boundaries of traditional homelands and various claims of preferential treatment as questions of fact. Setting the factual record straight, scholars believed, could settle competing claims and presumably quell nationalist passions. Attention to the discursive constitution of claims and purported facts is relatively recent. Attention to the play of gender in nationalist discourse too is relatively recent.

This chapter concerns Sinhala nationalism and those Sri Lankans who identify themselves as Sinhala people, a group that accounts for 74 percent of Sri Lanka's population and that dominates the political and cultural life of the island. It looks beyond Daniel's "unholy alliance" of race, religion and language to gender as another constituent of identity. More specifically, I ask about the representations of women produced in the discourses and practices associated with Sinhala nationalism. First, I consider the narratives of Sinhala identity that emerged in the anti-colonial movements that began at the turn of the century. Then I shift to the present to ask how persisting residues of those narratives sometimes constrain and sometimes enable certain activities and identities of women in contemporary Sri Lanka. The chapter focuses exclusively on women and images of womanhood. It remains for future researchers to probe representations of manhood and masculinity, as well as homoeroticism and homosociality, in Sri Lankan discourses of national identity.[3] Moreover, my focus does not include Tamil peoples (who account for 18 percent) and Muslim peoples (7 percent), peoples who struggle for political representation, justice and safety within the Sinhala-dominated state.[4] I hope that even if it does not answer fully the questions that it raises, this chapter will spur further inquiry into the constructions of identity of all ethnic groups in Sri Lanka, and the ideologies of gender and sexuality embedded within them.

Partha Chatterjee's (1990, 1994) analysis of nationalism in India offers a framework for opening inquiry into gender and nationalism in Sri Lanka. The cultural and political histories of Sri Lanka and India (especially its southern portions) overlap considerably, and so we should expect correspondences between Indian and Ceylonese nationalisms as they took shape in the early part of the twentieth century. Sri Lanka was an integral part of a polity comprising a number of kingdoms of the South Indian region from the fourth century B.C. onward. Moreover, Sri Lanka and India shared a history of British colonial rule; the British administered both territories with policies and regulations that were inflected with English ideas about men and women, sexual morality and family relations.

Speaking about India, Chatterjee identified the central problematic in the quest for sovereign nationhood as follows: how to modernize the nation along Western lines while simultaneously establishing an essential local identity as the basis for political claims to nationhood. In Chatterjee's analysis, gender operates as a key element of the symbolic code for organizing the social world in such a way that these dual goals both can be achieved and the contradictions between them minimized or concealed. Gender is one of a linked set of binary oppositions: outer/inner, material/spiritual, modern/traditional, men/women, public/private and West/East. When the social world is bifurcated along gender lines, the activities demanded by modernization become part of the domain of men, while practices that create and sustain a national identity that is continuous with the past are assigned to women.

In Ceylon, as in India, "one of the predominant nationalist responses and resistance to British colonialism . . . took the form of a gendered separation of spheres: a feminized 'private' and a masculinized 'public'" (de Alwis 1995–96: 17). Different roles, different spaces and different orientations to past and future were associated with men and women. In the masculine public sphere, the demands and possibilities of modernization could be given free play. It was in the feminine private sphere of domestic life that cultural traditions were to be revived and preserved. Representations of ideal womanhood sustained a specific Sinhala identity anchored in the ancient past. As we shall see later, this ideal woman was modest, chaste and ignorant of the world beyond home. Positioned in the confines of the home, she was obedient to her husband and devoted to the needs of her children and other family members. These roles and traits were inscribed both as aspects of "Eastern" women's essential nature and as matters of patriotic duty. Nationalist ideologies about gender difference, gender relations and a separate feminine sphere were devised primarily by and for members of the urbanized middle class. As such, they defined the aspirations, obligations and moral character of middle-class "gentlewomen"; in contrast, women from the urban lower classes were deemed coarse, unclean and sexually immoral (cf. Mohanty 1991).

The next section turns to Ceylonese anti-colonial movements in the early decades of the twentieth century; it traces ideologies of gender and family relations embedded in assertions of Sinhala identity at that time. Ironically, the efforts to construct a local identity different from and superior to that of the foreign rulers remained heavily invested with the political values, moralities, and gender ideologies of those rulers.

Anti-colonialism and the construction of Sinhala identity

In Ceylon, resistance to colonial rule and demands for self-determination mounted in earnest at the end of the nineteenth century and continued until independence was granted in 1948. The national liberation movements coalesced around the Sinhala people, who were the majority of the population, and around Buddhism, the religion associated with the Sinhala majority. Thus, this wave of Ceylonese nationalism was Sinhala and Buddhist; its "Others" were the British, Christians (both local and foreign) and the West. Yet even when the anti-colonial movements were violently anti-Western, the ideologies they proffered reproduced the Western project of progress, liberalism and humanism in many ways. Indeed, the very project of constructing a sovereign Ceylonese state was grounded in Western models and ideals. In Ceylon, as in India, organized resistance to colonial rule was not in opposition to the culture of the rulers, but contained within it (cf. Spivak 1985). This should not be surprising: many of those who spoke out against colonial rule in Ceylon were English-speaking elites; they had been exposed to Western ideas in their education and some had been schooled in the West.

By the beginning of the twentieth century, Ceylon had experienced 400 years of European colonization; colonization began with the Portuguese in 1505, who were followed by the Dutch (1640–1796) and the British (1796–1948) in unbroken succession. The prolonged colonial occupation had created changes in all aspects of social and economic life, including gender arrangements. The successive waves of colonizers had imposed changes in marriage, gender relations and sexual practices, bringing them more into line with Western norms. European missionaries brought religions that demanded monogamy, forbade sexual unions and childbirth outside of marriage, and insisted that marriage was lifelong. Practices in each of those areas had been more relaxed prior to colonization (Yalman 1967); moreover, Christianization transformed sexual behavior into a moral issue. Furthermore, under British rule, sexual behavior and family life came under state regulation. For instance, the colonial administration required registration of marriages and births, declared infanticide a crime, outlawed bigamy, limited the grounds for divorce and required it to be regulated by the courts (Risseeuw 1988, 1996).

The regulation of marital and sexual practices facilitated the transition to an economy based on private ownership and individual initiative. As Carla Risseeuw (1992) details, it also had the effect of transforming gender relations in ways inimical to women's interests. Defining marriage as a lifelong monogamous union concentrated land ownership and economic power in the male members of a family. Moreover, such unions made women more dependent on their husbands and diminished their connection to their families of origin. More generally, Western ideals of womanhood and family were imposed upon local conceptions of marriage, divorce, adoption, offspring, rights of widows, women's work and sexual morality. Ironically, it was these conceptions, practices and values – not pre-colonial ones – that figured in the anti-colonialists' ideologies of righteous Sinhala womanhood and their vision of an indigenous and specifically Sinhala civilization.

Dharmapala and Sinhala Buddhist nationalism

Anagarika Dharmapala (1864–1933) was a central figure in the national liberation movement and in the Buddhist revival that took place in the early decades of the twentieth century. Dharmapala is still revered among Sinhala people as a national hero and a folk historian. Ethnic and religious chauvinism, as well as gross inaccuracies, are palpable in Dharmapala's fabrications of the ancient past. Yet, his writings are frequently reprinted in major newspapers, especially on Buddhist religious holidays. They are offered – and consumed – as accurate accounts of Sinhala history.

Dharmapala's thought embodied both a quest for modern sovereign nationhood, conceived along Western lines, and the claim of an essential local identity on which demands for sovereignty could be based. In a turn against his family background and education, Dharmapala had become adamantly opposed to Western culture, which he characterized as "barbaric." Despite his repudiation of Western cultural values, he nonetheless favored Western science, technology and industrial development. At the same time, Dharmapala preached the separateness and difference of the East from the West, and the Ceylonese (for him, co-terminus with Sinhalese) from Europeans. He proclaimed the Sinhala people to be descendants of the "Aryan race" and the people chosen to be the defenders and preservers of Buddhism. Several volatile ideas weave through his writings: the revival of a glorious and righteous ancient Aryan civilization, the racial superiority of the Sinhala people, the deep connection between the Buddhist faith and Sinhala ethnicity, and the island of Ceylon as the chosen province of Buddhism. All these ideas remain prominent today. Buddhism is a potent political force as well as a key social institution drawing large-scale financial support and other resources from the state. Although freedom to practice any religion is guaranteed in the constitution, Buddhism remains the "first among equals" (Bartholomeusz 1995). Although not all Sinhalese are Buddhists, the notion of a special relationship of Buddhism to the Sinhala people and to the island of Sri Lanka remains a rallying cry for Sinhala nationalist movements (Obeyesekere 1992). Moreover, substantial numbers of Buddhist clergy, from prominent religious leaders to ordinary monks, are vehement proponents of Sinhala hegemony; some clergy members support militarization and even violence to achieve that end (Tambiah 1992).

It is Dharmapala's views on women and gender relations that are of interest to us here. Dharmapala unhesitatingly accepted the support and patronage of wealthy European women, among them feminists, suffragists and political activists. Nonetheless, he insisted that Sinhala women repudiate European ways and feminist values. He admonished Sinhala women to "return" to the purported ideals of Aryan womanhood: they should be chaste and modest, subordinate their needs to those of the members of their households, and submit obediently to their husbands. According to Dharmapala, only by embracing such ideals could women earn respect and find true liberation (Jayawardena 1986).

Dharmapala endeavored to make women the guardians and visible representatives of Sinhala culture, assigning them the task of enacting cultural superiority and separateness. He urged a number of reforms on Sinhala women that would serve to set them apart from women of other faiths and ethnic backgrounds. According to Dharmapala, a Sinhala woman should not allow herself to become Europeanized; she should not wear European clothing, smoke or drink. He urged Sinhala women to don the *sari*, a garment that he believed signified modesty and tradition. (Ironically, the *sari* was not traditional Sinhala dress, but an import from South India.) Dressing in *sari* was not only a practice of modesty and tradition, but also marked a woman as Sinhala by her outward appearance. Tamil women, Muslim women and European women all dressed differently.[5] Thus, for Dharmapala, women's dress and demeanor would serve to demarcate the boundaries of Sinhala identity. Moreover, by practicing a superior form of femininity, Sinhala women would embody and ensure the cultural ascendancy of the Sinhala "race."

Dharmapala's nationalism imposed strictures on women that constrained their behavior, freedom of movement and self-determination. Not all efforts to inculcate a distinctive and separate Sinhala identity had constraining effects on women, however. The nationalist project of promoting Buddhist education had emancipatory potential for women. By the end of the nineteenth century, opposition to foreign rule prompted a rejection of Christian missionary education and a drive for Buddhist education. This drive was propelled by a number of intertwined forces: the anti-colonialists' desire to undermine the dominance of Christianity; a reaction against the British rulers' preferential treatment of Christians in employment; and the impact of the worldwide Buddhist revival. As part of this drive, wealthy families funded the establishment of Buddhist schools in several provincial towns as well as in the capital city of Colombo (Jayawardena 1995a).

The drive for Buddhist education involved educating girls as well as boys and thus it occasioned a shift in educational practice that had a dramatic effect on women. Prior to that drive, Sinhala families had resisted education for their daughters, a fact remarked on by foreign observers throughout during the 1800s (Harris 1994). Rates of literacy among women were extremely low. As late as 1881, the census had estimated that only 2.5 percent of all women in the country could read, compared to 30–40 percent of men; rates of literacy for Sinhala women in particular were even lower. By 1911, however, literacy rates for women increased to 10.6 percent. An even more dramatic improvement was registered among Sinhala women, whose rates of literacy increased from 1.4 percent to 9.1 percent (Denham 1912). This embrace of girls' education has continued unabated; Sri Lanka presently claims the highest female literacy rates (88 percent in a 1986 assessment) in the South Asian region (Jayaweera 1995a).

The drive for Buddhist education was first and foremost a reaction against colonial rule. Yet the initiative to give girls a Buddhist education encompassed a diverse array of motives and interests. For some educators, a Buddhist education was a means of emancipating women, a project crucial to the political and social advancement of the nation. For others, the goal of educating girls was to produce Buddhist wives who would cultivate the Buddhist faith in their homes and secure their husbands' faith against pressures to convert to Christianity. For others, providing girls with a Buddhist education was part of the project of constructing a Sinhala national identity. Educated to understand and appreciate their own cultural and religious heritage, women would carry that heritage forward to future generations. Yet even when schooling was designed to train girls only for domestic roles, it equipped them for activities in the public sphere as well. And no matter what goals educators held for girls, schooling served to widen female students' horizons.

The legacies of Sinhala Buddhist womanhood

Ceylon was granted independence in 1948. In the fifty years leading to independence, anti-colonial nationalism had become thoroughly enmeshed with Sinhala chauvinism. Following Independence, the identification of the state of Ceylon with the Sinhala majority had a poisonous influence on communal relations. The Sinhala-dominated government instituted policies and practices that favored Sinhala and Buddhist interests, exacerbating differences and divisions among ethnic groups. The passage of the Sinhala Only Act in 1956 marked a watershed in the new nation's history. This piece of legislation made Sinhala the single official language, thus marginalizing Tamil speakers and English speakers (whatever their ethnic group membership – Sinhala, Tamil, Muslim or Eurasian) economically, culturally and politically. The politics of language, which persisted throughout the 1950s, created a milieu in which English-educated people were assigned "the cultural role of national scapegoat" (Gooneratne 1992: 16). Many English speakers – even those who were bilingual – were sufficiently discouraged by the situation and fearful of the future that they elected to emigrate. Moreover, tensions between ethnic groups escalated too, flaring on occasion into outbursts of serious violence.

The anti-Tamil riots of 1983, the worst in the island's history, marked another watershed in post-Independence history. The violence lasted for several days, destroying lives, businesses and homes. The Sinhala-dominated government delayed in taking measures adequate to stem the violence; moreover, many believe that some state officials and police actively abetted the Sinhala mobs. In the aftermath of the 1983 violence, some Tamil people fled the country and others sought safety in the northern and eastern parts of the island, areas in which Tamil people predominated. Furthermore, what had been sporadic guerrilla activity by groups of Tamil militants in the jungles of the north and east consolidated into a full-scale war waged by the Liberation Tigers of Tamil *Elam* (hereinafter L.T.T.E.) against the state. As the war has persisted, the L.T.T.E.'s demands for greater representation and more political autonomy have transformed into a secessionist call for Tamil *Elam*, that is, a sovereign Tamil state carved out of the northern and eastern territory of the island.

The civil war in Sri Lanka has produced a crisis of unprecedented proportions. As the war continued into its fifteenth year, the public, in Neloufer de Mel's words, "have adopted ferocious communal positions" (1996: 168). Claims about the distinctiveness and cultural superiority of Sinhala people, as well as their natural right to political supremacy and to ownership of the island, are asserted by various groups, including such extremist groups as the Front for the Protection of the Sinhalese, the J.V.P. (*Janatha Vimukthi Peramuna* or People's Liberation Front) and the *Jathika Chinthanaya* (more or less, Nationalist Way of Thinking) movement. Indeed, there are presently some 56 Sinhala and Buddhist organizations united under an umbrella organization called the Sinhala Commission. Among its activities, the Sinhala Commission has issued public reports detailing injustices suffered by Sinhala people, staged public hearings and organized large-scale protest demonstrations. It has been working to prevent a negotiated settlement of the civil war, seeking instead a military victory. The Sinhala Commission also has registered strong opposition to proposed constitutional reforms that would transform Sri Lanka into a federation of states wherein predominantly Tamil regions would gain a modicum of autonomy and protection from Sinhala majoritarianism. Whereas earlier forms of Sinhala nationalism were primarily anti-colonial and anti-British, today it is asserted *vis-à-vis* local Others, primarily Tamils.

Assertions that the Sinhala people constitute a "race" distinct from other groups on the island and from peoples of "the West" continue to be made not just by ideologues but also by officials at the highest levels of government. Claims of historical continuity with "Aryan civilization" are often asserted as well. Women remain central figures in contemporary discourses of Sinhala nationalism. As Kumari Jayawardena and Malathi de Alwis have put it:

> While the basis for identity has shifted within each country [in South Asia], each ethnic group still has its myths of origin (Aryan, Dravidian, Semitic) and its Golden Age in which women were "free." They were situated within a "glorious" mythical past in which social divisions were absent. . . . In such a context, minorities were represented as invaders, outsiders, or aliens who corrupted the pristine purity of the majority. They were accused of destroying ancient traditions and the old idyllic way of life and disrupting the political hegemony of a "united" polity; most of all, they represented the threat of rape and thereby the possible "pollution" of the "daughters of the soil." . . . As property of the national collective, the woman-mother symbolises the sacred, inviolable borders of the nation.
>
> (Jayawardena and de Alwis 1996: x)

In the remainder of this chapter, I focus on three groups of contemporary women who have stepped outside the entailments of Sinhala womanhood: women elected to political office; housemaids and garment workers (forms of wage work newly available to women); and women engaged in feminist activism and women's mobilization. Women in each category are in some ways refusing the ideology of authentic Sinhala/Sri Lankan womanhood, and thus have aroused more than their share of animosity and anxiety. The projects that these women are engaged in have little in common with one another. Yet, as we shall see, the attacks on them in popular discourse and mass media reiterate similar themes, themes that hark back to those promulgated by Dharmapala and other early nationalists.

Women in politics

Whenever norms of propriety exclude women from the public sphere and demand modesty and submission, political leadership by women remains fraught with contradiction. Yet Sri Lanka – like India, Pakistan and Bangladesh – has a strong record of female leadership at the highest level of government. In the fifty years since Independence, two women have been elected to the office of Prime Minister in Sri Lanka: Srimavo Dias Bandaranaike (elected in 1960) and her daughter, Chandrika Bandaranaike Kumaratunga (elected in 1995). Presently, Mrs. Bandaranaike holds the office of Prime Minister and Mrs. Kumaratunga, the office of Executive President. At lower levels of government, however, the record of female leadership remains bleak; indeed, women rarely even contest elections. Moreover, virtually all women who have held high political office in Sri Lanka since Independence – whether President, Prime Minister, Member of Parliament or head of a political party – have come from elite families with strong political backgrounds, often extending over many generations (de Silva 1995). Most such women first entered politics by stepping into a void created by the death (often by assassination) of a close male relative.

This pattern of women's political participation might suggest that women can safely enter the political arena if they come from an elite background or if they cast themselves as fulfilling the unfinished mission of a dead father or husband. However, Malathi de Alwis's (1995) reading of the careers of recent female political figures in Sri Lanka suggests that flouting the norms of womanly conduct is not so easy. Respectability is indeed an essential attribute for a woman in public life, but respectability, at least as far as women are concerned, is also highly unstable, as de Alwis points out.

The early years of Chandrika Kumaratunga's presidency offer a paradigmatic case of this instability. Mrs. Kumaratunga comes from an old aristocratic Sinhala family, one that has a distinguished tradition of political leadership. Her father was assassinated while serving as Prime Minister and her husband was assassinated while standing for election; in her campaign and afterward, she frequently alluded to their deaths and to their ideals and visions for the country. Mrs. Kumaratunga also holds an array of educational credentials earned abroad. On grounds of class, caste and family connections, her respectability is beyond reproach. Furthermore, Mrs. Kumaratunga's campaign platform seemed constructed to embrace many traditional virtues of Sinhala womanhood. She pledged to bring peace and unity to the country. She promised to end the civil war through political negotiation rather than military domination and conquest. She vowed to uncover and rectify the human rights abuses of the previous administration, to bring perpetrators to justice, and to recompense victims. She also vowed to "clean house" by abolishing the corruption identified with the past administration. In her campaign speeches, she invoked eloquent and emotion-laden images of herself as a mother and as a woman who had lost loved ones to violence, thus stressing her common cause with ordinary women.

Mrs. Kumaratunga's first years in office illustrate how entering public life can undo a woman's claim to respectability, no matter how carefully constructed and seemingly impeccable that claim is. Despite her aristocratic heritage and her use of the symbols of motherhood, widowhood and filial respect for her deceased father and her mother, Mrs. Kumaratunga's respectability came under siege as soon as she stepped into the public sphere. Neither her aristocratic background nor her effort to portray herself as a maternal figure insulated her from moralistic attacks by both mainstream newspapers and the tabloid press. Simply being in public life and fulfilling the ordinary social obligations of a head of state was recast as immoral conduct by the mainstream press. Drinking wine at formal state dinners, eating meals in five-star hotels, staying out "too late" and traveling without her mother's permission were described as improprieties in Mrs. Kumaratunga's personal conduct. Tabloids and gossip columns circulated inflammatory rumors about inebriation, alcoholism, love affairs and even casual sexual liaisons with men of lower standing. In short, those opposed to Mrs. Kumaratunga sought to undercut her political leadership not by criticizing her politics but by impugning her sexual morality and womanly virtue. Such attacks reverberated in the popular imagination, undermining Mrs. Kumaratunga's stature as a political leader.

Mrs. Hema Premadasa, whose husband served as President from 1989 until his assassination in 1993, shared a similar onslaught of public degradation once she entered the public arena, as Malathi de Alwis (1995) carefully documented. As the President's wife, Mrs. Premadasa labored assiduously in a series of widely publicized social service projects, "village upliftment" schemes, and charitable giveaways. These projects can be seen as attempts by Mrs. Premadasa to position herself as "a nurturing and beneficent 'Mother of the nation'" (de Alwis 1995: 146). Her efforts, however, were undercut by a continual stream of sexual jokes and rumors of sexual improprieties in the tabloids and gossip networks. Mrs. Premadasa's claims to respectability were further undercut by pointed allusions to her lower-class background, such as remarks about her "gaudy" *saris* and "vulgar" deportment (de Alwis

1995). Yet when Mrs. Premadasa dressed in styles associated with the upper-class Sinhala aristocracy, she was ridiculed for presuming to reach above her caste/class background, behavior regarded as distasteful and even morally offensive. One middle-class man, superintendent of a tea plantation, summarized to me what he regarded as Mrs. Premadasa's penchant for self-aggrandizement as follows: "What to do? These are the things that happen when nobodies try to be somebodies." Mrs. Premadasa was left sexualized, morally degraded and with her dubious class and caste origins exposed to public scrutiny.

By stepping into the public arena, Mrs. Kumaratunga and Mrs. Premadasa each defied long-standing ideals of Sinhala womanhood; Mrs. Premadasa and her husband further challenged the hegemony of the class- and caste-based aristocracy. The vicious character assassinations directed at them demonstrate the Catch-22 that women in political office face: respectability is a prerequisite for a woman to enter the public sphere, yet a woman cannot retain her respectability once she is there. Malathi de Alwis astutely interprets this Catch-22 in terms of the link between male dominance and nationalism: "the patriarchal gaze of the nation . . . constantly seeks to scrutinize the underbelly of respectability" (de Alwis 1995: 150). The relentless mudslinging that Mrs. Kumaratunga and Mrs. Premadasa faced served not only to discredit them but also to deter other women who might seek political office. Character assassinations that focus on women's sexual behavior and sexual morality not only impugn women themselves but also discredit the honor of their families of origin and their offspring. It is not surprising that women, though well represented in the civil service, have avoided running for elected office. Only a minute proportion (less than 4 percent) of candidates for elected offices at all levels of government (national, provincial and local) have been women (de Silva 1995).

Working women and wage labor

Economic necessity has long demanded that poor women in Sri Lanka contribute to the economic support of their families. Until recently, however, the wage work available to women consisted mainly of small-scale cottage industries, seasonal agricultural work and domestic service. Only among plantation laborers, a group composed mostly of Hill Country Tamils, were large numbers of women employed in an organized sector of the economy. In the late 1970s, however, two new forms of wage work for women emerged; both were available to women from all ethnic groups: domestic service in the Middle East and work in urban garment factories. Although both occupations involve low wages, no advancement and exploitative conditions, these jobs represent the best-paying (if not the only) employment for lower-class women irrespective of ethnic group identity. Many women have been recruited and many more desperately seek them out. According to the statistics compiled by the Sri Lanka Bureau of Foreign

Employment, 710,000 Sri Lankans were working in the Middle East as of 1997. Most Middle East workers are female housemaids; most are married and mothers. As of 1992, nearly 105,000 women were employed in garment factories in the Export Processing Zones (E.P.Z.s); most were young and single (Jayaweera 1995b).

In itself, the visible presence of women in the paid workforce violates the doctrine of separate spheres and female domesticity on which respectable Sinhala womanhood is premised. For housemaids and garment workers, the conditions of work further violate the norms of feminine sexual propriety, motherhood and subordination to male authority in the family. Housemaids live in the households of their foreign employers; most garment workers in the E.P.Z.s live in dormitories maintained by the factories. Both groups of workers thus live apart from their families; their behavior is not under the direct supervision of their fathers or husbands. Moreover, they have a degree of financial autonomy that is unusual. Housemaids in the Middle East further violate local norms by leaving their children in the care of others, most often their female relatives. In these ways, domestic service abroad and urban factory work place women workers outside the imperatives of respectable Sinhala womanhood. In consequence, these women have been targets of moralizing attacks.

The sexual morality of garment workers is a common target of popular suspicion. Factory girls whose virtue has been compromised, whether by seduction or coercion, are stock characters in tele-dramas (serialized television shows) and Sinhala-language cinema. Sensational rumors circulate freely through gossip networks about promiscuity, abortions (which are illegal in Sri Lanka), rape, H.I.V. infection and prostitution among garment workers in the E.P.Z.s. Indeed, the pejorative nickname for garment workers, "*Juki* girls" (after the brand name of the industrial sewing machines used in the factories), connotes sexual promiscuity (Lynch 1996).

In recent years, the number of rural women (as opposed to lower-class urban women) seeking wage work has escalated in response to adverse economic conditions. The entry of these women into the paid workforce has compounded anxieties about workers' sexual morality. It is in the rural countryside – a place of such mythic significance that most Sinhalese simply refer to it as "The Village" – that Sinhala culture is believed to persist in its original untainted form. Women of "The Village" are figured as innocent and simple custodians of modesty and virtue; they are hailed as continuing to embrace the tradition that "woman, as wife and mother, has her own preordained place" (Jayawardena n.d.). In a report of her fieldwork with garment workers, Caitrin Lynch (1996) quotes one informant, a factory supervisor:

> These young women are people who have grown up under the shade of their parents, in the village, preserving Sinhala customs and traditions. [They] are girls who consider ethics and culture as their lives.
>
> (Lynch 1996: 6–7)

Young women from "The Village" come in for special scrutiny and criticism from which urban women are exempted. For instance, garment workers are castigated for purchasing jewelry, perfume, handbags and Western-style clothing, items that signal a departure from feminine virtues of self-abnegation and modesty; young women working in city office jobs are free to buy such items (Lynch 1996). Indeed, a sexual double standard seems to operate with reference to middle-class "city girls" and lower-class "village girls." As one "village girl" said:

> Actually office staff in Colombo [the capital city] must be misbehaving too. But since they are dressed properly and they come in a vehicle and because of their family background, they cover-up themselves.
>
> (Lynch 1996: 3)

So strong are anxieties about the sexual morality of "village girls" that one plank in the campaign platform of President Ranasinghe Premadasa was the "200 Garment Factories Program," an initiative to build factories throughout the countryside expressly to keep women in the village and away from corrupting urban centers (Lynch 1996).

Virtually all Sri Lankan women in domestic service abroad are married and with children (Dias and Weerakoon 1995). This demographic pattern reflects the dual feminine imperatives of sexual purity and pronatalism embedded in discourses of national identity. The preoccupation with women's premarital chastity prevents women who are unmarried from seeking work that requires them to live in the household of an unrelated male. Moreover, women are expected to bear a child as soon after marriage as possible. Ignoring these imperatives places women at risk of social censure. But paradoxically, housemaids, who have conformed to them, still face charges of immorality. Though they are married, they are still accused of sexual promiscuity. Furthermore, as mothers who have entrusted their children to others' care, they endure further disapprobation.

In the spring of 1996, I interviewed three rural-based community health workers about their work. Without prompting, the two male workers named women's employment in the Middle East as the single root cause of problems they encountered: men's alcoholism, gambling and marital infidelity; fathers' neglect and sexual abuse of their children; children's truancy, suicides and drug addiction; separations, desertions and divorce; jealousies and conflicts among extended family members; and soaring rates of suicide. Yet, a case-by-case inventory of their current caseloads did not turn up a single situation that actually involved a housemaid working in the Middle East. Moreover, when I asked if they could recall *any* specific cases involving a mother working in the Middle East, they could not. When I asked whether men might bear some responsibility for their behavior, the third worker – a young woman – took over the floor. The blaming tone continued but the content shifted to housemaids' sexual morality. Housemaids in the Middle

East, she informed me, enjoyed such cosmopolitan lifestyles and lavish living standards that they could not readjust to village life on their return. Moreover, she asserted, many housemaids working abroad engaged in such morally objectionable practices as promiscuity, prostitution and producing babies to sell to European couples.

When women take up wage work away from their families, their actions challenge the gender dualisms and power relations on which the distinctiveness and superiority of Sinhala (and, by extension, Sri Lankan and even Eastern) identity has partly been predicated. Women who are wage workers and breadwinners challenge traditional conceptions of both femininity and masculinity, as well as traditional gender differentials in status and power in their households and in the community. Moreover, with large numbers of rural women desperate for jobs in the Middle East and in garment factories, the idealized images of rural life are hard to sustain. For most housemaids, working abroad is a last resort for the survival of their families; their husbands are unemployed, have deserted the family or otherwise do not contribute money to the support of their families (CENWOR 1987). Some go to the Middle East to escape from alcoholic husbands or abusive marriages (Gamburd 1995). Rural villages are beset by grinding poverty, rampant alcohol abuse among men, high levels of domestic violence, marital dissolution and desertions, and rates of suicide that are the highest in the world (Ratnayeke 1996). To confront the realities of village life is to explode an idealization that props up collective Sinhala identity. "The Village" is far from a haven of bucolic tranquillity and harmonious relations rooted in timeless Sinhala culture.

Women's activism

Women played a significant part in the anti-colonial resistance in Ceylon during the early decades of the twentieth century, whether as feminists, suffragists or participants in trade unions, political parties and peasant agitations. But Dharmapala and many other nationalists of that era did not acknowledge women's activism and contribution to public life. Instead, the proper model of righteous and respectable womanhood was the woman who served her family's needs, who practiced modesty and subservience, and who confined herself to the home. In contemporary Sri Lanka, women's activism still sits uneasily alongside such definitions of righteous womanhood. The vocal presence of women in public life, whatever causes they espouse, transgresses the ideal that women belong in the domestic sphere.

Kumari Jayawardena (1995b: 404), a leading feminist scholar in Sri Lanka, has traced reactions to feminist efforts to improve women's status by contrasting editorials published in Sri Lanka's leading national newspaper to mark International Women's Day. In the English-language edition (whose readership is limited to the Westernized, urban elite), the editors praised women's achievements, declared that "the much flogged male chauvinist pig is as dead as a dodo" and hoped that "women's libbers" would continue their work to "improve the status of women." In contrast, the Sinhala-language edition (with a readership that is more diverse economically, educationally and regionally) carried a dramatically different message. Its editorial mounted a scathing attack on feminists asserting that their "slogans and drumbeats" were borrowed from the West. "Feminists," it asserted, were financed by "foreign money" in order "to promote sexual promiscuity and cultural degeneration." Further editorial commentary repeated the theme of feminism as a despised Western import:

> The feminine consciousness as it obtains today is . . . ideological baggage borrowed from the West.
>
> Like most other fads to which our alienated elite genuflect, this too is a concept hatched in a West riddled by . . . problems.
>
> (Jayawardena n.d.: 2)
>
> The fashionable women's lib . . . is merely a chic posture devoid of any meaning to the large bulk of Sri Lanka's women.

The assertion that feminism is a faddish Western import alien to the true culture of Sri Lanka conjures up memories of the immediate post-Independence period (the 1950s) in Ceylon, when the linguistic nationalists turned against the so-called "brown Sahibs," the English-speaking, Westernized elite, driving many out of the country. It positions feminists as a tiny group of alienated, degenerate elites, setting them against "the large bulk of Sri Lanka's women." This latter group is presumed to share a commitment to a uniform indigenous culture, to uphold traditional sexual morality (as opposed to promiscuity) and to be content with the status quo. One can guess that the dramatic difference in editorial content between the English-language edition and the Sinhala-language edition reflects the editors' assessment of their readership's stance on both gender equality and questions of national identity. The scathing indictment of feminism as a "concept hatched in the West" serves to promote the idea that the Sinhala-speaking majority share a unitary, unique and superior identity. Imagery that portrays Sri Lankan feminists as a deracinated and morally suspect elite exerts a powerful constraining effect on many of them. Some activists have elected to position themselves as public mothers and as guardians of the nation's morality; they have directed their energies toward such issues as childhood prostitution, sex tourism and foreign pedophiles. Other activists are privately critical of this imagery, but choose to craft their public stances to sidestep the charges that they are outsiders to the "true" culture and purveyors of Western degeneracy. Thus, they have made strategic choices not to bring certain issues into public debate, for fear of backlash and a greater repression than already exists. Many of these issues center on women's sexual agency, for example, the preoccupation with female premarital chastity; the use of gynecological tests of virginity prior to marriage and the inspection of bed sheets after the wedding night; the ban on legal abortion; the expectation that all women marry and bear children; the silence around lesbian sexuality; and the sexual abuse of children and servants by male members of the household. More generally, feminists have steered away from challenging gender relations in the domestic sphere. For instance, even with the rising number of women who work outside the home, there has been no public call for men to shoulder a share of the burden of housework and child care (Kiribamune 1993). Thus, the association of women with sexual purity, innocence and modesty stands largely unchallenged, as does the double standard of sexual morality. Moreover, the unfair division of labor in the family goes unremarked.

Women's peace activism and nationalism

Peace activism is an arena in which some women in Sri Lanka have mobilized to confront nationalism directly. Women For Peace, a movement for women of all ethnic communities, formed in the late 1980s. In the fall of 1995, Women For Peace promulgated a statement on nationalism and the war between the Sinhala-dominated government and the secessionist L.T.T.E. This document, entitled "Through the eyes of women: a new way of seeing and knowing our reality," deserves a close reading, both because of its wide dissemination within Sri Lanka and because of what it reveals about the complexities and difficulties of positioning women *vis-à-vis* nationalist struggles.

The opening paragraphs of the document frame the war as something visited on women ("us") without their knowledge, agreement or participation:

> Our country is being torn apart by war. The politicians, the television, and newspapers tell us this is the way things are and the way things must be. . . . We did not start it, but we suffer, as daughters, sisters, wives, and mothers, from the devastating consequences of war. . . . We are told that war is being fought on behalf of the nation. The Tamil nation against the Sinhala nation. But whose nation is this anyway?
>
> (Women For Peace 1995–96: 37)

The remainder of the document consists of a series of four paragraphs, each one detailing some consequences of the war for a specific group of women. Each paragraph ends with the refrain, "Shouldn't we challenge this denigration of women's lives?" For "Sinhala mothers," the first group of women mentioned, the denigration is sending sons to the battlefield. For "Mothers of the North" (i.e. Tamil women living in the battle zones of the northern province), the document points to the forcible conscription of their sons and daughters by the L.T.T.E., and life in a "militarised, socially repressive environment." For "Tamil women in Colombo," the document points to the vulnerability of their husbands and sons to abduction and torture at the hands of state security personnel; it also points to the vulnerability of these women themselves to ethnic prejudice and state terrorism. For "Muslim women of the East" (i.e. those living in the eastern province of Sri Lanka, another zone of armed conflict), the document enumerates impoverishment, restriction of movement, the disintegration of community life and the ongoing risk of rape and violent abuse by men fighting on one or the other side of the conflict.

"Through the eyes of women" offers important counterpoints to the insistent pro-war rhetoric of Sinhala nationalist groups. The document asserts that there is such a thing as a women's perspective on questions of national identity and on nationalist-inspired violence, one that involves seeing and understanding differently than men; this feminine perspective necessarily is both anti-war and anti-nationalist. Moreover, the Women For Peace refuse the metaphor of woman as nation, waiting to be protected by her sons and avenged in war. As the document asserts, women are victimized by this "spurious justification" for "endless, meaningless war." Furthermore, the document embodies a call to women to set aside national interests in favor of female solidarity and women-centered interests.

The courage of the Women For Peace should not be minimized. Yet a closer reading of the document reveals that although it challenges some elements of nationalist discourses about women, it accepts others without question. For example, the document embraces and reiterates the premises that nationalism is a masculinist ideology and that "aggressive militancy and revolutionary violence" are solely the province of "male nationalists." Indeed, the text is explicitly built upon familiar binary oppositions of men and women, agents and victims, aggressors and pacifists. The text figures women exclusively as non-combatants, victims and horrified witnesses to the devastation of war. Even the women soldiers of the L.T.T.E. are recast as "children" "snatched away from their mothers' domestic embrace" and "forced" to take part in suicide missions. Rhetorically appealing as such portrayals may be, they constitute a rewriting of history. In all ethnic groups in Sri Lanka, some women have embraced nationalist causes and some have been active combatants. Perhaps the most publicized examples are the numerous female cadres of the L.T.T.E. who have functioned as assassins and suicide bombers. (Two of the most visible are the female suicide bombers who assassinated Rajiv Gandhi in India and President Ranasinghe Premadasa in Sri Lanka.) Sinhala women serve in the government armed forces, albeit not in combat roles. Indeed, as the war has dragged on, the participation of Sinhala women in the government armed forces and in ancillary occupation has expanded greatly. For instance, in the fall of 1997, women were recruited as police constables so that more male constables could be posted in dangerous border zones. In that same time period, the Sri Lanka Air Force set out to recruit and train women to pilot surveillance and transport aircraft. Moreover, as Daniel (1996) documents, Sinhala and Muslim women have functioned as interrogators in the prison camps of the Sinhala-dominated government;

in that role, they have not stepped back from using torture on prisoners. In village massacres in border territories, women and children of all ethnic groups have participated in the carnage alongside men (Pieris and Marecek 1991). Less dramatic but not to be disregarded are the legions of women who support the nationalist projects of their ethnic communities. In summary, in their text, Women For Peace attempt to dissociate women from nationalist ambitions and projects. But this is a fiction that gives way to the reality of women's complicity and active engagement in those projects. Women in Sri Lanka have not stood apart from nationalist projects but have been deeply implicated in them.

"Through the eyes of women" strives to create a unified category of "women," in order to promote a woman-centered antiwar politics. Thus, the opening lines of the document assert the common suffering of women *qua* women: "we suffer as daughters, sisters, wives, and mothers." But this assertion is problematized by the subsequent enumeration of the experiences of women from different ethnic groups. Although this enumeration underscores the extent to which women suffered, it inevitably points out that women have not all suffered equally. Singling out the specific ways in which women from different ethnic groups experience ethnic violence invites invidious distinctions among them. Moreover, not all suffering can be attributed to generic "men"; each ethnic group has suffered because of the acts of members (men and women) of other groups. Indeed, the suffering of some women (e.g. the conscription of their children) is instrumental to the suffering of other women. Thus, dichotomies such as women versus men, victims versus agents, domination versus freedom, and perhaps even peace versus war are too simplistic. Instead of exposing the complexities, ambiguities, and crosscurrents in the historical record of ethnic relations in Sri Lanka, such dichotomies blanket them over. We must question whether initiatives for conflict resolution framed in such dichotomies can work.

"Through the eyes of women" reveals just how difficult it is to move beyond the gender dualisms that are embedded in discourses of Sinhala identity. Yet peace activists must interrupt these dualisms. One reason, as I have noted, is the importance of acknowledging that, in all ethnic groups, many women in Sri Lanka are not for peace; indeed, some have taken active roles in injuring others. Another reason is that claiming an essentialized female identity as peacemakers positions men as war-makers. As "Through the eyes of women" puts it, "If men are for war, can't we women be for peace?" But not all men are for war. Yet if war is insistently linked with masculinity, those men who are for peace are effectively marginalized.

Shortcomings aside, "Through the eyes of women" succeeds in claiming the right of women from all ethnic communities in Sri Lanka to insert their voices into ongoing public debates on ethnic relations, nationhood and war. Moreover, it forcefully asserts that women's experiences and perspectives are not necessarily congruent with those of men; it demands that those experiences and perspectives be publicly registered. The document poses two powerful questions in succession: "But whose nation is this anyway?" and "Shouldn't we challenge this denigration of women's lives?" It challenges the exclusion of women from full and equal participation in the nation, linking that exclusion to material practices that jeopardize women's physical safety and security, dignity, and autonomy.

Summary and conclusion

At the turn of the century, nationalist movements in Ceylon focused on opposition to colonial rule and the right to sovereign nationhood. Instead of embracing all ethnic groups, however, those movements coalesced around a single ethnic group, the Sinhala people, and around Buddhism. In producing a Sinhala Buddhist identity, a cultural heritage harking back to a glorious and ancient civilization was constructed. This construction served both to establish the difference and separateness of the Sinhala from other groups, and to justify their political and cultural hegemony. This heritage included an elaborate gender ideology, which specified the attributes of righteous womanhood and the relations of power between women and men. As I have indicated, the ideologies of womanhood that prevail in present-day Sri Lanka bear the direct imprint of this gender ideology. Thus, women who conform to the ideals of traditional womanhood are extolled as custodians of the cultural heritage; women who step outside them come under attack. The strength of the link between traditional ideals of womanhood and national identity is apparent in the media attacks on feminists who have challenged traditional gender ideologies. They are labeled as inauthentic, alienated from their own culture, and deracinated. Their ideas are discredited as borrowings from the "outside" (specifically from the "degenerate" West), not as *bona fide* expressions of dissent from within.

At the anniversary of the first fifty years of Independence, Sri Lanka faces the formidable task of crafting a state in which the needs and interests of different ethnic groups are protected and respected. That task is made all the harder because the post-Independence record thus far has been marred by crude majoritarianism, mistrust among ethnic groups, violence and a protracted civil war. Thus far, women too have not been included fully as equal and respected participants in political life. As we have seen, the gender ideologies arising from Sinhala nationalism have functioned to keep women in a silenced and subordinate status. To be respectable, women must remain in the confines of the domestic sphere, under the protective authority of men and without a public voice. To do otherwise risks moral and sexual degradation. People do not, as Rosa Luxemburg (1976) observed, "homogeneously inhabit any given piece of territory." Women's interests and needs are not always congruent with men's, even when they have a common ethnic identity. In order for those needs to be recognized and respected, women must be included in the nation not as symbols, but as speaking subjects.

Notes

1 The quotation is a line from a poem composed by a Buddhist nun in the first century B.C., collected and translated by Caroline Rhys-Davids (1909: 45–46). The stanza reads:

> How should a woman's nature hinder us?
> When hearts are firmly set, who ever moves
> With growing knowledge onward in the path? . . .
> Am I a woman in these matters, or
> Am I a man, or what am I then?

2 For example, individuals might identify themselves as Jaffna Tamils, Batticaloa Tamils, Colombo Tamils or Hill Country Tamils. The last group have also been called Estate Tamils because they or their ancestors worked on tea estates and Indian Tamils because their ancestors migrated from India to labor on British-operated estates.

3 Masculinities and male sexuality have been subjects of much fruitful theorizing about the British Empire and Indian anti-imperialism. Given the cultural and historical parallels between India and Sri Lanka, there is good reason to believe that they will be significant in the Sri Lankan case.

4 Many Tamils and Muslims live in areas that are battlegrounds in the ongoing civil war between the L.T.T.E., a militant Tamil nationalist group, and the state; they live in conditions of extreme hardship and terror. On Tamil national identity, the reader is referred to Balasingham (1993), Cheran (1992), Coomaraswamy (1987), Daniel (1996), Hoole *et al.* (1990), Lawrence (1997), Maunaguru (1995) and Pfaffenberger and Chelvadurai (1994). On Muslim identity, see Ismail (1995).

5 Tamil women wear *saris* but also mark their foreheads with a *pottu*, a small red circle.

References cited

Balasingham, A. A. (1993) *Women Fighters of the Liberation Tigers.* Jaffna, Sri Lanka: Thasan Printers.

Bartholomeusz, T. (1995) "The clash of 'selves' and 'secularisms' in contemporary Sri Lanka," paper presented at the Twenty-Fourth Annual Conference on South Asia, Madison, Wisconsin, October.

CENWOR (1987) *Women's Work and Family Strategies, Sri Lanka.* Colombo, Sri Lanka: Centre for Women's Research.

Chatterjee, P. (1990) "The nationalist resolution of the women's question," in K. Sangari and S. Vaid (eds.) *Recasting Women: Essays in Indian Colonial History.* New Brunswick, N.J.: Rutgers University Press, pp. 233–253.

—— (1994) *The Nation and its Fragments: Colonial and Post-colonial Histories.* Delhi: Oxford University Press.

Cheran, R. (1992) "Cultural politics of Tamil nationalism," *South Asia Bulletin* 12,1: 42–56.

Coomaraswamy, R. (1987) "Myths without conscience: Sinhala and Tamil nationalist writings in the 1980s," in C. Abeyesekera and N. Gunasinghe (eds.) *Facets of Ethnicity.* Colombo, Sri Lanka: Social Scientists' Association.

Daniel, E. V. (1996) *Charred Lullabies.* Princeton, N.J.: Princeton University Press.

de Alwis, M. (1995) "Gender, politics, and the 'respectable' lady," in P. Jeganathan and Q. Ishmail (eds.) *Unmaking the Nation: The Politics of Identity and History in Modern Sri Lanka.* Colombo, Sri Lanka: Social Scientists' Association, pp. 137–157.

—— (1995–96) "Notes toward a discussion of female portraits as texts," *Pravda* December–January: 16–21.

de Mel, N. (1996) "Metaphors of women in Sri Lankan war poetry," in K. Jayawardena and M. de Alwis (eds.) *Embodied Violence: Communalizing Women's Sexuality in South Asia.* New Delhi, India: Kali for Women, pp. 168–189.

Denham, E. B. (1912) *Ceylon at the Census of 1911.* Colombo, Ceylon: Government Press.

de Silva, W. (1995) "The political participation of women in Sri Lanka 1985–1995," in CENWOR, *Facets of Change: Women in Sri Lanka 1985–1995.* Colombo, Sri Lanka: CENWOR, pp. 229–263.

Dias, M. and N. Weerakoon (1995) "Migrant women domestic workers from Sri Lanka: trends and issues," in CENWOR, *Facets of Change: Women in Sri Lanka 1985–1995.* Colombo, Sri Lanka: CENWOR, pp. 195–228.

Gamburd, M. (1995) "Sri Lanka's army of housemaids: control of remittances and gender transformations," *Anthropologica* 37, 1: 49–88.

Gooneratne, Y. (1992) "The English-educated in Sri Lanka: an assessment of their cultural role," *South Asia Bulletin* 12, 1: 2–33.

Harris, E. J. (1994) *The Gaze of the Coloniser: British Views on Local Women in 19th Century Sri Lanka.* Colombo, Sri Lanka: Social Scientists' Association.

Hoole, R., D. Somasunderan, K. Sritharan and R. Thiranagama (1990) *The Broken Palmyra: The Tamil Crisis in Sri Lanka – An Inside Account.* Claremont, CA: SLSI.

Ismail, Q. (1995) "Unmooring identity: the antimonies of elite Muslim self-representation in modern Sri Lanka," in P. Jeganathan and Q. Ismail (eds.) *Unmaking the Nation: The Politics of Identity and History in Modern Sri Lanka.* Colombo, Sri Lanka: Social Scientists' Association, pp. 55–105.

Jayawardena, K. (1986) *Feminism and Nationalism in the Third World.* London: Zed.

—— (1995a) *The White Woman's Other Burden: Western Women and South Asia during British Rule.* London: Routledge.

—— (1995b) "The women's movement in Sri Lanka 1985–1995: a glance back over ten years," in CENWOR, *Facets of Change: Women in Sri Lanka 1985–1995.* Colombo, Sri Lanka: CENWOR, pp. 396–407.

—— (n.d.) *Feminism in Sri Lanka.* New York: WIRE.

Jayawardena, K. and M. de Alwis (1996) "Introduction," in K. Jayawardena and M. de Alwis (eds.) *Embodied Violence: Communalizing Women's Sexuality in South Asia.* New Delhi, India: Kali for Women, pp. ix–xxiv.

Jayaweera, S. (1995a) "Women and education," in CENWOR, *Facets of Change: Women in Sri Lanka 1985–1995.* Colombo, Sri Lanka: CENWOR, pp. 96–130.

—— (1995b) "Women and employment," in CENWOR, *Facets of Change: Women in Sri Lanka 1985–1995.* Colombo, Sri Lanka: CENWOR, pp. 156–194.

Kiribamune, S. (1993) "Introduction," in S. Kiribamune (ed.) *Women, Work, and Role Reconciliation in Sri Lanka.* Delhi: Navrang, pp. xi–lix.

Lawrence, P. (1997) "Work of oracles, silence of terror: notes on the injury of war in eastern Sri Lanka," unpublished dissertation, Department of Anthropology, University of Colorado.

Luxemburg, R. S. (1976) "The rights of nations to self-determination," in R. S. Luxemburg, *The National Question: Selected Writings by Rosa Luxemburg.* New York: Monthly Review Press.

Lynch, C. (1996) "The production of newly traditional women in rural Sri Lankan garment factories," paper presented at the Twenty-Fifth Annual Conference on South Asia, Madison, Wisconsin, October.

Maunaguru, S. (1995) "Gendering Tamil nationalism: the construct of woman in projects of protest and control," in P. Jeganathan and Q. Ishmail (eds.) *Unmaking the Nation: The Politics of Identity and History in Modern Sri Lanka*. Colombo, Sri Lanka: Social Scientists' Association, pp. 158–175.

Mohanty, C. T. (1991) "Cartographies of struggle," in C. T. Mohanty, A. Russo and L. Torres (eds.) *Third World Women and the Politics of Feminism*. Bloomington, IN: University of Indiana Press, pp. 12–13.

Obeyesekere, R. (1992) "The Sinhala literary tradition: polemics and debate," *South Asia Bulletin* 12, 1: 34–41.

Pfaffenberger, B. and M. Chelvadurai (eds.) (1994) *The Sri Lanka Tamils: Ethnicity and Identity*. Boulder, CO: Westview Press.

Pieris, S. J. R. and J. Marecek (1991) *The Welikanda Massacre*. Colombo, Sri Lanka: International Centre for Ethnic Studies.

Ratnayeke, L. (1996) "Suicide and crisis intervention in rural communities in Sri Lanka," *Crisis* 17, 4: 149–151, 154.

Rhys-Davids, C. A. F. (1909) *Psalms of the Early Buddhists, vol. 1, Psalms of the Sisters*. London: Pali Text Society.

Risseeuw, C. (1988) *The Fish Don't Talk about the Water: Gender Transformation, Power, and Assistance among Women in Sri Lanka*. Leiden: E. J. Brill.

—— (1992) "Gender, kinship, and state formation: the case of Sri Lanka under colonial rule," *Economic and Political Weekly* (October 24–31): WS-46–WS-53.

—— (1996) "State formation and transformation in gender relations and kinship in colonial Sri Lanka," in R. Palriwala and C. Risseeuw (eds.) *Shifting Circles of Support, Contextualizing Gender and Kinship in South Asia and Sub-Saharan Africa*. Delhi and London: Sage, pp. 79–109.

Spivak, G. C. (1985) "Three women's texts and a critique of imperialism," *Critical Inquiry* 12 (autumn): 243–261.

Tambiah, S. J. (1992) *Buddhism Betrayed: Religion, Politics, and Violence in Sri Lanka*. Chicago: University of Chicago Press.

Women For Peace (1995–96) "Through the eyes of women: a new way of seeing and knowing our reality," *Pravda* December–January: 37.

Yalman, N. (1967) *Under the Bo Tree: Studies of Caste, Kinship, and Marriage in the Interior of Ceylon*. Berkeley, CA: University of California Press.

Arafura Sea
Darwin
Gulf of Carpentaria
Indian Ocean
Coral Sea
Wyndham
Derby
Northern Territory
Cooktown
Cairns
Townsville
Mount Isa
Mackay
Queensland
Alice Springs
Western Australia
Rockhampton
Ayers Rock
A U S T R A L I A
Charleville
South Australia
Brisbane
Geraldton
Bourke
Kalgoorlie
Broken Hill
Port Augusta
New South Wales
Perth
Newcastle
Albany
Adelaide
Sydney
Canberra
Victoria
Melbourne
Indian Ocean
Tasmania
Hobart
0 km 500

Australia

7 Native sex

Sex rites, land rights and the making of Aboriginal civic culture

Elizabeth A. Povinelli

> The phenomena which early societies present us with are not easy at first to understand, but the difficulty of grappling with them bears no proportion to the perplexities which beset us in considering the baffling entanglement of modern social organization. It is a difficulty arising from their strangeness and uncouthness, not from their number and complexity.
>
> (Sir Henry Maine, *Ancient Law*, 1986 [1864]: 115–116)

On June 11, 1936, the Darwin Administrator of the Northern Territory of Australia sent an urgent telegram to the federal Department of Interior in Canberra.

> CONSTABLE PRYOR FROM DALY RIVER ARRIVING ON 12TH JUNE WITH 6 ABORIGINES CHARGED WITH RAPING LUBRA STOP UNDERSTAND OFFENCE IS MORE OR LESS TRIBAL CUSTOM WHEN FEMALE ABORIGINAL WALKS ON SACRED GROUND RESERVED ONLY FOR MALES SUCH FEMALE BEING REQUIRED [TO] HAVE SEXUAL INTERCOURSE WITH ALL AND SUNDRY STOP MORE NATIVES WOULD HAVE BEEN IMPLICATED BUT FOR THE TIMELY ARRIVAL OF PEANUT FARMER HARKINS[1]

This telegram raced across continental wires and delivered to southern federal administrators what many Australians considered the scandalous open secret of "native" sacred rites. In public and semi-public spaces, in regional newspapers, in administrative meetings and hearings, in beer halls and across campfires, white Australians talked about the sex acts performed in Aboriginal men's rituals. By 1936, sexological, anthropological, popular and religious writings reported that genital operations, bestiality, masturbation, group sex and same sex were integral components of some of the men's sacred rites. For instance, in *Across Australia* (1912), a book written for a general audience, the ethnologists W. B. Spencer and F. J. Gillen, wrote that indigenous ceremonies consisted of "naked, howling savages" engaged in bodily acts that were "crude in the extreme."[2] For the Australian bourgeois public these extremes were certainly reached in instances of so-called ritual rape – the means by which indigenous men punished women who violated their ritual spaces and knowledges.

Aboriginal men and women intimately engaged each other during their ceremonies in ways that white Australians did not consider to be sexual. In the rituals of many northern indigenous groups, sweat and blood were rubbed into or washed over human bodies, ritual objects and ancestral sites. Words and sounds, spoken and sung, communicated to and penetrated into initiates and ancestral sites. Ancestral spirits entered, possessed and spoke to and through initiates and ritual leaders. But these ritualized intimacies did not provoke critical public debate. Nor did these other intimate, corporeal exchanges cause the Australian public to pause and wonder whether acts of ritual sex were *sex*. What the public saw was sex: public and semi-public sex, wild savage sex wholly divorced from the heterosexual couple. And what they saw and talked about was the shocking fact that their government, liberal men in Canberra who knew little of real life in the frontier, protected these saturnalian customs from police and public interference. Perhaps, more shocking still, the Australian state protected these sex acts, even though, unlike the United States, Canada and New Zealand, it had never recognized native sovereignty or native customary law.

The Australian public did not simply look at indigenous sex, imagine it, discuss it. They felt implicated in acts like the one above. It was as if each Australian citizen had to choose to participate or not in the actual rape of the Aboriginal female by deciding whether or not the police should be allowed to interfere in "more or less tribal customs." And, in so choosing, it was as if each Australian citizen had to weigh the cost of transgressing either the moral or the juridical foundations of the nation. If the police punished indigenous men for, paradoxically, the most public of their private affairs, they would do so in ways that seemed unjust to many average white Australians. How could the state punish "more or less tribal" men and women for a crime that had no local name? But not to punish these men seemed grossly immoral. Average Australian citizens began to ask what were, after all, the principles of rightness and fairness on which "the English speaking race" based their interference in their tribal affairs.[3] The lack of a common language or shared moral universe between settler and indigenous groups revealed to many white Australians the function of force in liberal nation-building. Force was not a distortion of a civil conversation between indigenous and settler groups, it was the conditions of this conversation's possibility.[4]

Thus, ritual sex did not simply make the nation talk. Whether the state protected or punished the ritual actors, ritual sex terrorized hegemonic understandings of Australian nationalism in the early twentieth century. Ritual sex tore the Australian public's attachment to an ideal image of the liberal democratic state as the administrative arm of the moral nation. It tore the notion of a moral Australian nation from the notion of a just Australian nation as the public was forced to choose between a commitment to the principles of moral sense and sensibility and a commitment to the principles of legal fairness. And, finally, ritual sex tore the imaginary shared fabric of the Australian normative white public's collective will as it was forced to recognize the depth of its civil disagreements. Who were *We the People* of Australia in relation to a minority indigenous population who did not share the white majority's most basic understandings of the privacy and sanctity of sex and intimacy? How could a modern nation condone a potentially permanent space of sexual immorality, perversion and violence? But, equally, how could the indigenous population be integrated into the nation and be given equal citizenship rights and responsibilities while they maintained customs antithetical to civil society? Should and, if it should, how could the state punish indigenous men and women for their sex acts in a civil and civilizing way?

This chapter examines how indigenous ritual sex called into question the foundations of Australian nationalism, helping to build a new language of national unity based on a more flexible form of cultural difference. But this chapter is not a tale of liberalism's progress. In the second section of the chapter I suggest how anthropologists, in particular, developed a modern language of cultural diversity that allowed the nation to retain its moral foundation and its fantasy of a normative division of society into a rational critical public sphere and intimate, sexual domestic sphere even as it experienced itself as supporting Aboriginal customary life. I argue that to accomplish this feat, anthropologists had, first, to create sex and, then, distribute this sex into its proper social domains. In the last section of the chapter, I change time frames rather dramatically. I discuss a contemporary Aboriginal women's menstruation ritual in order to suggest the legacy of national anxieties about ritual sexuality in the (post)modern multicultural Australian nation.

Sex rites and state rights

Let me return to the telegram. If the state were literally an apparatus, a horrid grating sound would have been heard when J. A. Carrodus, the Secretary of the Department of the Interior, responded to the Darwin Administrator.

> YOUR TELEGRAM 11TH JUNE WAS CHIEF PROTECTOR CONSULTED BEFORE APPROVAL GIVEN FOR ARREST OF NATIVES STOP ON INFORMATION GIVEN BY YOU MATTER APPEARS ONE IN WHICH THERE SHOULD HAVE BEEN NO POLICE INTERFERENCE[5]

Rather than praise the investigatory promptness and moral sense of northern police, Carrodus reprimanded them for their actions and implicitly threatened their careers. In a strongly worded letter sent six days after this telegram, Carrodus wrote, "It is considered that the action of Constable Pryor in arresting the natives and bringing them to Darwin was at variance with the spirit of the instruction given by the Minister and conveyed to you in my memorandum of 10th February, 1936."[6] The spirit of administrative law was clear. Neither settlers nor police should interfere with Aboriginal practices where "tribal laws only are concerned and where no white person is involved."[7] In practice this meant that northern police had to suspend their moral sense at the moment when it was being most severely tested. And they had to protect what they considered to be immoral practices and to be insults to masculinity and femininity and to society more generally. State policy demanded that Daly River Constable Pryor hold back the hands of Peanut Farmer Harkins and that the Constable and Farmer do no more than look on as the "female [was] required [to] have sexual intercourse with all and sundry." Why had the Department of Interior formulated a policy that so clearly violated the commonsense morality and civil sensibilities of the majority of the Australian public including state employees like Constable Pryor?

State policy barring police and settler interference in "tribal affairs" reflected, in part, the state's broader role in regulating the economy. By 1936, social theorists had fostered a commonsense view that, in primitive societies, men's rituals lay at the heart of the society's customary law, social regulation and social reproduction. And, by the same time, the long *durée* of imperialism had demonstrated to the state the administrative and economic usefulness of supporting the "internal stability" of "wild natives," those groups who lived "more or less permanently in remote areas."[8] Protecting "the free exercise" of "native customs" provided agriculturists and pastoralists with a self-regulating, self-reproducing, disposable labor pool, even as it helped insure peace in the frontier.[9] The more indigenous groups maintained the traditions necessary for a self-regulating society the less state and private capital had to be expended to maintain and reproduce the Aboriginal labor critical to the appropriation of Aboriginal lands and resources. Aboriginal laborers were paid in minimal rations, employers arguing that Aborigines could forage for the extra provisions that they needed during the work season and during periods when no work was available. This all but free labor was critical to economic enterprise in the north where profit margins were thin as best. As the frontier was absorbed into the nation, Aboriginal customs were said to lose their "wildness." No longer wild, they were no longer recognized as "more or less tribal" and the full force of state law bore down on their practices.[10] No little irony is found here.

State policy was not simply a matter of economic expediency. It also reflected the majority white population's aspirations to leave behind its violent frontier image and, instead, to consider itself and to be considered a modern liberal nation as the discursive foundations of liberalism shifted to a looser, more tolerant view of cultural difference. Australians were forced to question the baffling entanglement of the nation in the sex acts of indigenous groups in part because of the late settlement of the Australian frontier and the cultural integrity of indigenous groups living there. While Australian settlers and indigenous groups struggled to control the economic, moral and social terrain of the nation, representations of "primitive society" were slowly changing. By the 1930s, the relativization and pluralization of culture had gained hegemonic status in liberal bourgeois public sphere.[11] Justifying the extermination or gross mistreatment of colonial subjects on evolutionary or religious grounds grew more difficult. Instead, the Australian state's treatment of the indigenous population, its ability to maintain peaceful relations between indigenous and settler groups, and its promotion of civil values gradually became key indices of the nation's achievement of modernity. When its police and juridical institutions disciplined indigenous women and men, did it do so based on reasoned and impartial law and toward the end of the making of modern citizens? Or was the nation still mired in a frontier mentality, its native policy based on force rather than reason, on punishment rather than reform, on physical extermination rather than social advancement?

Prohibiting white interference in tribal customary practices and protecting "tribal laws . . . where no white person is involved" meshed well with the national aspirations of white Australians when the customary practices protected were minimally dissonant with normative values and practices. It cast the nation in the role of ward, protector, pater. And it reinforced public identifications with and idealizations of the white Australian nation as a good, tolerant, reasonable "subject" (Povinelli 1997, Balibar 1991) and of the state as the nation's political-administrative prosthesis.[12] In these instances, the state could be seen as the apparatus that the national body needs to institutionalize itself and its ideals while remaining a sublime material, an indestructible and immutable body which persists beyond the corruption of the state.[13] The state's tolerance, even narcissism, of minor cultural differences reinforced the imperial fantasy that colonial appropriation was a form of paternalistic recognition and gift-bestowal.

But in cases in which the state protected cultural practices that seemed morally incomprehensible, even evil, to ordinary people, state policy threatened to rive national subjects from their ideal images of themselves, their nation and its hyphenated relation to state institutions. The internal stability of "wild natives" might depend on "the free exercise of their native customs." But, for the majority of white Australians, sacred acts of bestiality, masturbation, same sex and group sex seemed, at the least, incommensurate with a modern civil society's understanding of sex and intimacy as a private, normatively heterosexual affair. Aborigines' public and semi-public sacramental sexuality conjured images of profound depravity, an absolute limit of a civilized nation's tolerance of cultural difference.

We should not underestimate what was being asked of ordinary national subjects; by that I mean, persons holding government jobs, designing and administering state policy. If state workers were to protect "more or less" customary practices, they had to make critical cultural and moral judgments which might have broad social ramifications. People like Constable Pryor had to decide quickly whether Peanut Farmer Harkins collided into indigenous culture or stumbled upon the perverting effects, "more or less," of a Christian colonialism gone terrible awry. Was an indigenous practice part of the group's precolonial traditions or was it a response to the "type of whiteman" living in the frontier, to Christian missionaries proselytizing across the outback, or to the very *laissez-faire* entrepreneurs the policy was, in part, designed to protect and support? The scandalous nature of indigenous sexuality was not the only news making national headlines. The sexual relations between white men and Aboriginal women shocked and troubled the nation. Underscoring the uncanny intercalation of settler and indigenous culture, spokespersons of indigenous and national culture reported that many northern tribes dubbed one of the ceremonial complexes in which sex occurred as "Big Sunday" services (Freud 1959). Was this moniker a clue to the ceremonies' origins? Was it an uncanny commentary on the corrosive, degenerative effects of civilization in its frontiers? Some public officials and anthropologists thought so. In an administrative submission about Daly River men's ceremonies, a government worker in Native Affairs, W. E. Harney, argued that although the sexual law of men's rituals predated settlement, its application had of late dramatically increased because "contact with civilization tends to make the native women disobey the laws and taboos of the tribe."[14] Some anthropologists disagreed with Harney about the origin of ritual sex acts, but agreed that they were related to the civilization process. Anthropologists argued that Aboriginal men had borrowed these rites from Asia in order to compensate for their loss of power and status in the colonial nation.[15] The worry that colonial contact caused Aborigines' ritual sex acts made anxious the nation's claims to modernity. What type of civilization produces rather than civilizes primitive sexuality?

The state did more than hone the cultural skills of its employees. In asking its employees and the public more generally to act against their deepest moral convictions, the state opened a space between the ideal image of the moral nation and the ideal image of the state as the apparatus for upholding and promoting the nation's values. Whose moral values was the state upholding, encouraging? How were indigenous men taking advantage of the sexual and cultural naïvety of liberal cultural relativism? Another Daly River Constable, J. T. Turner, argued not only that state-sanctioned "black" rituals violated deeply held national "white" values, but also that these so-called tribal customs were nothing more than grotesque ceremonial masks, cunning masquerades of cultural difference, mocking the state's inability to discern culture from connivance (Bhabha 1994, Fanon 1967). According to Turner, indigenous male rituals were simply the diabolical deceits of "Blood lustful Aboriginals," their "excuse" for rape and murder; and male ritual was on the rise not because of the functional integrity of indigenous society at the frontier, but because of Aboriginal men "understanding nothing will be done to them" in instances such as the Daly River "rape" because these rituals were "considered a Tribal Affair."[16]

For white Australians like Turner an interior social pocket of sexual violence and immorality anticipated what the contemporary liberal theorist of multiculturalism, Charles Taylor (1994), described as the "awkwardness" of the nation bearing "substantial numbers of people who are citizens" of the nation but also "belong to the culture that calls into question" its "philosophical boundaries." The awkwardness of ritual sex led many persons in the state and public to argue for the selective suppression of native sexual customs. The Acting Chief Protector of Aboriginals, W. B. Kirkland, argued just this in the Daly River case. According to Kirkland "ritual rape" should be suppressed even if the only way of doing so was the application of a law that so-called tribal natives did not understand.[17] For Kirkland principled and just law must give way to the moral foundation of the nation; and the moral perspective of the natives must give way to the moral perspective of the nation.

But many anthropologists and popular writers were making the Aboriginal perspective a compelling part of public discourse. They reported that, from an intracultural perspective, ritual sex was not a monstrous aberration of the social order, but the proper method of organizing, expressing, regulating, and disciplining that order. In a widely read ethnography of Aboriginal society published two years after the Daly River case, A. P. Elkin reported to the reading public that Aboriginal women did not object to the "practices" to which their bodies were put even though their treatment was supposedly to men's benefit.[18] According to Elkin (1938), "tribal" women and men considered ritual sex acts to be a legitimate part of religious services, the law, community bonding, cosmological expression, and conflict resolution.

While aboriginal sexual practices defied the hegemonic privilege settlers gave sex acts as opposed to other social, physical and psychical fluids, acts, exchanges and identities, arguments that white law should be used to suppress certain Aboriginal customs in the absence of a common language opened a disturbing question. Was a moment of injustice, blind and unprincipled force, essential to the civilization process (Derrida 1990)? Were the sexual morals of white Australians relevant to the social organization, regulation and reproduction of Aboriginal society? The Australian ethical public sphere was not alone in asking these questions. The critical international public sphere asked how could "full justice . . . be done";[19] and what did "legal justice" mean in a colonial context where there was no common set of rules and definitive criteria between indigenous and settler communities? As far back as 1837, the British House of Commons Select Committee on Aborigines observed that for wild natives "the observation of our laws would be absurd and to punish the non-observation of them by severe penalties would be palpably unjust."[20] In 1933, 60 jurists from the Northern Territory echoed this parliamentarian report. They called on the Darwin Supreme Court to try Aborigines according to their customary laws where the offense was of a "more or less tribal" nature. To do otherwise would expose "white" law to shameful charges of savagery, injustice, and inhumanity: by what right try and potentially condemn to death an Aboriginal man for a crime he has no language for or understanding of? The absence of a common language and practice blurred the line between force and justice, trial and torture, civil and savage law, the frontier and modern nation.

What I am arguing is that the line of antagonism Aboriginal ritual sex marked forced national subjects to recognize and confront a series of separations that the ideology of an Australian national modernity masked: the modern Australian nation did not consist of a "popular collective will"; the nation's commitment to "a superior, total, form of modern civilization" was often at odds with state and capital objectives and often resulted in the degeneration of indigenous society rather than its advancement; and, finally, civil law might demand a moment of unfair, brutal, force (Zizek 1994, Gramsci 1992). If, as Etienne Balibar (1991) has argued, the nation is "presented to us in the form of a narrative which attributes to . . . [the nation] the continuity of a subject," radical sexual practices revealed the incoherence, the partiality and fragmentation of national and individual subjects and the role national ideology plays in fabricating a "totality set on effacing the traces of its own impossibility" (Zizek 1989). While ceremonial masks might have hidden the fact of sexual predations, they unmasked the fiction of a collective Australian national will and of an unproblematic Australian modernity. Ritual sex forced the nation to experience itself not only as out of sorts, but also as out of its collective mind. But ritual sex did something else. It prompted the Australian public to consider a new foundation for its national aspirations, a more diverse, more tolerant mode of national unification.

Many state administrators and public spokespersons turned to anthropology to provide a language for this new form of national integration. J. A. Carrodus, for instance, instructed the Administrator of the Northern Territory that "the question of whether ritual rape and certain other rites and customs should be suppressed is one which should be determined in the light of the advice given by anthropologists or officers trained in anthropology."[21] Likewise, the Darwin jurists who called on the Darwin Supreme Court to try Aborigines according to their customary laws suggested that the decision of whether an offense was more or less tribal should be left up to "men who have studied their laws and customs from a scientific point of view, and . . . men who are genuinely sympathetically interested in the Aborigines. . . . Leaving the matter in the hands of those who have no knowledge of the Aboriginal would only result in a remedy worse than the disease."[22]

Men of sympathy: anthropology and the technology of difference

Anthropology was able to influence the Australian state's indigenous policy for a number of historical reasons. On the one hand, by the 1930s, anthropology had established itself globally as the liberal science of primitive societies. State and public anxiety over how to administer justly indigenous sexual customs found a discipline waiting to provide answers. On the other hand, the second Chair of Anthropology in Australia, A. P. Elkin, set a particularly broad liberal agenda for the discipline.[23] Elkin had a clear vision of anthropology's role in the making of a modern tolerant Australian nation. He sought to overcome ethnological amateurism and establish the field as a modern scientific discipline ruled by analysis rather than anecdote, by fact rather than fear, by understanding rather than abjection. And he sought to make anthropology relevant to national consciousness about and governance of the indigenous population. He and some of his students and colleagues sought to extend the field's influence from the narrow and hermetically sealed confines of the academy into the corridors of federal, state and territory governments and into the public conversations about Aboriginal society occurring in lecture halls, newspapers and radio broadcasts through the Australian nation.

But if this broad liberal program was to succeed, anthropology had first to catch and captivate government and public attention, compel them to turn toward the new science. Anthropologists had to meet, at least half way, public, law, and state obsessions. "Learned theorizing" had to merge with "concrete phantas(ies)" to organize and cohere the nation's collective will (Gramsci 1992: 126). To do so, Elkin and his students moved among several different registers in their ethnographies and policy papers – citing the voices of neutral and objective science, of a horrified and traumatized public, and of their readers' intimate desires and shames for a modern Australian nationalism.

What, then, were administrators, publics, and jurists worried about? The answer to this question clustered around issues of sexuality, citizenship and nationalism: the strange sexual customs of Aborigines; whether and how Aborigines could be integrated into the nation; and, finally, in what way these sexual practices were related to white Australians' social and economic practices at the expanding national frontier. Whereas other disciplinary knowledges and practices had failed to provide principles for answering these questions, Elkin promised that anthropology would provide not only a rational explanation for Aborigines' strange sexual customs, but also a means for doing away with them, and thereby, doing away with a whole host of horrors threatening the Australian nation's claim to be a modern, civil and humane society. This was the deep seduction of Elkin's rhetoric: he promised to accommodate customary Aboriginal law to emergent liberal notions of cultural tolerance without altering the moral foundations of the nation.

The first task that anthropologists set themselves was, therefore, to establish the discipline's expertise on Aboriginal culture and morality. Among the various civic and state institutions intruding into Aborigines' intimate lives, the role of the church as moral arbitrator of Aboriginal ritual sex was an especially acute problem for Elkin. Along with being the Chair of Anthropology in Australia, Elkin was a practicing Anglican minister. Perhaps because of this, while Elkin argued for the governmental regulation of missionaries, he was not entirely critical of missionaries' role at the nation's frontiers. He noted that the very zealous, provincial missionary character that many upper-class, well-educated government administrators "sneered upon" provided the state a free economic and labor infrastructure for the administration of Aborigines in the north (see Elkin 1944: 74–75; see more generally Swain and Rose 1988).

The real trouble with missionaries, Elkin argued, was that too many of them lacked the respectful understanding of Aboriginal "cosmology" and "philosophy" that anthropology could provide. Much later, another anthropologist, W. E. H. Stanner, similarly argued that, rather than men of cosmology, missionaries saw promiscuous women and "men of sodomy, sinners exceedingly."[24] Anthropologists argued that this unfounded prejudice led to the unscientific suppression of native customs and thereby to the destruction both of the valuable scientific knowledge embedded in Aboriginal culture and practice and of the social fabric of Aboriginal life. Missionaries were right, Elkin argued, to see ritual sex acts and some other customary sexual practices as antithetical to the goal of Aboriginal citizenship. He himself made a similar argument in a short text, *Citizenship for the Aborigines* (Elkin 1944). But, Elkin argued, missionaries were wrong to confuse Aboriginal customs and cosmology with Aboriginal sexuality. Aboriginal sex acts and Aboriginal customary law and ritual were not identical. Elkin's anthropology promised to demonstrate to Aboriginal elders and to Australian publics, missionaries and state administrators the difference between sex acts and customary law. And Elkin promised to take over the troubling task of deciding what was and what was not a native custom, which of these customs should be suppressed, and how.

What anthropologists had that missionaries and governments did not, and thus what made arguments like Elkin's compelling, was the discursive power of science and scientific methods. These methods were thought to allow anthropologists to understand how sexuality fitted in native social and cultural systems. Although counter-intuitive, a willingness to understand the sense their sex made would provide the means for doing away with the sex troubling the nation. As Elkin was wont to remind his audience – to understand was not to fail to judge, was not to condone. Understanding sex's insinuation throughout the Aboriginal social body (kinship, ritual, marriage, warfare, economic practices) made anthropologists better social surgeons. They could disambiguate sex from customary law without destroying Aboriginal society and, in so doing, could produce: a sanitized sex-free "cosmology," a privatized native sexuality conversant with Aboriginal citizenship, and a morally reintegrated nation.

Along with its analytic methods, anthropology also had a spatial component, an injunction to travel between national and indigenous fields. Armed with their scientifically acquired understanding, anthropologists could engage in a critical-rational discourse with both "tribal elders" and the liberal national public. Jack-of-all-trades, social surgeons, grand mediators and master sanders, anthropologists would cut and smooth away the sharp edges of cultural alterity, dulling its radical challenge to liberal models of sex, civil society and publics (Habermas 1991: 79–88).

It was a fantastic vision, really. Anthropologists would sit with "tribal elders" and, through critical conversation rather than force, persuade them to alter those aspects of "social organization and custom" which were objectionable to civil society.[25] And anthropologists did try. With the help of missionaries and government officials they pursued sex, focusing their notes and queries on local practices and beliefs about sex and social organization. And anthropologists persuaded some Aboriginal men and women to devalue some of the "bad objects" of culture that, in critical ways, they themselves were valuing by showing such intense interest in them. The "good objects" of sexual difference would remain. In Australia as elsewhere, anthropologists pored over and presented to the public the puzzling maze of Aboriginal kinship and descent. Charts and graphs outlined how gender (sex of the body) and sexuality (descent) sorted Aboriginal individuals into local descent and ritual groups.

In these ways, anthropologists conversed with the national public, calming their fears, explaining away the "strange and puzzling" and the "degrading" and disgusting of Aboriginal social life, even while reassuring the public that these "bad objects" of Aboriginal social life were being systematically eliminated (Elkin 1938: 108). And because anthropologists were men of good will, men who could demonstrate a real knowledge and passion for Aboriginal society, they could reassure the public that whatever the discipline advocated for Aboriginal society was advocated humanely, tolerantly, and on its behalf. Sex is (soon to be) nothing (to fear): this is the ambivalent syntax that defined modern anthropological approaches to native ritual sex. In so convincing the public, anthropology not only salvaged indigenous culture, but salvaged the ideal of nationalism and dispassionate reason from its miscarriage at the nation's frontier (Elkin 1938: 108, 130–131).

But what was this "sex" that everyone was talking about? Were the genital acts horrifying the nation "sex" at all? And if not, how did they become "sex acts," to and for whom? Were anthropologists, like the public, misrecognizing the relation between a structured network of corporeal exchanges and one of its elements, genital exchanges, and, in the process, creating the very sex that they hoped to civilize?

In an essay, "Religion, totemism, and symbolism," the anthropologist W. E. H. Stanner (1979 [1965]) confronted and tried to counter public perceptions of Aboriginal ritual sexuality. His aim was like Elkin's, to distinguish between sex acts and customary law. To do so, Stanner differentiated the cosmological aspects of Aboriginal High Culture from what he called its "vehicle or symbolizing means" (ibid.: 235). According to Stanner, sex was simply a symbolic tool, a substance indigenous men chose to vehiculate, to communicate, their cosmological values. From Aboriginal men's perspective, sex was one of a number of powerful and transformative corporeal and noncorporeal substances. Semen, sweat, blood, songs and clays penetrated initiates' bodies and sacramentally reformed them into ancestral beings. All these substances created intimate corporeal relations between humans and landscapes, transferred ancestral powers and conveyed cosmological meaning. In other words, sex was merely equivalent to rather than the general equivalent of these other bodily exchanges. And in portraying sex as merely an equivalent, Stanner enunciated the grounds of a radical critique of the object status of sex in the sex acts riveting the nation and discipline.

But while Stanner de-privileged sex as the final interpretant of social exchange and other "intercourses" of the symbolic in indigenous society, he re-established sex as the final interpretant of bodily and social exchanges in civil society. According to Stanner, neither Aboriginal subjects nor Aboriginal cosmology ("the Dreaming") could take their rightful place in Australian civil society unless a more suitable vehicle for symbolic locomotion were found. Like Elkin, Stanner argued that until cosmology and sexuality were separated and properly relegated to their public and private domains citizenship would elude Aborigines. And, in so arguing, Stanner privileged sex in relation to other corporeal substances and exchanges and removed sex from public and semi-public semiotic circulations. Rather than pursuing the difference of indigenous sex, Stanner reconfirmed civil society's supposedly naturalized relationship to sex acts, publics, privates and sacramentalities and, thereby, contributed to the slow reformation of the place and function of sexuality in Aboriginal society. In the civil form of Aboriginal society, sex was separated from the semiotic, the cosmological, the public and semi-public transmission of meaning. Rather than just one of a number of public and semi-public cosmological and corporeal exchanges, sex became a privatized technology whose end was the biological reproduction of bodies and the formation of social groups. Sex created alliances and made family groups, clans and local descent groups. Ritual made meaning.

In sum, Australian anthropology did not explain radical cultural difference to the state and its national publics so much as create a common language and practice in indigenous society that dulled the antagonism of its sexual alterity. Anthropology did not simply sort sex into its proper domain, removing it from the sacramental and relegating it to the familial. Anthropology helped create sex for the nation and for indigenous women and men. In so fashioning sex and cosmology, anthropology created a form of cultural difference easily digested into the emergent multicultural Australian nation. And, in the process, they left unchallenged dominant models of sex and civil society, emergent discourses on sexual subjectivity (heterosexuality, homosexuality) and dominant hierarchies of bodily fluids and acts.

Multicultural sex

In 1991, ten women and I lounged around the verandah of the Belyuen Women's Center, a house on the edge of the Belyuen Community, northern Australia. The Belyuen community derives its name from a large fresh water spring located just below where we were sitting. In that spring resides the Rainbow Serpent Dreaming (*durlg*, *therrawen*), *Belyuen*, which provides Belyuen people with *maruy*, variously described as a person's conception Dreaming, their shadow and their essence (Povinelli 1994). When Belyuen women swim in the waterhole, the Belyuen Rainbow Serpent enters them, depositing a part of itself in the children that the women will one day bear.

The women and I had gathered at the Women's Center for a *pidjawagaidj* ceremony, a series of rituals held for a young woman's first menstruation. The young initiate was locked up inside the Center where she would stay for the next five days or so. Off in the grass to the left of us was a smoking fire. Next to the smoking fire lay other half-burnt logs, the remainder of a women's ceremony held months before. Under them were the ashes of numerous other fires made for the same purpose. When the women are conducting a *pidjawagaidj* ceremony, they keep a fire constantly burning to cover up the initiate's strong smell. But they also use the fire to signal to the Rainbow Serpent in the Belyuen waterhole that one of its children is becoming a reproductive woman (*djïpel*).

As we sat talking, one of the women heard Agnes Lippo's footsteps ruffling the grass around the east side of the building. Agnes Lippo is dead, but her *nyuidj* (dead spirit) often visits the Women's Center when we go there to discuss women's business or to hold women's ceremonies. She was the ceremonial leader of the women's business before she died of kidney failure, a cause of death too common in the community. As in other interiors of the first world, at Belyuen poverty is embodied, mottling most people with the cicatricose leftovers of endemic streptococcal sores and exhausting some people with diarrhea, diabetes and other degenerative diseases. But not only are the material conditions of national life embodied at Belyuen, Belyuen bodies become part of the material conditions of the landscape. Agnes's work, her sweat (*ngungbudj*), attached her to this place. A specificity to her identity will persist while those of us who knew her remain alive. As we die her specificity will slowly dissipate, but it will not vanish if others remain who, knowing nothing of Agnes, know, nevertheless, that a Belyuen *nyuidj* lives in the vicinity of the Women's Center.

Many details of the *pidjawagaidj* ceremony are secret. But the communication between people and ancestral beings and their mutual embodiment play an integral role in all parts of the ceremony. While the *pidjawagaidj* ceremony marks a young woman's sexual maturity, the making of bodies at

Belyuen is not simply a sexual affair. It necessitates the intimate penetration of the human body by substances most non-Aboriginal Australians would not consider sexual. Women sing to alert various ancestral beings of their ritual presence. Water, sweat and smoke are rubbed into the initiate's body. Blood and sweat are diffused in fresh waterholes and in the sea. Ritual leaders burn and bury in the ground objects saturated with bodily substances just as their mothers and grandmothers burned and buried objects. Burning, burying, soaking, singing, rubbing, sweating, smoking, being born from a place (*maruy*) and sinking back into it at death (*nyuidj*): in these ways this land and these women have come to share a corporeal substance, come to be for and from each other.

The meaning and value of Belyuen women's rituals do not simply derive from local notions of bodies, intimacies, substances and landscapes. Like their ancestors, their most intimate ritual relations are caught up in national dramas and aspirations. By 1991, the Australian state had renounced the ideal of "a unitary culture and tradition" and instead "recognized" the value and worth of "cultural diversity within it, as the basis of, a more differentiated mode of national cohesion" (see Frow and Morris 1993, Povinelli 1997). Rather than outside the nation, Aboriginal traditional culture is now represented as a vital part of the nation's social, economic and cultural fabric. The Australian juridical, state and public commitment to supporting Aboriginal traditional culture is testified by a range of land rights legislation that provide Aboriginal groups an opportunity to regain their traditional lands.

The Belyuen, for instance, have petitioned to be recognized by the state as the "traditional Aboriginal owners" of land surrounding the community under the aegis of the Aboriginal Land Rights (Northern Territory) Act 1976. As we sat on the verandah, we discussed the claim. I explained the statutory requirements of traditional Aboriginal ownership. The Belyuen had to demonstrate that they were members of a "local descent group" which had "common spiritual affiliations" and "primary spiritual responsibility" for the land. I explained that land claim legislation was intended to support what most non-Aboriginal Australians considered the three elements of traditional Aboriginal ownership: family ties, religious ties and customary economic rights. In other words, I explained how the state had encoded anthropology's separation of sexuality and spirituality into legislation supposedly supporting traditional Aboriginal culture. The state represented itself as recognizing that Aborigines used sex as a technology of descent, as a means generating social groups, in a somewhat different way than the normative white public.

This was hardly news for these women. They were members of various local descent groups for land south of the claim and received exploration, pastoral and agricultural royalties as members of these local descent groups. The Belyuen, however, are not a local descent group in any immediate or obvious sense for the land surrounding their community. Instead, they are widely viewed as a collection of densely intermarried migrant groups displaced from their traditional country during the colonial period. During the late

1930s, they and their parents were forcibly interned at Belyuen. In 1991, many Belyuen considered themselves to be, most immediately, for and from the community and surrounding landscape as a result of their marriage and kinship affiliations *and* as a result of the corporeal practices like those summarized above. Sexual substances did not make the Belyuen into a group, a community, a family; sweat, blood, language, *maruy* did.

But this structural network of corporeal exchanges does not make a local descent group. And so, if these women are to gain the material security that the state offers in the form of land rights, they must find a way of fashioning the cultural bonds that tie them to each other and to the country into a model of heterosexual descent. As we waited for tea to boil, the women meditated on the consequences of failing, of "being wrong," of "not fitting the law," of they and their parents having made "mistakes." Women asked each other whether, in the event that they failed to convince the Land Commissioner that they were a local descent group, the entire community would be sent back to their respective southern countries. From these women's historical perspective, this seemingly fantastic communal apocalypse is not so far fetched. Soon after the Japanese bombing of Darwin in 1942, the war government transported the entire community to war camps in Katherine. Closer to the present, these women have watched other communities displaced in the wake of lost or disputed land claims.

Terrified and fascinated, the women and I outlined various ways of joining Belyuen families by marriage and descent. We turned to and were transfixed by this thing, the local descent group. At this moment, we were not simply engaged in a language game. They did not feel in conflict with the state or with normative national notions of sex and sexuality. They wanted to be this traditional thing that would give them more power over their lives, give them and their children material security. And I wanted to give it to them. In other words, Belyuen women's optimisms and desires, their sense of what kinds of intimate exchanges have value, were trapped by a national tradition: reforming local corporeal practices into a language of sex and sexuality.

Of course, land claim legislation is not the only state resource channeled through normative notions of sex and sexuality. Our conversation on the verandah did not stay on land claims. We also talked about whether various people had gotten or cashed various types of social welfare checks: single mothers' benefits, employment benefits for married and single persons, widows' benefits.

Even as the law provides an incentive for these women to fold their community into a familiar form of marriage and descent, public discourse about ritual sex and other forms of non-normative sexuality place a confining and conforming interpretative and identificatory pressure on Aboriginal women and men when they practice their rituals. It is important, in this context, to remember that although early anthropologists may have persuaded some Aboriginal men and women to substitute supposedly "proper" symbolic vehicles for improper ones, they did so without altering many public and semi-public discussions about the sexual "truth" of Aboriginal customary law and without challenging common or statutory laws outlawing certain sex practices. Ritual "sex," even where no longer practiced, continues to haunt Aboriginal cosmology as its real, true and final interpretant. Disciplinary and legal archives remain. Anthropologists pass on the facts, fantasies, and mythologies of the past to their students. Semi-public and private tales continue to circulate. Indeed, in the multicultural nation, Aboriginal men and women are caught between the multiply directed force of the law and public opinion: the state and public injunction to be culturally authentic, but not to be sexually perverted. We catch the disjunctions of normative Australian law and public opinion in the voice of a Darwin taxicab driver.

I met the taxicab driver on my way from a large card-game at the Darwin Marina to the ferry at Cullen Bay in 1992. I am not the only white woman who frequents the park "casino" but the large majority of people who play there are Aboriginal women. As the white cab man and I pulled away, he asked "You getting a little Top End Abo culture, ei?" "No," I said, "many of the women are old friends of mine, I am just going home." "And where is that? You're American aren't you?" he asked. "Yes, but I am living across the harbor at Belyuen." "You a school teacher?" he asked. "No," I said, "I am an anthropologist." Perhaps because my identity switched from American tourist to American academic, the cab driver's comments turned from cultural consumerism to cultural politics and expertise. My cab driver argued that "Blacks' problems" would never be solved by giving "natives" land, because "real Blacks don't need land, they already have more than they need to live in the bush which most don't really want to do anyway." The problem, he surmised, was lack of labor-discipline. All the talk about protecting "Abo culture" was "for the birds." "Have you heard of this Peppi thing? That's Abo culture around here."

"This Peppi thing." It is likely my cab driver heard about the Peppimenarti trial from local radio broadcasts and the local paper, the *Northern Territory News.* As reported in the mass-media, several men from Peppimenarti, an inland Aboriginal community south of Belyuen, were on trial for and eventually found guilty of manslaughter. The group had beaten to death another Peppimenarti man who they believed was sexually harassing a white nurse then working in the community. They were afraid the community would lose its medical service if the nurse was forced to leave. The publicity of the Peppimenarti death derived in large part from the defense argument that the beating (which was not meant to have killed the man) was part of customary men's business and thus not subject to the Australian penal code (Watt 1992a). The beating was reported to be a new application of a precolonial ceremonial law wherein sexual misconduct is physically punished (Watt 1992b).

This semi-public conversation continues older ones I described above. For my taxi driver the "Peppi thing" was paradigmatic of the public secret of Aboriginal culture in the north, especially Aboriginal ceremony and ritual.

Echoing Constable Turner, my taxicab driver argued that "Blacks" were taking advantage of the good intentions of liberals by excusing their social dysfunction as a traditional legal function. Government support of Aboriginal ceremonial practices betrayed politicians' "lack of guts," their refusal to "own up" in these "politically correct times" to the sex, violence and social chaos rampant in Aboriginal communities and hidden behind ceremonial masks. For this taxi driver and many other northerners, state support of Aboriginal culture heightened the contradictory nature of multiculturalism.[26]

The taxicab driver was not simply wrong. Whatever "sex" was in the precolonial period, today sex and sexuality are hardly absent from *pidjawagaidj* ceremonies. For example, when the young initiate is considered ready, she is taken from Belyuen to a women's ceremonial ground on the northern shore of the Cox Peninsula. Here women "play fun (*erere*) with each other, following or playfully citing and breaking kinship roles. For example, *meinggen* (cross-cousins, "wives" in the local creole) are encouraged to physically and sexually "tease" one another. And aunts are encouraged to "kill" (to hit hard with mud, sticks, rocks or clay) their nieces, the future wives of their sons. In other words at specific segments of the ceremony, sexual teasing between various kin is mandatory. These roles will be important during a "hard" ritual held later in the day. Like the penetration of material substances into the human body, these erotic games and narratives build intimate attachments between people and places. They form communities from biologically and socially connected persons. And they are part of the process by which human and land corporealities, intimacies, and desires become for, from, and identified with each other.

But these ritual and everyday intimate and erotic interactions build other things in the contemporary multicultural nation. They build sexual identities and subjectivities. For example, in the hot afternoon as we waited for another segment of the *pidjawagaidj* ceremony to begin, my younger sister (*edje*), my *meinggen* and I were sent to a local pub to buy some food and drinks for everyone. I drove. As is proper, our *meinggen* sat between my sister and me. As we bumped along the dirt road in my large Toyota Landcruiser, my sister drew our cousin into her arms and began "swearing" at me, saying that I was trying to steal her wife. Really, I wasn't. To our *meinggen*, my *edje* said, "You don't love Beth. You love me. You're married to me." And to me she said, "I'll kill you dead, you steal my wife!" Just as I began my oral defense, we hit a huge bump in the road, the stick shift banged against the inner thigh of my *meinggen*'s leg, and I flew against her side. "*Yagarra*," my sister said, "*kaidjek mungga!*" ("Oh Lord, she fucked you!"). Just at this point we reached the road leading to the pub. My cousin finally spoke, "Shut up you two, white people might hear us and think we're lesbians."

Others might think so too. At other ceremonies attended by white lawyers, the wives of Land Claim Commissioners and linguists, women reform their bodies, intimacies and desires around normative sexual frameworks.

Women discipline each other. They say, "Don't get up (and tease), maybe this *mitjitj* (white women) will think we are hungry for men (sexually promiscuous)" or "that we are faggots." A host of national sexualities insinuate themselves in the attachments Aboriginal women and men form with their ritual practices at least in territorial proximity to "white" interpretation. Proscribed kinship relations are sorted out into sexual identities: homo- and heterosexuality, Black and white sexuality, licit and illicit sexuality. The focus on the relationship between sex acts, race and sexual identities obscures the function of other fluid exchanges between human and mythic bodies.

Conclusion: native sex, from antagonism to difference

This chapter has examined the implication of two forms of Australian nationalism in the ritual sexuality of indigenous subjects. And I have suggested how, in both forms, the nation encloses subaltern practices and languages of intimacy, erotics and corporeality in the very language of sex and sexuality that both structures and haunts the nation.

The torsions Australian national ideology exhibited as public spokespersons shifted the discursive grounds of what makes the Australian nation *a* nation rather than a collection of divergent people raises questions about sexuality and modern liberal ideology more generally.

Long ago Marx (1963) argued that democratic-republican institutions function to weaken the antagonism between capital and wage labor, harmonizing the two without doing away with their material oppositions. In the mid-twentieth century liberal democratic nations like Australia began to understand that social harmonics were not the only means of doing away with the sexual alterities threatening national ideology, identity and identifications. While points of sexual alterity threaten social harmony and the fiction of a collective national will, liberal democracies slowly began to understand how a milder form of social and sexual difference provided the ideology of nationalism flexibility. In Australia a mild form of difference dulled both dominant and subaltern experiences of their mutually conflicting desires, needs, practices and discourses and mitigated the possibility that either group would experience in these extremes the potential for radically reconstituted social relations, intimacies and practices, that is, the social stuff of nations and capital relations. National and international queers become marriage partners. Mormons become monogamous. Female same-sex Aboriginal cross-cousins become lesbians. As they do, alternative models of community-building, of erotics and of corporeality are excised from critical debate about what makes someone for and from a place and a people.

Thus, this chapter is not simply about the nation and its troubled relation to non-normative sexual modalities. It is also about what are the grounds for deciding who is a member of a family, a community, a land and a nation. Belyuen women struggle to form intimate communities based not on marriage and descent simply, but on other penetratrive and reproductive intimacies, if intimacy is the right word to use. And the Australian state and public continue to enclose these attempts in a language of sex and sexuality. But it is useful to remember that the incoherencies of national, state and capital needs and desires around Aboriginal culture and sexuality more generally also open up possibilities for new social languages and practices. The liberal public's very desire to understand themselves and their nations as good, tolerant and reasonable means that even as some subaltern discourses are modified, other discursive and imaginary zones are created and opened for new enunciations and experimentations.

Acknowledgments

Research for this chapter was supported by the National Science Foundation (SBR-9630155) and the Wenner Gren Foundation. I would like to thank Susan Edmunds, Lauren Berlant and Tamar Mayer for their critical, intellectual and editorial contribution to this chapter.

Notes

1 Australian Archives CRS F1 Item 36/592, June 11, 1936.
2 See also Stocking (1995: 94).
3 Suggested, for instance, in a letter to the editor of the *Northern Territory Times and Gazette* published six years before the Daly river "incident," from a well-known northern public figure, Joe Croft.

> The Old People and Their Tribal Ways. I maintain we should not interfere with them. If we do we should start in Darwin. I can take the authorities to the Compound and charge fourteen of the aboriginals with bigamy. Then I can take them to the Daly River and charge aboriginals with cutting their fingers off. Then we can go down to the McArthur and Roper River and charge the natives with inflicting torture on young girls. This horrible rite is practiced for the sole purpose of forcing maturity on girl children so the old men of the tribes are kept supplied with wives. Just after this ordeal 10 and 21 year old girls can be seen newly operated on like calves in a branding yard. Then I can take the authorities further out to the Hubert River where natives perform the surgical operation on the males which prevents any possibility of propagation. Now I maintain that if these outrageous offenses are permitted by the aboriginal in their tribal affairs the fights between them should be allowed and that we of the English speaking race have not the right to interfere in their tribal affairs.
>
> (*Northern Territory Times and Gazette* May 16, 1930)

4 See Calhoun (1995: 52).
5 Australian Archives CRS F1 Item 36/592, June 12, 1936.
6 Letter from J. A. Carrodus, Secretary, Department of the Interior, to the Administrator of the Northern Territory, June 18, 1936. Australian Archives CRS F1 Item 36/592.
7 Memorandum sent to the Administrator in Darwin, February 10, 1936. Australian Archives CRS F1 Item 36/592, June 18, 1936. See also the Administrator's response, June 12, 1936, Australian Archives CRS F1 Item 36/592.

8 Australian Archives CRS Fl Item 36/592, March 8, 1940. J. A. Carrodus, Secretary, Department of the Interior, to Administrator of the Northern Territory.
9 Australian Archives CRS Fl Item 36/592, March 8, 1940. J. A. Carrodus, Secretary, Department of the Interior, to Administrator of the Northern Territory.
10 Four years after sending the above memorandum, J. A. Carrodus reiterates this de facto recognition:

> The Minister has not agreed to rescind the ruling referred to in its general application. In the future, the direction will only apply in the case of relatively uncivilised natives who live more or less permanently in remote areas, who are not under any form of permanent European control, assistance or supervision, and who depend for internal stability on the free exercise of their own native customs.
>
> (Australian Archives CRS F3 Item 20/32, March 8, 1940)

11 See, for instance, Stocking (1995) and Kuper (1993).
12 On the prosthetic person, see Berlant (1993).
13 On the sublime body, see Zizek (1989).
14 Bill Harney was a supervisor of Aboriginal settlements and, later, a popular writer, radio personality and collaborator with the anthropologist, A. P. Elkin. In this case Harney wrote:

> Contact with civilization tends to make the native women disobey the laws and taboos of the tribe, and they would pass over or near these taboo sports knowing they are protected by the law, or the white people of that part, and so the natives seeing their greatest weapon for law and order (increase, regeneration and clearing up of tribal disputes) becoming useless by these women, become annoyed and use force.
>
> (Australian Archives NT Medical Service File of Papers 20/103, September 6, 1940)

15 Stanner notes, for instance, that when Aborigines' lives "came under pressure – from falling numbers and the ageing of leaders – and when the objective circumstances of life were about their worst (during the post-World War I period) an effort was made to reconstitute the High Culture." The old All-Father Cult was replaced by the new fertility All-Mother Cult: "The All-Mother, Karwadi, who is The Old Woman of the Kunabibi cult." He goes on to note:

> Before I had heard a word of Kunabibi I had been told that Angamunggi [All Father] had "gone away." Many evidences were cited that he no longer "looked after" the people: the infertility of the women (they were in fact riddled by gonorrhea), the spread of sickness, the dwindling of game among them. The cult assumed the local form of a cult of Karwadi, by which the bullroarer, the symbol of the All-Mother, had been known in the days of the All-Father. Karwadi became the provenance of the mixed but connected elements which I term the new High Culture.
>
> (Stanner 1979 [1965])

See also Swain (1993).
16 Australia Archives, CRS F3, Item 20/100, August 8, 1940.
17 "Certain rites and customs of the aboriginals should be suppressed and it may be argued that the only method of suppression justifiable is the application of the white law. Notwithstanding the opinion expressed in the Department's memorandum of 18th June, it is respectfully submitted that ritual rape is such a custom" (Australian Archives CRS Fl Item 36/592, July 10, 1936).

18 "The fact that women may often not object does not justify the custom. But those who would modify or prohibit them must do so through the elders and help them find other means of expressing friendship, symbolizing a readiness to negotiate for peace and such like" (Elkin 1938).
19 *Seventy Fifth Annual Report* (1933) of Aboriginal Friends' Association, cited in Law Reform Commission (1986). Also cited by Elkin (1944).
20 Law Reform Commission (1986).
21 Australian Archives CRS F1 Item 36/592, August 16, 1936.
22 *Daily Telegraph*, Sydney April 13, 1933, cited in Elkin (1944).
23 For a biography of Elkin, see Wise (1985).
24 W. E. H. Stanner quoting "the amiable Mr. Dredge," the early-nineteenth-century protector who so described his wards (Stanner 1979 [1965]: 235).

25

> We must also recognise if changes of belief and custom are to be made amongst the Aborigines without disaster, they must be made through the elders, and not just by external authority. As we have seen in the preceding chapters, many important changes have been made in social organization and custom, and also in the ceremonial life; but they have been made after consideration by the elders.
>
> (Elkin 1938: 109)

26 Like Negt and Kluge they point to a certain "oscillation between exclusion and intensified incorporation" of nonnormative counter-publics whose actual practices and beliefs are incommensurate with core national values. See Negt and Kluge (1993: 14).

References cited

Balibar, E. (1991) "The nation form: history and ideology," in E. Balibar and I. Wallerstein (eds.) *Race, Nation, Class, Ambiguous Identities*. New York: Verso, pp. 86–106.

Berlant, L. (1993) "National brands/national body: *imitation of life*," in B. Robbins (ed.) *The Phantom Sphere*. Minneapolis: University of Minnesota Press, pp. 173–208.

Bhabha, H. (1994) "Of mimicry and man: the ambivalence of colonial discourse," in H. Bhabha (ed.) *The Location of Culture*. New York: Routledge, pp. 85–92.

Calhoun, C. (1995) *Critical Social Theory*. Cambridge, MA: Blackwell.

Derrida, J. (1990) "Force of the law: the 'mystical foundation of authority'," *Cardozo Law Review* 5–6: 920–1045.

Elkin, A. P. (1938) *The Australian Aborigines: How to Understand Them*. Sydney: Angus & Robertson.

—— (1944) *Citizenship for the Aborigines: A National Aboriginal Policy*. Sydney: Australasian Publishing Company.

Fanon, F. (1967) *Black Skin, White Masks*. New York: Grove Weidenfeld.

Freud, S. (1959) "The uncanny," in *Collected Papers, Volume 4*, edited by J. Riviere. New York: Basic Books, pp. 368–407.

Frow, J. and M. Morris (1993) "Introduction," in J. Frow and M. Morris (eds.) *Australian Cultural Studies: A Reader*. Urbana and Chicago: University of Illinois Press.

Gramsci, A. (1992) "The modern prince," in *Selections from the Prison Notebooks*. New York: International Publishers, pp. 123–205.

Habermas, J. (1991) *The Structural Transformation of the Public Sphere*. Cambridge, MA: MIT Press.

Kuper, A. (1993) *The Invention of Primitive Society*. London: Routledge.

Law Reform Commission (1986) *The Recognition of Aboriginal Customary Laws*, vol. 1, Report no. 31. Canberra: Australian Government Publishing Service.

Maine, Sir Henry (1986 [1864]) *Ancient Law: Its Connection with the Early History of Society, and its Relation to Modern Ideas*. Tucson, AZ: University of Arizona Press.

Marx, K. (1963) *The 18th Brumaire of Louis Bonaparte*. New York: International Publishers.

Negt, O. and A. Kluge (1993) *Public Sphere and Experience: Toward an Analysis of the Bourgeois and Proletarian Public Sphere*. Minneapolis: University of Minnesota Press.

Povinelli, E. A. (1994) *Labor's Lot: The Power, History and Culture of Aboriginal Action*. Chicago: University of Chicago Press.

—— (1997) 'The state of shame: Australian multiculturalism and the crisis of indigenous citizenship,' *Critical Inquiry* 24, 2: 575–610.

Spencer, W. B. and F. J. Gillen (1912) *Across Australia*. London: Macmillan.

Stanner, W. E. H. (1979 [1965]) *White Man Got No Dreaming: Essays 1938–1973*. Canberra: Australian National University Press.

Stocking, G. (1995) *After Tylor: British Social Anthropology 1888–1951*. Madison: University of Wisconsin Press.

Swain, T. (1993) *A Place for Strangers*. Cambridge: Cambridge University Press.

Swain, T. and D. B. Rose (eds.) (1988) *Aboriginal Australians and Christian Missions: Ethnographic and Historical Studies*. Bedford Park, South Australia: Australian Association for the Study of Religions.

Taylor, C. (1994) *Multiculturalism*, edited by A. Gutmann. Princeton, N.J.: Princeton University Press.

Watt, B. (1992a) "Flogging a custom, court told," *Northern Territory News* July 28: 3.

—— (1992b) "Flogging outside the law," *Northern Territory News* August 6: 3.

Wise, T. (1985) *The Self Made Anthropologist: A Life of A. P. Elkin*. Sydney: George Allen & Unwin.

Zizek, S. (1989) "How did Marx invent the symptoms?," in S. Zizek (ed.) *The Sublime Object of Ideology*. New York: Verso, pp. 11–54.

—— (1994) "The spectre of ideology," in S. Zizek (ed.) *Mapping Ideology*. New York: Verso, pp. 1–33.

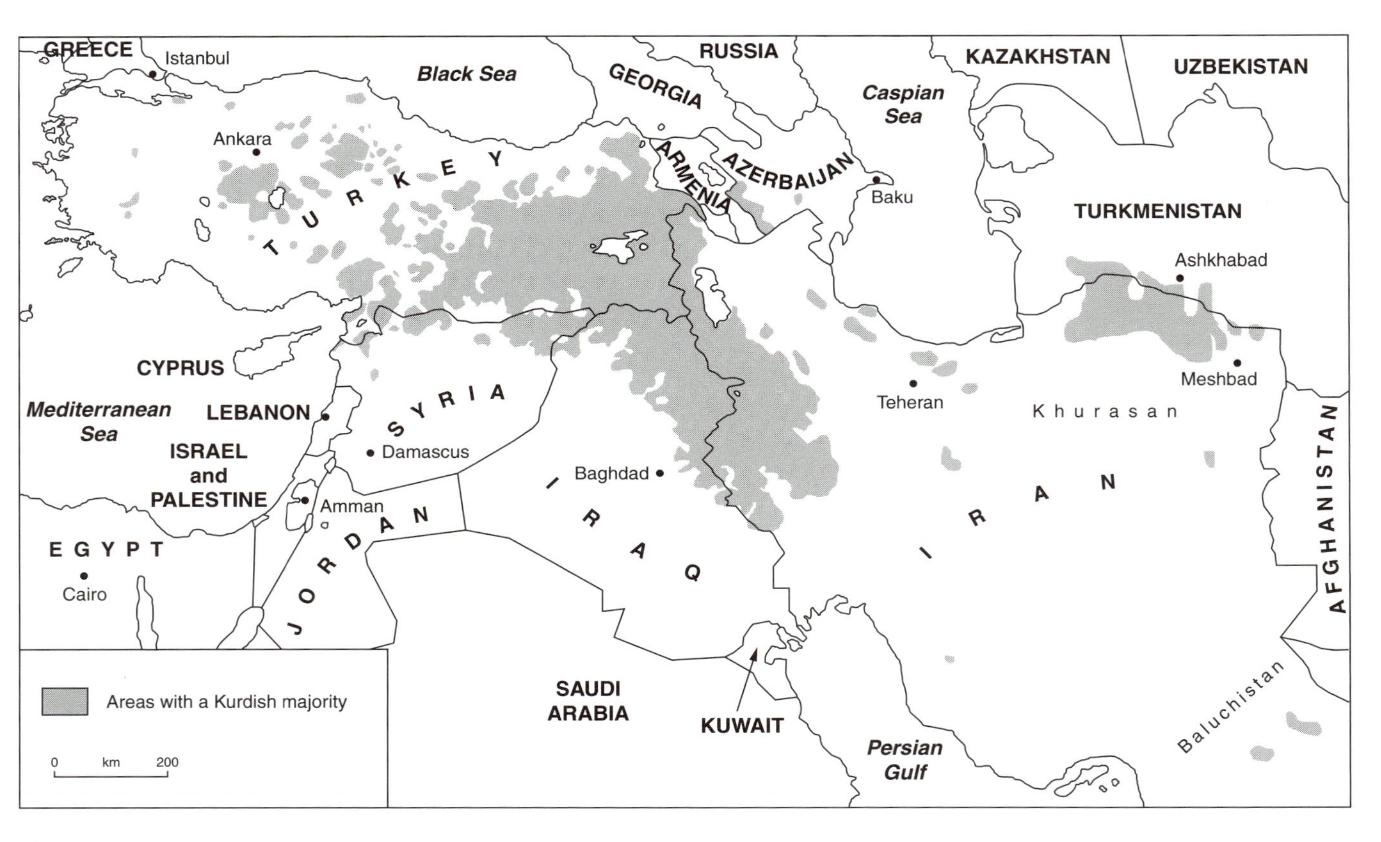

GREECE
Istanbul
Black Sea
RUSSIA
GEORGIA
KAZAKHSTAN
UZBEKISTAN
Caspian Sea
Ankara
TURKEY
ARMENIA
AZERBAIJAN
Baku
TURKMENISTAN
Ashkhabad
Meshbad
CYPRUS
SYRIA
Teheran
Khurasan
Mediterranean Sea
LEBANON
Damascus
ISRAEL and PALESTINE
Baghdad
IRAQ
IRAN
AFGHANISTAN
Amman
JORDAN
EGYPT
Cairo
Areas with a Kurdish majority
0 km 200
SAUDI ARABIA
KUWAIT
Persian Gulf
Baluchistan

Kurdistan

8 Kurdish nationalism in Turkey and the role of peasant Kurdish women

Cihan Ahmetbeyzade

Throughout the twentieth century, the public image of the Kurds has been obscured through fabrication and manipulation by the Turkish state.[1] The Kurdish Question is not considered in ideological, intellectual or political discourse by the policy makers of the Turkish nation-state. Instead, the issue of Kurdish nationalism has continued to be defined within the realm of military policies which have, since the 1923 formation of the Turkish nation-state, consisted of aggressive imperialism manifested by a brutal level of killing, village burning and sexual intimidation of Kurdish populations.[2]

My purpose in this chapter is to restore focus on Kurdish nationalism by critiquing the role which the ideologies of Turkish nationalism played in creating an "imagined community" (Anderson 1991) of the Turkish nation-state and in creating an illusion of "a new Turkish woman" after the formation of the state. The discourse of nationalism developed by the Turkish elite created an interdependent relationship between the state and a "new female" figure. As a result, Turkish feminist scholars began postulating sets of "relations of ruling" (D. Smith 1987: 2); ideological constructs which I argue subsume Kurdish women under the Turkish nationalist project, and misrepresent them as oppressed and marginalized. The absence of a nationalistic critique of Turkish feminist scholarship has left the mistaken impression that Kurdish women constitute a homogeneous group.

I argue that Kurdish women have critical perspectives of their own situation. They interpret, act and react against oppression, as well as negotiating with various forms of patriarchies. Although these women may be oppressed by diverse patriarchal systems, they are not simply victims as portrayed by various Turkish feminist scholars – but subjects with choices. I offer a new and necessary analysis of Kurdish peasant women in which they are seen as having fluid and multiple identities; actively involved at both local and national levels in the creation of new meanings of "peasant household," "family," and "tribal loyalties." Kurdish peasant women contribute to Kurdish nationalism in ways that appear to run counter to the Turkish state's discourses and nationalistic ideologies.

The building of the Turkish nation-state and the construction of Kurdish identity

Situated primarily in the heart of Asia Minor, Kurdistan presently is divided by the internationally acknowledged borders of Iran, Iraq, Syria and Turkey.[3] To fully understand Kurdish nationalism, we must first consider the distinct political and economic position of Kurdistan within the Ottoman Empire (1299–1918), Kurdish economic and socio-political organizations and contemporary development policies within the region.

Excluding the territories in Iran, Kurdistan remained the only region within the core area of the Ottoman Empire which was exempt from Ottoman agrarian structure. The Land Code of the Ottoman Empire made special arrangements for the Kurdish region, guaranteeing the reproduction of the Kurdish tribal structure. The empire did not systematically assume complete control over either land titles or the collection of tribute for agricultural surplus because the tribal system of the Kurds, based on descent and kinship, real or putative, had a strong internal hierarchy that allowed the tribal chief to internally control economic resources as well as political and social affairs. This system also maintained the survival of Kurds as a community of people with a shared identity.

The policies of the Turkish nation-state toward its Kurdish subjects and Kurdish region continued Ottoman policies. That is, the Turkish state was similarly interested in preserving the socio-economic and political structure of the Kurds in order to thwart Kurdish nationalistic aims. For that reason, the Turkish state eliminated the community leaders who waged a series of rebellions against the Turkish state during 1921–38. In contrast, Kurdish tribal leaders who were not interested in representing the articulated needs of their communities were invited to share political power with the state, and thus permitted to accumulate land and wealth. The Turkish state fostered the power of the landlord class, thus undercutting the possibility of serious land reform and guaranteeing the continuity of the tribal system and agrarian structure within the region.[4] Furthermore, the lack of industrial and infrastructural development perpetuated within the region the conditions of underdevelopment – inadequate education, inadequate transportation facilities, deteriorating infrastructure and inadequate medical facilities – further strengthening the power of the tribal chiefs/landlords. Under these conditions, the degree of dependence of Kurdish peasantries on tribal chiefs/landlords is heightened, resulting in increased acceptance of the *protection* they offered. In other words, the Turkish state aimed to curtail the political mobilization of the Kurds against the state by restraining their economic development and by declaring Kurdish populations unfit to participate in national politics as Kurds. The Turkish state legally forbade them from creating their own national vision and contesting the vision of the Turkish state.

Turkey's move toward capitalism during the 1950s made significant inroads in the Kurdish region. The arrival of agricultural mechanization displaced established sharecropping arrangements and devastated the daily life of peasants in the Kurdish region. The resulting geographic and social mobility introduced proletarianization and class-consciousness, as well as education possibilities, for Kurdish peasants who migrated to the cities. As they traveled between the cities and their villages as seasonal workers, these peasants developed new ideologies and practices.[5] The resulting social, political and economic divisions among the Kurds eliminated the possibility of mobilization for a nationalist movement. These changes also created harsh experiences for peasant Kurdish women, and their experiences played a major role for them in connecting the policies of the Turkish state to the impoverishment of their families and children. Simultaneously, through their collective resistance, a growing number of villages are challenging and condemning dominant tribal power. Peasant Kurdish women's contributions to this resistance movement and to the efforts of breaking down the feudal system are undeniable. The political and economic upheavals in the region affect the formally institutionalized roles for women within their families and communities, providing them with new possibilities for establishing networks necessary for dealing with their daily struggles. These networks create the potential for peasant Kurdish women, as members of the resisting class and ethnic group, to contribute individually and collectively to the development of Kurdish nationalism.

Since the mid-nineteenth century, nationalisms in Turkey have been intertwined in complex ways.[6] While Kurdish nationalists struggled throughout this period, Turkish nationalists achieved a victory. The basic outline of this history is well known. The collapse of the Ottoman Empire at the end of World War I resulted in the dismemberment of the empire and the occupation of the core area by the Entente Powers.[7] According to Turkish official history, the nation-state of Turkey was first imagined by the Ottoman elites. But the fact is, in order to gain legitimacy and the support of the populations, the elite of the Turkish national revolution successfully incorporated popular resistance among peasantries, by manipulating various cultural, social and religious images. While various forms of sovereign states have long been imagined by many inhabitants of Anatolia who had cultures and languages that were different from the Turks, the victorious politically centralized power minimized the future contributions of peasants in the formation of the Turkish nation-state.

Turkey's deliberately one-sided official historiography tells us that the successful formation of the Turkish nation-state was the result of effective calculations by the Ottoman urban male elite, who planned and carried out their nationalistic goals. Their exclusive imagination of a nation involved fundamental changes in "cultural conceptions" (Anderson 1991: 36). A new community in Turkey was imagined, with the erasure of the sacred – the Sultan, the Caliph and Islam – in order for the sovereign state to be claimed in defined territories. As Anderson (1991: 45) argues, print capitalism created a language of power. A linguistic unification was envisioned among the elite intellectuals, establishing Turkish as the language of power. At the discursive level, the elite of the state assumed cultural homogeneity, in opposition to the ease of Ottoman cultural heterogeneity.

Just as Turkish independence helped to articulate emerging discourses of democracy, the newly formed nation-state created new national dynamics among its people. A new category – citizenship – based on the equality of all people, eliminated in legal terms all other categories such as gender, linguistic distinctions and ethnicity. Attempts were made by the elite to create a more democratic form of nationalism than the European version, using a fully inclusive model. Thus, in theory, no one was barred from access to citizenship. As a result, individuals – all now Turkish citizens under the new law – gained a collective identity as citizens, becoming politically homogeneous. Yet gender and ethnic hierarchies continued to function in socio-political and economic practices. Especially for Kurdish populations, the basic contradictions between the promise and practice of democracy challenged the meanings of these newly articulated discourses.

Historically, the resistance of Kurdish peasantries against power and authority has been a constant factor in their daily struggles. Maintaining the Ottoman Empire required the militarization and mobilization of armed peasants.[8] Yet, the same militarization of Kurdish peasantries also turned against the empire, making the Kurdish region a site for peasant rebellions throughout Ottoman history.[9] This historical dynamic – the simultaneous reproduction and the contestation of state institutions – provided the Kurds with power, meaning formation and grounds for challenging hegemonic processes. Hence, the establishment of the Kurdish national struggle. Following its formation in 1923, the state instituted policies aimed to silence Kurdish groups that had enthusiastically embraced the establishment of a multinational sovereign state.[10] Significantly, the existing literature about the Kurdish rebellions that developed after the formation of the Turkish nation-state indicates that Kurdish rebellions were nationalistic.[11]

The concept of democratic patriarchy that Judith Stacey (1983: 116–117, 155–157) has used as a paradigm for the communist revolution in China is also useful in analyzing the role that the Turkish state plays as a public patriarch, constructing a centralized nationalist ideology in order to control populations. The newly formed Turkish nation-state – with its articulated ideas of nationalism, democracy, modernization and secularism – promised improvements in the living standards of the peasants, especially those in the western regions. As time passed, however, the peasants realized that the changes which had occurred did not benefit the Kurds. Although these new models generated new opportunities for Turkish-speaking populations and helped to develop a national consensus, these gains came at the expense of the Kurdish populations. Stacey's conceptualization of de facto hierarchical order illustrates how relationships and shifts between the multiple processes of state power and the formulations of gender and ethnic group equity are established. Her concept forces us to reconsider the concrete historical and political forms of ruling in order to understand resistance movements developed against dominance. The concept of democratic patriarchy also allows us to recognize state policies anchored in eliminating, transforming and excavating institutions, simultaneously producing new forms of knowledge and constructing notions of the present that are different from the past. Many historical accounts have shown that Kurdish populations jointly ventured with Turks into the Independence War against Anatolian occupation with an understanding that the newly founded state would have a multinational character. However, after the war was won, the powerful Turkish democratic patriarchy eliminated the past and present claims of Kurdish populations within the borders of the nation-state. It offered them full Turkish citizenship at the expense of their own cultural identities – for the Kurds were forced to adopt Turkish history and language.

The "war zone" and Kurdish peasant women

The most recent armed nationalistic uprising of Kurds living in Turkey began in 1984. It is led by the Kurdistan Workers' Party (Partiya Karkarena Kurdistan or P.K.K.), which has appealed to underprivileged Kurds in order to create a social base for the organization.[12] Theoretically sophisticated leaders of the P.K.K. belonged to the generation born after the suppression of Kurdish rebellions in 1938. With their background rooted within the lower and middle classes, as well as in the peasantries, they were able to establish the necessary connections between the real needs of the Kurdish peasant populations and the nationalistic goals of the P.K.K. As a result, Kurdish peasants easily identified their daily struggles with the nationalistic goals of the P.K.K.[13]

Since 1984, the Turkish state has insisted on military solutions, despite an estimated twenty thousand deaths on both sides, including Kurdish civilians within the Kurdish region. At present, the living conditions of the Kurds inhabiting ten cities of the Kurdish region are very difficult under the State Emergency legislation.[14] Turkish military forces also direct their activities against Kurdish civilians in order to curtail their real or alleged economic contributions to the P.K.K.[15] Since 1993, over three thousand villages and hamlets have been burned by Turkish military forces, leaving hundreds of thousands of Kurdish peasants unsheltered.[16] Fundamental human rights are significantly restricted in the Kurdish region, as are health services (*Kurdistan News* 1995, 10: 8, 15–16: 3–4). Thousands of schools have been closed down,[17] Kurdish women have been sexually abused,[18] and people have been tortured or killed during detention.[19]

The Turkish state's militaristic attempts to suppress the Kurdish uprising have exacerbated the hardships faced by Kurdish peasant women. State repression forces peasant Kurdish women to respond to these interventions without the support of their husbands, who have most often migrated abroad as workers. Some male family members have been jailed or killed because they have dared to fight with the P.K.K.; some have been summoned by the Turkish state to join the Turkish military for their obligatory military duty or as "village guards" to fight against the P.K.K.[20] As a result, peasant communities are largely represented by women in the Kurdish region. In addition to exacerbating the deterioration of health and education services, the impoverishment of living conditions, lack of access to land, water, deficiencies in rural infrastructure, electricity, drainage and transportation in the rural areas, the guerrilla warfare has forced these women to develop individual survival strategies. These strategies are based on traditional gender roles that assign women the responsibility of shaping Kurdish social networks.[21] Women's maintenance of household tasks – including daily chores, participation in kin networks, and ceremonial occasions, medical provisions and psychological care, and meditation of family disputes and conflicts – enables them, in turn, to organize their individual survival strategies into collective approaches. As a result, in the "war zone," these women take action in accordance with their gender roles, to participate in the new national Kurdish movement.[22]

Information about the Kurds living in the region is controlled and censured by the Turkish government. Some information is also released by various offices of human rights organizations. However, Kurdish women are largely absent from this information. Thus, reconstructing the ways in which Kurdish peasant women contribute to Kurdish nationalism means engaging in a process of identifying those issues that are represented officially. Chandra T. Mohanty maintains that "The fact that women are representationally absent from history does not mean that they are/were not significant social actors in history" (1987: 35). Evidence has shown that Kurdish peasant women are directly affected by the nationalist and military practices of the Turkish state and that they have resisted those policies in many ways. In addition to having their ethnic identity erased from official history books of the Turkish state, peasant Kurdish women are also forced to disown the language they speak since the establishment of the Turkish republic.

Teresa de Lauretis (1988) defines the state of displacement and self-displacement as a shift:

> leaving or giving up a place that is safe, that is "home" (physically, emotionally, linguistically, and epistemologically) for another place that is unknown and risky, that is not emotionally but conceptionally other, a place of discourse from which speaking and thinking are at best tentative, uncertain, unguaranteed. But the leaving is not a choice: one could not live there, in the first place.
>
> (de Lauretis 1988: 139)

While Kurdish peasant women are uprooted from their land by the agrarian policies of the state due to military practices, they are evicted from their burnt houses.[23] Their displacement results directly from the political, economic and military policies of the Turkish nation-state. The oppression they experience emanates from the authority of state power in their daily lives, from the police force, from prisons and from military measures. This experience and their engagement in a struggle to maintain their livelihood simultaneously produces a struggle against the state and a conscious experience of political struggle. Under these conditions, they acquire new information to act and resist state oppression. At the same time, the contradictions and oppositions embedded in their identities – the citizens of a democratic Turkish republic and as the members of an oppressed ethnic group and class – are recognized in terms of the relationship between the oppressor and oppressed.[24]

In the war zone Kurdish peasant mothers and wives create and re-create new oppositional gender-specific social roles as female heads of their households. Kurdish women's economic contributions to Kurdish nationalism have been tremendous. The military strategies of both Turkish forces and the P.K.K. have narrowed the definition of both the "homefront" and the "battlefront" as Kurdish women, children and elders are left in danger of military attacks of Turkish forces. Under these conditions, peasant Kurdish women are seen by both sides as combatants. The ongoing transformation of production relations in the war zone necessitates cooperation among Kurdish households and the development of extensive personal networks in order to meet daily consumption needs and to back up the developing guerrilla forces. Peasant Kurdish women also supply food for the P.K.K., even though the state limits the productive activities of women in certain areas and in some cases their crops are burned down. In addition, women use already established inter-household networks to pass vital information to combating units. Many of them fight against the Turkish state as guerrillas. Some of them wear bombs and enter police headquarters, blowing up their young bodies in order to protest the policies of the Turkish state (*Wall Street Journal* October 30, 1996: 1).

"Unemancipated" women and Turkish feminist scholarship

Mainstream Turkish feminist scholars have tended to group together middle-class and working-class women with peasant women of limited economic resources living in rural areas or on the outskirts of cities and in shantytowns. In her analysis of western feminist discourse, Chandra T. Mohanty (1991: 52) states that an analysis of the discursive construction of "third world women" indicates that such approaches emerge from the problem of the conception of a monolithic patriarchy. Western feminists have tended to establish a homogeneous category of patriarchy that Mohanty calls "'the third world difference' – that stable, ahistorical something that apparently oppresses most, if not all, women in these countries" (1991: 53–54). Turkish feminists' hegemonic analyses have, likewise, persisted in defining and grouping together these different women in terms of their "shared oppression" and their "total dependency on patriarchy." Similarly, Turkish feminist scholarship posits a female category whose "sameness" is assumed, a priori, in relation to a singular patriarchy. In Deniz Kandiyoti's (1977, 1987) work, for example, a Western intrinsically ahistorical feminist model is used which assumes women as the Other, already constituted within a homogeneous category of oppressed victims. This vision produces constructed ideas of women's behavior that are wrought with contradictions, especially in the cases of Kurdish peasant women, as it essentially dismisses individuals' experiences, agency and possible negotiations. The establishment of an ahistorically constituted group of unified women as "powerless victims" located in undifferentiated socio-economic, political, cultural, linguistic and religious contexts sustains the victimization of women by the very scholars whose work and actions are meant to uplift women. These analyses inevitably – and unfairly – locate women at the lowest level of the production system and define women's labor as unrecognized or undervalued by various forms of patriarchies. Furthermore, such analyses dismiss the strategies through which women negotiate their position by resisting familial, tribal and state patriarchy in order to gain access to resources and control their labor. My purpose in what follows is to analyze the theoretical problems embedded in such Turkish feminist scholarship.

Since the immediate concern of Western feminist scholarship has largely been to examine the oppression of female subjects, persistent contradictions tend to be overlooked. Descriptive attempts at establishing the power of familial patriarchy assumes with respect to so-called "traditional" life a static conception of familial patriarchy within which there is no room for women to initiate change. The analytical trap dooms feminist analysis to presenting rural women as "powerless victims." This analytical outcome of Turkish feminist scholarship on "unemancipated" women reflects their analyses of emancipated women. Turkish feminist scholars, like Kandiyoti (1991), had already accepted the view that the changes affecting their own position in Turkish society were not produced from within, but instead were imposed from outside by the state patriarchy according to the discourses of nationalism. The fundamental problem with this analysis is that it assumes the impossibility of changing the conditions of women unless there is an outside force, and insists that women who are "oppressed" and "submissive" remain choiceless because "there has been no shift from an unrecognized, underprivileged laborer status to that of free and emancipated one" (1977: 72).[25] Where change is observed it positions women as objects of male pleasure, turning them into "a conspicuous consumption item for males" and "not into a productive member of the community" (ibid.). Within this framework, subaltern rural women's identity is conceived as "dependent, psychologically deprived" and "sexually and socially segregated" from, "subordinated" to and "oppressed" by men (*passim*).

Turkish feminist scholarship has simply focused on examining the unemancipated women's oppression – assuming, as Kandiyoti does, that they are unable to recognize their true interests – rather than on locating the coping strategies that women use to deal with their oppression or considering the empowerment of women even under difficult conditions (Kandiyoti 1977: 60). The oppression of women remains a priori to their argument. In short, scholarly work dividing women into two separate but homogeneous categories – i.e. secular, Western, educated, urban and middle and upper-middle class as opposed to women who are not yet affected by nationalistic ideologies – sets up problematic binaries; as Marnia Lazreg brilliantly remarked, such assumptions "subsume other women under one's own experiences" (1988: 99).

In her comparative study of "social settings" Kandiyoti describes the nomadic tribe, the traditional peasant village, the changing rural environment, the small town and the large urban center to establish dichotomies in terms of rural–urban and traditional–modern settings. The term "social settings," as defined, deflects the description of differences, for her aim is to construct two categories which are assumed to be coherent within themselves for a feminist analysis. Unemancipated women in rural areas and on the fringes of the cities and emancipated women in modern cities, then, are fixed in their respective settings. Yet, in reality, each "social setting" is a socio-economic and political category containing within itself cultural, ethnic/racial, linguistic and religious differences. For example, the Yoruk group and the Alikan tribe are described by Kandiyoti as nomadic tribal organizations, within which, she claims, "striking uniformities in women's roles do exist" (1977: 58). What is missing from her description is the fact that the Yoruks are Turkish, while the Alikan is a Kurdish tribe. Therefore, the racial-ethnic, cultural, linguistic, religious, socio-economic and political characteristics of these two tribes are actually radically different.

Conflating women into "sameness" limits the analysis not only of gender identity but also of their multiple identities as constructed within different contexts. A conceptualization of alternative forms of nationalism developed by Kurds in Turkey is essential to feminist analyses that seek to include women's struggles with and resistance to state patriarchy (Strathern 1988). Moreover, because the position of many Turkish feminists does not challenge the notion of citizenship created by the hegemonic Turkish state patriarchy, their work actually supports the ideologies of the nation-state and thus silences women who do not accept the state definition of Turkish national identity. The lack of historically grounded feminist analysis of the intersection between ethnicity and nationality in effect erases peasant Kurdish women's past and present participation in revolutionary and nationalist struggles.

The oppression of Kurds occurs through multiple layers of domination, including both the Turkish state and Kurdish tribal authority. Kurdish women, however, face three-fold oppression: that of the Turkish state, the feudal/tribal system and patriarchal familial authority. Their position within Turkish and Kurdish society produces specific and sometimes conflicting experiences. Yet, this three-fold oppression does not simply act either on powerless victims or on a homogeneous group of women under oppression. If Turkish feminist analysis is not to become complicit with state ideologies of nationalism it must confront class and status distinctions, and local and regional variations, resulting in both multi-layered exploitative practices and diverse modes of resistance.

Turkish feminism

State nationalism in Turkey has failed to bring significant solutions to the problems of the masses. The experiences of the elite intellectuals has remained far removed from the experiences of the Muslim subjects of the Ottoman Empire. Newly established categories which complement the ideologies of nationalism – such as nation-state, the undifferentiated category of citizenship, the secular state apparatus, the imposition of cultural homogeneity and linguistic unity – have not had meaningful effect on the socio-economic and political structures in the countryside. Moreover, growing barriers among the historically differentiated masses have followed the state's failure to bridge promise and practice with the theoretical formulations of nationalism. Contrary to the discourse of nationalism which assumed homogeneity, the elite intellectuals and bourgeoisie were well aware of the differences between themselves and the subaltern classes. Turkish feminist scholars must be included in their failure to decenter the discourses and institutions of nationalist thought. As a result, Turkish feminist scholarship, influenced by dominant nationalist ideologies, has also silenced female subjects of the nation-state, especially Kurdish women.

After the formation of the nation-state, the Grand National Assembly of Turkey announced the deposing of the Caliph and the abolition of the Caliphate (1924). The end of the Caliphate was followed by legislative changes and the adaptation of the Swiss Civil Code.[26] The last constitutional blow to the traditional way of living and thinking came in 1926, with the decision to break the official ties between the Turkish state and Islam. In the eyes of nationalist, secular legislators, the declaration of a secular state marked the end of the sovereignty of God and the beginning of the sovereignty of the nation. In turn, the monolithic category of "the new Turkish woman" – secular, public, educated and Westernized – constituted the antithesis of Islam.

The concept of "citizenship" around which the Turkish nation-state structured its power supported the position of women as fully participating members in the modern society. A series of new legislations were passed which meant to emancipate women and to transform the power relations within the patriarchal household. In theory, the new civil code guaranteed the freedom of women by outlawing polygyny, and by giving women the right to choose their marriage partners, to divorce their husbands if necessary, to obtain education, to work, to vote and to be elected to political positions. However, the patriarchal state, capable of silencing oppositional voices against legal emancipation, also claimed the power to control the political, economic, social and sexual practices through which "emancipated" Turkish women were to experience these new rights.[27]

Turkish feminist scholarship proliferated after the 1970s, paying much needed attention to issues of gender and patriarchal oppression in Turkey.[28] These feminist scholars have seemed primarily interested in positioning "Turkish women" in "Turkish society" in relation to the hegemonic state patriarchy (Tekeli 1981: 293–310, 1982). It is, for example, a generally accepted view among feminist scholars that the laws passed by 1937 declaring the "Turkish woman" emancipated were legislated from top down by men (Kandiyoti 1991).[29] However, the category of "Turkish women" is not usually clearly examined in these arguments. In spite of the contradiction between the discourses of emancipation and the patriarchal state's control of the rights of "the new Turkish women," feminist scholarship also claims that educated, urban, middle-class and upper-class women exercised their democratic rights and enjoyed a new degree of freedom in social, political and economic life as equals to men, with the ability to control their production and reproduction choices (Kandiyoti 1991, 1987).

Most contemporary Turkish feminist scholars were educated in accordance with ideologies of nationalism within the secular education system, along with other emancipated women. Because these emerging feminist scholars created distinctions between emancipated "Turkish women" (whose experiences were enriched by their freedom to make choices) and "Middle Eastern women," they have also distinguished themselves from white, middle-class, Western feminists' view of them as weak and oppressed, due to underdevelopment, high illiteracy, the oppressive practices of Islam and poverty. By differentiating their own experiences from those of such homogeneously categorized "Middle Eastern women," Turkish feminist scholars have thus allied themselves with Western feminism (see e.g. Kandiyoti 1977, Toprak 1981).

Turkish feminists have also been concerned with the oppression of emancipated Turkish women as female subjects of the state, a category in which they saw themselves included. In this regard, they have recognized the contradictory relationships within "the new concept of women." On the one hand, "the new women" were emancipated and protected from familial patriarchy by the state; on the other hand, the cultural experiences of "the new women" were publicly shaped by dominant state propaganda. For that reason, Turkish feminists did mount a critique of the institutions of the nation-state that defined their gender identity. Without a clearly defined political agenda, however, their vision and ability to act upon changes were limited, and their struggle remained positioned within the same silencing discourses which were created by the ideologies of nationalism.

The "new Turkish women" were invited to share power and to join in all prescribed political, economic and social activities as full members of the "new Turkish nation." Yet, in the context of the development of political consciousness among Turkish feminists, this apparent enlargement of power actually produced limits to their agency: their right to power was valid only as long as they complied with the terms of "nationalism" set by the monolithic category of "citizens," who recognize and contribute to secular and Western discourses of the nation-state. At the level of discourse, the state eliminated traditional hierarchical differences in social, economic and political spheres, while claiming that race, ethnicity, class and gender differences have been eradicated in Turkey by the category of citizenship. Yet, in practice, the nation-state privileged certain groups, including Turkish feminist scholars, who were invited to share power by representing the interests of the illiterate, backward masses – women and peasantries. And unfortunately, Turkish feminist scholars worked uncritically within this framework of nationalist ideologies. Their limited vision restricted their examination of Kurdish women who had different historical experiences from their own.

Their unwillingness to deconstruct the discourses and practices of nationalism and to decenter the concept of citizenship resulted in Turkish feminists' uncritical acceptance of the privileges bestowed upon them by the state patriarchy and determined, as well, the forms of critiques that they produced. Their lack of interest in developing a radical political feminist movement with a commitment to problematizing the ideologies of nationalism, patriarchal discourses and practices at all levels impeded their consideration of the possibility that identities are complex. Hence, in their ideological complicity with the policies of the nation-state, Turkish feminists failed to consider the possibility that Kurdish women did not feel represented by them solely on the basis of shared gender.

Peasant Kurdish women and the patriarchal household: local considerations

In what follows, I argue that alternative analytical strategies need to be used in examining the social agency of Kurdish women. Using the limited written sources about Kurdish women, I present the diverse ways in which Kurdish women deal with oppression and resist it. Ultimately this approach demonstrates how Kurdish peasant women contribute to Kurdish nationalism through their resistance to the ideologies of Turkish nationalism.

The ethnic and the national identities of Kurdish women are intertwined in complex ways. Over the years, due largely to earlier anthropological interests in tribal structures of the Kurds, scholarly consensus has established the social, political and economic activities of Kurdish women as the direct product of tribal, lineage and/or traditional household structures. Women's complex roles and their contributions to various institutions, organizations and ideas have not entered the literature. Earlier accounts simply presented the activities of Kurdish women as determined by certain male-dominated structures, and regulated by male elders and seniors through hierarchical relations (van Bruinessen 1992, Barth 1953, Leach 1940). This ahistorical approach to the social fabric of Kurds tended to disregard power networks among women, and between women, and men. As a result, the ways in which Kurdish women acted and reacted to historical changes taking place in the region, the tactics they developed to secure the survival of Kurdish life, and the means by which they negotiated their position within their own societies are hardly considered.

We need, instead, to ask a different set of questions: how do women negotiate their positions and collectively organize themselves to challenge and change patriarchal processes aimed at controlling them? What networks exist among women, that both establish closer ties among them and enlarge the grounds of their negotiation? How do women construct their identities as plural, collective, and contradictory in order to deal with state control? What are the contradictory elements constitutive of women's consciousness?

I shall therefore address the specific problems of Kurdish peasant women, who experience the effects of state dominance while living under tribal and patriarchal authority. These women dwell in extended and nuclear families and tend to have close relations with relatives living in adjacent houses, in small towns or in house clusters (*mezra*); they possess little or no land. This group of rural women differs from other Kurdish women living in rural villages, in the metropolis, and in wealthy households.

Most Kurds have been peasants throughout history.[30] They have been involved in both nomadic and sedentary agricultural production. While remaining rural, they have tended to maintain close relations with urban centers. Multi-occupation enterprises and multi-production sites – including petty-commodity production, artisan activity, trade, even wage labor – have complemented their agricultural activities.

In most peasant households control of resources is exercised by both female and male heads of the household in their respective work spheres. The labor of the children and the labor of the sons' spouses is often regulated by the female head of the household as part of the household economy. The Kurdish household (*mal*) is a unit of production and consumption where all the involved parties constantly negotiate their productive power.[31] Households that consist of three generations are always units in transition, as they break down into smaller units with the separation of grown offspring. This transitional pattern of peasant households is further intensified among semi-nomadic and nomadic peasants during seasonal movement.

The physical division of the *mal* and the usage of space in the contemporary Kurdish peasant setting are described in detail by Yalcin-Heckmann (1991: 137–148). According to Yalcin-Heckmann, many Kurdish women claim rights over the physical space within the household, limiting the mobility of men inside. Hence, Kurdish women exercise power within their homes and actively negotiate their power with the other members of the household. The living/family room is a space allocated both for the needs of family members and for receiving guests. Women's access into this central room is not restricted. Women, as well as men, share this space in socializing, eating or hosting guests as they please. The usage of this social space as a living/family room available to all for social events, regardless of gender, is unique to peasant Kurds. Yalcin-Heckmann remarks that the "male–female or age segregation is not as strict as has been reported in some Middle Eastern societies" (1991: 139). Yet, according to Yalcin-Heckmann (1991: 142), Kurdish peasant women also insist on using or deliberately refusing to use the main room. Kurdish women exploit this space in negotiating their power. Therefore, the Kurdish peasant household is also a site in which women's agency is exercised. Moreover, when it is financially possible to add new rooms, often the secondary and tertiary rooms are designated for the sole usage of women and children during the day. Men, including the male head of the household, are barred from entering the women's space (ibid.). Males' mobility within their home often is thus limited by women's demands in creating their own universe.

The above example of the ways in which peasant Kurdish women challenge domination and negotiate their power illustrates their understanding of their oppositional agency as members of an oppressed group. By refusing the dominant forms of oppression, Kurdish women both individually and collectively create strategies and tactics to resist their oppression. As Mohanty (1987: 37) stresses, shared historical experiences are important in defining politics. The shared experiences of Kurdish women as tribal, landless peasant women enable them to decenter the power of domination and to subvert the oppressive practices of patriarchy by creating a universe of their own. Reclaiming and recreating their identities as subjects of their own is similar to the position which bell hooks (1984) defines as "both looking from outside in and from inside out." Peasant Kurdish women's ability to move from margin to center with knowledge of their historical and present realities creates grounds on which they can construct and negotiate their own space.

The female head of the Kurdish peasant household, the *kabani*, is the eldest woman of the household. She holds the position of power and thus is hierarchically superior to the other women of the household. She regulates all the aspects of labor relations and daily chores, and she controls consumption. However, her power is occasionally contested by individual women who attempt to change their social position within the household hierarchy (Yalcin-Heckmann 1991: 159–162). Peasant Kurdish women's ability to contest the power of domination forces the *kabani* to consider the needs of every woman in the household. In the peasant Kurdish society, a peasant woman resists to challenge existing knowledge and patriarchal dominance with an agency of her own. Yalcin-Heckmann's explication of the contested terrains within the household allows us to recognize the individual agency and conceptualize collective agency. Evidently both individual and collective agency is produced by women who consider themselves having the power to resist.

Like certain labor processes – including carpet weaving, spinning, dyeing, leather tanning and gardening for household use or sale – daily chores, processing milk products such as butter and cheese for consumption and marketing, are jointly organized by women within the household network. The necessity for a cooperating workforce, especially for agricultural production, gives rise to intra-household networks among women. Resulting social networks, organized between relatives and peers, create legitimate terrains on which they can challenge the idealized roles of women within the traditional family structure. Although these networks are embedded in patriarchal relations of power, peasant Kurdish women's ability to constantly negotiate for networking is recognized because of women's potential power to mobilize strategic resources when needed, especially during Kurdish uprisings.

Kurdish women's activities are not limited by the physical boundaries of the household. Certain productive activities of women are strictly performed outside.[32] At this time, there is an absence of scholarship dealing with the cultivation work done by these women. We do know, however, that when children's contributions to production cannot be expected, the production capacity of poor, landless peasant families working in various share-cropping arrangements often depends on the labor of the wives. Kurdish women's contributions toward sustaining access to limited resources, essential to their families' survival, gives their labor legitimacy.

According to Ismail Besikci (1969: 175), peasant Kurdish women in the pastoral Alikan tribe have authority equal to men in both social and household relations. He insists that even in the patriarchal Alikan tribe, women are able to negotiate their power, because they contribute extensively to economic production. Networks among the women of neighboring households are essential, for allocating the time necessary to finish daily jobs and seasonal work. These inter-household networks are an integral component of Kurdish women's labor, providing the basis for more extensive networks at the local and regional level. It is difficult for Kurdish peasant women to mobilize their resources in order to connect local level networks with the regional Kurdish institutions – indeed, some institutions find it difficult to function as part of the Kurdish civil society under present conditions in the war zone. However, Kurdish peasant women contribute to democratic popular organizations, such as Kurdish Women's Commission of Human Rights in Kurdistan (I.H.D.), and Kurdish Women's Commission of Popular Culture (H.K.D.).[33] Both of these organizations have a heterogeneous base and both include and expect contributions from peasant women, from laborers, from white-collar workers and from housewives. Furthermore, Kurdish peasant women have contributed to the Independent Kurdish Women's Organization since its formation in 1992, in the Kurdish cities of Agri, Batman, Diyarbakir, Van and Urfa. Peasant Kurdish women immediately respond to the forced migrations of their families to the larger districts and cities by establishing social and economic networks for incoming families. These newly established networks guarantee solidarity among them and further intensify the possibility of their contributions to Kurdish women's institutions, such as Jiyan Kurt Kadin Kultur Evi and Kurt Kadinlari Dayanisma Vakfi.

As previously stated, the female and male elders' authority in controlling labor and resources is always subject to negotiation and contestation through the household consensus process. Though such apparent contradictions, challenges to dominant practices persist. Peasant Kurdish women's recognition of the value of their own labor allows them to mobilize their limited resources and to challenge patriarchal dominance. Within intra- and inter-household networks, where women's productive power is allocated and negotiated by women themselves, daughters grow up acquiring experience which contradicts dominant Kurdish practices, and learn both to contribute to the labor process and to challenge authority. Peasant Kurdish women's participation in the development of *halk koyleri* is evident and their contribution to this strategic move – challenging and condemning dominant tribal power with their collective resistance – is undeniable. This collective move involves the cooperation of both female and male members of the resisting class. As they develop their identities and experiences within these women's networks women also develop political consciousness through which they contribute to Kurdish nationalistic commitments. Although their heroic deeds resisting the state hegemony and fighting against Ottomans and Turks are generally not recorded,[34] networks established among women contribute to the development of Kurdish nationalism in various ways.[35] Since these networks provide the seeds of political infrastructure and alternative ways for political mobilization, it is not surprising that a growing number of Kurdish women are becoming active political agents by joining P.K.K.

Western feminist approaches have the tendency to dissolve quickly into narrow discussions of the oppression of women, as determined by patriarchal practices. But feminist thinking which presents third world women as victims of their own cultures, colonial administrations, nation-states, capitalist practices, and multinational corporations is now, increasingly, being challenged by feminists of color and third world women (Anzaldua 1992). These new theoretical approaches allow us to re-analyze the oppression and the silencing of Kurdish women by examining the discourses they create and by paying more critical attention to the practices of Turkish nationalism as expressed by the Turkish nation-state and/or Turkish feminist scholars. Attempts to explain the oppression of Kurdish women, through simplistic formulations of patriarchal practices such as arranged marriages, sister exchanges, discriminating inheritance and different forms of exploitative demands on women's labor, reduces Kurdish peasant women to victims and ignores the dynamic strategies through which they combat oppression. "Third wave" feminists' awareness of the contradictory nature of many of their experiences has, instead, helped them frame a model of *difference* (Alarcon 1990), which recognizes the ways in which the location of women's identity within multiple axes of patriarchy naturally establishes different experiences for each woman (Alarcon 1990).

The economic and political policies of the Turkish state aim to oppress peasant Kurdish women for the benefit of the state, while also empowering the landowning Kurdish class. As productive members of their families under these strenuous conditions, Kurdish women recognize that their complex experience with multiple forms of patriarchies, necessitates multiple forms of resistance to patriarchal structures. Therefore, "third wave" feminists have insisted that we must listen to women's voices coming from multiple and fluid positions in order to understand the totality of the "master's tools" (Lorde 1981). In effect, the daily struggles of peasant Kurdish women force them to acknowledge the relationship between the laws of the nation-state and the political and economic oppression of the Kurds. Under the present conditions in Kurdistan, the oppressive dynamic between the policies of the nation-state and Kurdish nationalistic goals demands peasant Kurdish women's contributions to Kurdish nationalism. While resisting familial and tribal patriarchal oppression, they give new meanings to traditional conceptions of peasant households, family and tribal loyalties. Moreover, on the new terrain, where they have been positioned by the state patriarchy, they must also negotiate and construct their identities as members of a non-recognized ethnic group within the borders of the Turkish nation-state. In claiming their fundamental rights – to speak their native language; to express their cultural identity; to defend their historical claims to land and resources – Kurdish peasant women's activities within networks also confront on the state policies that oppress their culture. Their national struggle in Kurdistan is thus a continuation of a long historical conflict. Presently, peasant Kurdish women not only participate at all levels of that conflict but also they and their children become casualties within the war zone when Turkish military destroy their fields and burn their villages, and when they take to the mountains to fight against Turkish military. Ultimately, the role that peasant Kurdish women play in the Kurdish national struggle will be certain in determining the future of Kurds and solutions to "the Kurdish question" within the borders of Turkish nation-state.

Conclusion

All citizens of the Turkish nation-state have been considered equals by law, with full political rights, as long as their legitimate claims remain within the boundaries of "Turkish national consciousness." Turkish feminist academicians, educated within state schools, have either refused or been unable to grasp the multinational and diverse ethnic origins within the nation-state and their implications for women. Within Turkish feminist scholarly discourse, women who are members of marginalized economic classes, women who have not had access to higher education, and women who do not speak Turkish are seen as homogeneous victims of patriarchal structures, who lack the opportunity to resist patriarchal authority and negotiate their own rights. I have looked, instead, to other forms of feminist theory, specifically the theories of multiple positioning, to reveal significantly more complex layers of oppression within and beyond axes of gender, ethnicity and class. The notion of *difference* makes possible the recognition of women, such as Kurdish peasant women, who are situated in marginal location, but who are agents in their own rights, especially in the local/domestic and national power. This shift in feminist consciousness necessitates an understanding of both the "temporality of struggles" and the location of power as well as the complexities of the processes of patriarchal dynamics both at local and national level.

Acknowledgments

I am especially grateful to Tamar Mayer, who has creatively critiqued earlier drafts, fed me informative ideas and helped to make this chapter more readable and cohesive. I take this opportunity to express my utmost gratitude to various Turkish and Kurdish scholars, who must remain unnamed at this time. I should like to acknowledge the invaluable institutional support and intellectual and political guidance that I received from all contributions of the following institutions: JIYAN-Kurt Kadin Kultur Evi, Istanbul, Kurdistan News-International Association for Human Rights in Kurdistan, Bonn, and KOMJIN-Yekitiya Jinen Kurdistan, Cologne. Finally, Jennifer Burrell, Kamari C. Clarke, Saloni Mathur and Ted Kafala, through fruitful discussions in graduate seminars and outside, have contributed immensely to my intellectual growth. For their understanding, caring, encouragement and just for being there, I thank these friends and fellow scholars.

Notes

1 Different forms of Kurdish nationalism exist within the region. The historical processes during the Ottoman Empire and after the disintegration of the empire took different directions for Kurds living in Iraq, Syria and Turkey. While Kurds living in Iraq and Syria had to go through a mandate period under both French and British occupation, Kurds living in Turkey seemed to face no Western interference, but were suppressed and persecuted by the newly established Turkish Republic. These historical processes differently affected the experiences of distinct groups of Kurds and resulted in contradictory forms of nationalisms within the borders of the nation-states of Iran, Iraq, Syria and Turkey. Any discussions about Kurdish nationalism must account for the ways in which these historically diverse experiences variously affect the trajectories of that nationalism. Therefore, this chapter does not attempt to analyze "Kurdish nationalism" as a whole and concentrates only on the ways in which the particular experiences of Kurds have contributed to the development of Kurdish nationalism in Turkey.

2 In his book *Imperialism*, J. A. Hobson writes:

> Aggressive imperialism, as our investigation has shown, is virtually confined to the coercion by stronger or better-armed nations of nations which are, or seem to be, weaker, and incapable of effective resistance; everywhere some definite economic or political gain is sought by Imperial aggressor.
>
> (Hobson 1965: 200)

3 The term " Kurdistan" refers to the geographic areas historically inhabited by the Kurds. This term has never been used by mapmakers. Ottoman politicians used the term only to refer to the Diyarbakir Governorship. Kurdish populations also live in former Soviet Caucasus and in the western Diaspora. See the map of Kurdistan in the context of the Middle East at the beginning of this chapter.
4 Accordingly, the 1945 and 1973 Land and Agrarian Reforms were not meant to bring a fundamental change in the Kurdish agrarian structure. While Kurdish peasantries gained very little, especially in the southeast, because the rich landed class was able to seize the land distributed to Kurdish peasants (Keyder 1987), tribal chiefs were able to maximize their accumulation of surplus.
5 Under these conditions, developing resistance against tribal/feudal authority caused a breakdown in traditional relations of reproduction within tribal/feudal systems. For example, landless peasants who traditionally worked for tribal chiefs/feudal lords and lived in *aga koyleri* (house clusters owned by landlords) developed a form of resistance and constructed their own villages, "*halk koyleri*," known as the people's villages (Besikci 1969: 129).
6 The emerging national consciousness of the "Young Turks" enabled them to construct numerous national discourses, such as pan-Islamism, pan-Turanism, pan-Turkism and Ottomanism, within the original multi-ethnic and multilinguistic face of the *Ittihad-i Terakki* (Committee of Union and Progress). Within the *Ittihad-i Terakki* "Ottomanness" versus Turkish nationalism and Kurdish nationalism, as well as other national-isms, developed.
7 The occupation of Anatolia was not controlled by a single army. It was rather a regional occupation by the armies of England, France, Greece and Italy.
8 For example, *Derbend* was a special gendarmerie force that protected border peasants, long-distance caravan trade and mail service, and defended and built strategic resources, roads, mines, bridges, waterways and *kervansarays*.
9 For detailed information about the Kurdish rebellions during the Ottoman Empire see van Bruinessen (1992: 159, 168, 169, 186–190) and also Olson (1989: 205, n. 9).
10 It took the Turkish military seventeen years of killing and destruction to suppress approximately seventeen rebellions and demilitarize the Kurdish population between the years 1921 and 1938. For the list of rebellions including locations and dates see Olson (1989: 205, n. 39).
11 The nation-state of Turkey insistently characterized Kurdish rebellions as religious rebellions directed against secular reforms of the state. Both van Bruinessen (1992) and Olson (1989) argue for the clear nationalist aspirations of the rebellions aimed at the establishment of an independent Kurdistan.
12 The P.K.K. was formally established in 1973 by its leader Abdullah Ocalan and his close friends.
13 It has been alleged that the P.K.K. has between 10,000 and 20,000 active full-time guerrillas, and over half a million supporters and sympathizers in Turkey. See Imset (1996) for further details on such figures.
14 The available information about Kurds living in these cities is obtained from the *Kurdistan News International Association for Human Rights in Kurdistan*, a publication that is made possible by the European Human Rights Foundation.
15 For example, a Turkish official confirms the necessity of the food embargo as preventing the smuggling of "excess commodities" to the P.K.K. (*Kurdistan News* 1996, 30: 4).
16 The former Minister of Human Rights in Turkey reported that in the Kurdish Province of Tuncelli (Dersim) the food embargo further exacerbated the suffering of the peasants. In addition to evacuations, houses, stables and grain stores had been burned down. Access to pastures was prohibited; the peasants were forbidden

access to their land. Crops, fruit trees, beehives, woods and vegetable gardens were either confiscated or destroyed (*Kurdistan News* 1996, 30: 4).

17 *Kurdistan News* (1997, 21: 19) reports that 24 Kurdish provinces of Turkey have, in theory, 11,592 schools. Of these 3,288 (about 27 percent) are at present closed, 1,859 for "security reasons," 1,248 for lack of teachers and 181 because there are not enough pupils.

18 To break down their resistance and alienate women who help the P.K.K., the state forces women to undergo a gynecological check up to determine if they are visited by their husbands during the night (*Cumhuriyet* September 3, 1996). In other words, the Turkish state rapes Kurdish peasant women with scientifically forced entries, to confirm whether or not their husbands ejaculated the night before. The rape and sexual harassment of Kurdish women as a part of physical torture by "security forces" continues in Kurdistan (Il 1997, Kitis 1997, *Jiyan* 1997, 20: 14–16).

19 Every month, *Kurdistan News* reserves its first two pages for the reports of tortures in detentions and extrajudicial executions.

20 See Imset (1996: 79) for detailed information about the "village guards," the government-paid paramilitary forces.

21 The exchange of reciprocal and non-reciprocal labor, goods and visits are detailed by Yalcin-Heckmann. Her examination highlights the inter- and intra-household networks (1991: 169–183).

22 The term "war zone" refers to the Kurdish territories where "State of Emergency Legislation" is in effect. In this region, the definitions of combatants and non-combatants or civilians are blurred contrary to the Geneva Convention of 1949 which guarantees the protection of innocent civilians. The Hague Protocol of 1954, which endorses the protection of cultural property, is also ignored by the Turkish military.

23 Continuing since 1994 the armed forces and the "special forces" are systematically burning all the houses of designated villages after forcing evacuations by the official pretext to eradicate the P.K.K. guerrillas and cut off their logistic support (*Kurdistan News* 1996, 21: 3, 22: 2).

24 The assimilationist strategies that the Turkish state applies to the Kurds are formulated in the legal and educational system and are identified with the official language, Turkish. The legal system in Turkey records the defendants' Kurdish response to questioning in court minutes as "[t]he defendant made use of his right to be silent" (*Kurdistan News* 1995, 14: 6). Kurdish women are aware of the silencing policies of the Turkish state. For that reason, following the military coup in 1981, in Diyarbakir, they took to the streets to protest against suppression of Kurds in jail. In 1989, they silently joined their relatives on hunger strikes, refusing to speak Turkish outside the walls of the penitentiaries.

25 Kandiyoti (1987: 62–63) examines unrecognized women's labor.

26 The Islamic Hicri year was abolished and the Gregorian Calendar adopted. The Arabic alphabet was replaced by Latin script in 1928. In 1925, wearing *Fez*, the symbol of Islam, was considered a criminal offense by the modernizing state. In 1926, the Shari'a was abolished and the new Civil Code was effective in changing laws of marriage, divorce, and inheritance. Polygyny and divorce (*talaq* or repudiation which is strictly a male prerogative) were outlawed. The elimination of inheritance according to Islamic Law and introduction of secular law gave equal rights to women to inherit. Turkish women obtained the right to vote in 1934 and right to be elected in 1937.

27 See Kandiyoti (1991: 41) regarding the elimination of the Turkish Women Federation in 1924 by the orders of the state. "The 'corporate' control of female sexuality" is examined in Kandiyoti (1987, 1988).

28 My analysis of Turkish academic feminism does not claim that there is only one form of feminism in Turkey. At the grassroots level, various feminist projects and activities concerning issues such as Islam, education, health and battered women are taken up by different groups that publish periodically. These groups are critical of the state policies that do not facilitate their objectives. For a historical examination of the development of the grassroots level organizations, especially after the 1980s, see Sirman (1989).

29 Although Kandiyoti (1993), in theory, stresses the heterogeneous and complex role and agency of women in their emancipation and mentions the ways in which the daily experiences of women contribute to their struggle, these complexities and experiences have yet to be fully analyzed. See also Kandiyoti (1994) on the oppression of man by patriarchy. For the counter-arguments about women's initiatives and contributions to their emancipation see Baykan (1992) and Cakir (1991).

30 The term "peasants" carries various definitional problems. See Roseberry (1991: 19–21) and G. Smith (1989: 18–28) for the definition and elaboration of the problem.

31 See van Bruinessen (1978: 55–56) and Yalcin-Heckmann (1991: 149–151) for other meanings of the term *mal.*

32 Milking, fetching water and food for animals as well as collecting wood are all done by women. Some women collect herbs and mushrooms for consumption. They carry hay from the mountains for winter feed and transport milk from summer pastures. Kurdish women and children also collect gum tragacanth to sell to traders (Yalcin-Heckmann 1991: 178–179).

33 The information about the peasant Kurdish women's contributions to various Kurdish organizations was provided by Jiyan-Kurt Kadin Kultur Evi, Istanbul.

34 We have only rare glimpses of women's historical contributions to rebellions. During the Dersim Rebellion in 1937, Aliser's wife Zarife's martyrdom by the enemy while fighting is mentioned by Goktas (1991: 135). See also van Bruinessen (1993: 25–39) about women rulers and their deeds in Kurdish history.

35 Personal conversations and correspondence with Kurdish women involved in various Kurdish women's organizations at various levels establish the fact that at the present there are an estimated fifteen hundred plus guerrilla women combatants within the P.K.K. Considering that Turkish official records underestimate the number of P.K.K. guerrillas, and the P.K.K. overstate their power in combat, this approximation of the number of women guerrillas represents only an informed estimate.

References cited

Alarcon, N. (1990) "The theoretical subject(s) of *This Bridge Called My Back* and *Anglo-American Feminism*," in G. Anzaldua (ed.) *Making Face, Making Soul Haciendo Caras*, San Francisco: an aunt lute foundation book, pp. 356–366.

Anderson, B. (1991) *Imagined Communities: Reflections on the Origin and Spread of Nationalism*, London and New York: Verso.

Anzaldua, G. (1987) *Borderlands/La Frontera*, San Francisco: spinsters/aunt lute.

Barth, F. (1953) *Principle of Social Organization in Southern Kurdistan*, Oslo: Brodene Jorgensen Boktr.

Baykan, A. (1992) "The Turkish women: an adventure in feminist historiography," submitted to the Third Biennial Symposium on New Feminist Scholarship, State University of New York at Buffalo.

Besikci, I. (1969) *Gocebe Alikan Asireti* (Nomadic Alikan Tribe), Ankara: Dogan Yayinevi.

Cakir, S. (1991) "II. Mesrutiyet'te Osmanli Kadin Hareketi ve Kadinlar Dunyasi Dergisi" (Woman's movement during II Mesrutiyet and the magazine of Women's World), unpublished Ph.D. dissertation, Istanbul Universitesi Siyasal Bilgiler Fakultesi, Siyaset Bilimi, Anabilim Dali. Istanbul.

de Lauretis, T. (1988) "Displacing hegemonic discourses: reflections on feminist theory in the 1980's," *Inscriptions* 3/4: 127–141.

Goktas, H. (1991) *Kurt, Isyan, Tenkil* (Kurd, Rebellion, Repression), Istanbul: Alan Yayincilik.

Hobson, J. A. (1965) *Imperialism*, Ann Arbor: University of Michigan Press.

hooks, b. (1984) *Feminist Theory from Margin to Center*, Boston, MA: South End Press.

Il, N. (1997) "Militarizmin soguk yuzunu yasayan kadinlar" (Women who live the cold face of militarism), *Jiyan* 20: 12.

Imset, I. (1996) "The P.K.K.: terrorists or freedom fighters?" *International Journal of Kurdish Studies* 10, 1 and 2: 45–101.

Kandiyoti, D. (1977) "Sex roles and social change: a comparative appraisal of Turkey's women," *Signs: Journal of Women in Culture and Society* 3, 1: 57–73.

—— (1987) "Emancipated but unliberated? Reflections on the Turkish case," *Feminist Studies* 13: 317–338.

—— (1988) "Slave girls, temptresses, and comrades: images of women in the Turkish novel," *Feminist Issues* 8, 1: 35–49.

—— (1991) "End of empire: Islam, nationalism and women in Turkey," in D. Kandiyoti (ed.) *Women, Islam and State*, Philadelphia, PA: Temple University Press.

—— (1993) "Strategies for feminist scholarship in the Middle East," paper presented at the Twenty-Seventh Annual Meeting of the Middle East Studies Association, North Carolina.

—— (1994) "The paradoxes of masculinity: some thoughts on segregated societies," in A. Cornwall and N. Lindisfarne (eds.) *Dislocating Masculinity: Comparative Ethnographies*, London and New York: Routledge, pp. 197–230.

Keyder, C. (1987) *State and Class in Turkey: A Study in Capitalist Development*, London and New York: Verso.

Kitis, G. (1997) "Tecavuz magduru Zeynep Avci: Tecavuze ugradigimi feodal nedenlerden dolayi sakladim" (A rape victim Zeynep Avci: I disclosed my rape, because of the feudal forms of oppression), *Jiyan* 20: 10.

Kurdistan News (1995) "News about human-rights violations in Kurdistan," *Kurdistan News*, Bonn, Germany.

Lazreg, M. (1988) "Feminism and difference: the perils of writing as a woman on women in Algeria," *Feminist Issues* 14, 1: 81–107.

Leach, E. R. (1940) "Social and economic organizations of the Rawanduz Kurds," LSE Monographs in Social Anthropology, no. 3, London: London School of Economics.

Lorde, A. (1981) "The master's tools will never dismantle the master house," in C. Moraga and G. Anzaldua (eds.) *This Bridge Called My Back: Writing By Radical Women of Color*, New York: Kitchen Table: Women of Color Press.

Mohanty, C. T. (1987) "Feminist encounters locating the politics of experience," in *Copyright, Fin de Siècle 2000*, Indiana: Indiana University Press, pp. 30–44.

—— (1991) "Under western eyes: feminist scholarship and colonial discourses," in C. T. Mohanty, A. Russo and L. Torres (eds.) *Third World Women and the Politics of Feminism*, Bloomington: Indiana University Press, pp. 51–80.

Mohanty, C. T., A. Russo and L. Torres (eds.) (1991) *Third World Women and the Politics of Feminism*, Bloomington: Indiana University Press.

Olson, R. (1989) *The Emergence of Kurdish Nationalism 1880–1925*, Austin: University of Texas Press.

Roseberry, W. (1991) *Anthropologies and Histories: Essays in Culture, History, and Political Economy*, New Brunswick, N.J.: Rutgers University Press.

Sirman, N. (1989) "Feminism in Turkey: a short history," *New Perspective in Turkey* 3, 1: 28–53.

Smith, D. (1987) *The Everyday World as Problematic: A Feminist Sociology*, Boston, MA: Northeastern University Press.

Smith, G. (1989) *Livelihood and Resistance: Peasants and the Politics of Land in Peru*, Berkeley and Oxford: University of California Press.

Stacey, J. (1983) *Patriarchy and Socialist Revolution in China*, Berkeley: University of California Press.

Strathern, M. (1988) *The Gender of the Gift*, Berkeley: University of California Press.

Tekeli, S. (1981) "Women in Turkish politics," in N. Abadan-Unat (ed.) *Women in Turkish Society*, Leiden: E. J. Brill.

—— (1982) *Kadinlar ve Siyasal Toplumsal Hayat* (Women and Socio-Political Life), Istanbul: Birikim Yayinlari.

Toprak, B. (1981) "Religion and Turkish women," in N. Abadan-Unat (ed.) *Women in Turkish Society*, Leiden: E. J. Brill.

van Bruinessen, M. (1978) "Agha, Shaikh and state: the social and political organizations of Kurdistan," published Ph.D. dissertation, Utrecht: Ryksuniversiteit.

—— (1992) *Agha, Shaikh and State: The Social and Political Structures of Kurdistan*, London and Atlantic Highlands, N.J.: Zed.

—— (1993) "Matriarachy in Kurdistan? Women rulers in Kurdish history," *International Journal of Kurdish Studies* 6, 1 and 2: 25–39.

Yalcin-Heckmann, L. (1991) *Tribes and Kinship among the Kurds*, New York: Peter Lang.

RUSSIA
Lake Baikal
KAZAKHSTAN
Lake Balkhash
MONGOLIA
Hailar
Heilongjiang
Qiqihar
Harbin
Karamay
Kulja
Urumqi
Changchun
Jilin
KYRGYZSTAN
Nei Mongol
Shenyang
Liaoning
Kashgar
Xinjiang
NORTH KOREA
Hohhot
Hebei
Beijing
Tianjin
Dalian
Yuman
SOUTH KOREA
Yinchuan
Taiyuan
Shijiazhuang
Yantai
Jinan
Ningxia
Qingdao
Golmud
Xining
Shanxi
Shandong
Qinghai
Lanzhou
Xi'an
Zhengzhou
Jiangsu
Shiqunhe
Gansu
Nanjing
Henan
Shaanxi
Shanghai
Hefei
Xizang
C H I N A
Hubei
Anhui
Hangzhou
Wuhan
NEPAL
Lhasa
Nanchang
Zhejiang
Sichuan
Changsha
INDIA
Jiangxi
BHUTAN
Hunan
Fuzhou
Guizhou
Fujian
Guiyang
Kunming
TAIWAN
BANGLADESH
Guangdong
Xiamen
Yunnan
Guangxi
Guangzhou
Hong Kong
MYANMAR
Nanning
South China Sea
VIETNAM
Haikou
LAOS
Bay of Bengal
Hainan
0 km 500 km

China

9 Calligraphy, gender and Chinese nationalism

Tamara Hamlish

One of the most powerful images of the political struggles marking the emergence of the Chinese nation resides in the Chinese term for the nation itself. The compound *guo jia* lays claim to the conflicting loyalties of Chinese imperial subjects for both empire (*guo*) and patriline (*jia*).[1] The term expresses the nation's appropriation of the processes of cultural production embodied in the "traditions" of ritual that order the empire and the processes of reproduction embodied in the patrilines' responsibility for and control over women's sexuality. The synthesis of these discrete categories in *guo jia* establishes an intimate link between production and reproduction, embedding the sexualization of women with respect to Chinese "tradition" in the construction of a modern Chinese nation.

In this chapter, I explore this relationship between nation, gender and sexuality with respect to the "traditional" practice of Chinese calligraphy. I maintain that the elite art of calligraphy serves as a particularly powerful emblem of the ways in which gender is sexualized in the construction of the modern Chinese nation, in large part because of the sense of mystery and awe that calligraphy often inspires in both Western and non-elite Chinese viewers. The seemingly esoteric and unique qualities of calligraphy often lead to depictions of this art as somehow inaccessible; for example, a "quaint survival" of a glorious, yet distant, past, or a powerful and mysterious symbol that embodies and preserves the distinctive and rich cultural history of the Chinese people. However, the sense of "inscrutability" that surrounds the art of calligraphy masks an ideology of gender and sexuality that lies at the center of the production and reproduction of cultural heritage in the modern Chinese nation.

In this chapter I investigate the attribution of gendered qualities to women calligraphers and their practices as a vehicle for the sexualization of women in the process of reproduction of China's national heritage. Following along the lines of much of the recent scholarship on Chinese women (Waltner 1996), I suggest that the category "women" should be read as political strategy rather than social reality, constructed in order, on the one hand, to insure the "purity and beauty" of Chinese tradition, and, on the other hand, to define and control the locus of (re)productive activity in the nation. I argue that constructions of women as sources of both continuity and rupture – tradition and modernity – can offer important insights into the sexualization of gender in the Chinese nation.

Gender and the construction of Chinese national heritage

Every nation, by definition, must possess a traditional and distinctive culture. This culture lends historical legitimacy to the nation – as distinct from the imperial, feudal, primitive or colonial past, yet continuous with a more ancient and primordial community. It further contributes to creating a singular, undifferentiated *national* heritage out of otherwise disparate communities and interests. This appropriation of tradition generally consists of (unsystematic) re-membrances of assorted elements of aristocratic, imperial, elite, urban or otherwise "high" culture that contrast favorably with other instances of national heritage and readily encompass the political realities of diversity, often characterized as folk, peasant, rural, ethnic or minority cultures.[2]

National heritage thus reconciles two seemingly irreconcilable categories: traditional culture and the modern nation. But it also obscures tensions between the two by reifying otherwise fluid processes of transformation and change, giving rise to extreme forms of cultural relativism, on the one hand, and a kind of theoretical essentialism, on the other hand. Intended to frame our investigations, these perspectives ultimately severely inhibit our understanding of the processes of nationalism and the significance of cultural heritage in this process. For cultural relativists, the rise of the modern Chinese nation heralds the demise of the exceptional and unique characteristics of Chinese culture. National heritage both salvages and preserves the "purity and beauty of tradition," thus reinforcing and legitimizing the historical and cultural authority attributed to these forms, transforming them into symbols of the enduring spirit of China's unique cultural heritage. Such celebrations of a singularly *unique* culture and history exhibit a particularism that severely limits generalizations and abstractions that might allow us to gain an understanding of practices such as calligraphy that goes beyond "Chinese tradition."

Essentialists, on the other hand, regard tradition as an obstacle to achieving the standards of modernity and development that define the modern nation. An evolutionary continuum from tradition to modernity allows us to plot the development of any particular nation according to the degree of culturally specific traditional "survivals" that impede the nation's progress. Abstract, universal, homogenizing categories framed in the language of scientific objectivity, development, evolution and progress render all nations and nationalisms familiar, and thus manageable under a single (culture-less) rubric. From an essentialist perspective, traditional practices such as calligraphy embody traditional values and beliefs that impede progress and slow the realization of a modern society and nation. Yet characterizations of traditional practices such as calligraphy as simply "antiquarian survivals" contribute little to an understanding of how these "traditions" contribute to the construction of modern nations and national identities.

Indeed, neither relativist nor essentialist approaches interrogate the distinction between tradition and modernity as part of the cultural process of production of a modern nation. Not surprisingly, the lived experiences of women often disappear in the chasm that emerges between these two approaches. Struggles over the representations of nations and nationalisms that either seek to preserve tradition as the core of the modern nation or to eliminate antiquarianism in the name of progress and development do not attribute much significance to gender. Instead, women become emblems – called upon to act as "guardians of national culture, indigenous religion and family tradition – in other words to be both 'modern' and 'traditional'" (Jayawardena 1994: 14), women find their lives and experiences readily reduced to instances of the successes and failures of modernity, or to quaint exotica in the culture gardens of tradition without the benefit of a critical understanding of how local constructions of tradition and modernity shape definitions of gender and sexuality. As I shall argue in more detail below, the radical distinction between tradition and modernity, reified in both relativist and essentialist approaches, obscures the use of tradition as a vehicle for inscribing gender and sexuality in a modern nation. Investigations into the gender of tradition, embodied in the encounter between Chinese women, Chinese tradition and China scholarship in the West, can thus offer powerful insights into the gendering of a nation (Chow 1991: 95).

A substantial body of scholarship on Chinese women and the Chinese nation provides us with excellent documentation on the changes in women's status and roles that took place amidst the radical social and political changes in China during the early twentieth century. These studies make clear the tremendous importance attached to gender in the construction of the Chinese nation. For example, in her book on the historical transformation of women's identities in twentieth-century China, Elisabeth Croll (1995) describes the elaborate effort expended to construct an identity for Chinese women consistent with the social and political transformations taking place at the time. She suggests that, especially during the early years of Communism, the nation's effort to address issues of gender are "perhaps unmatched by any other government at any time" (Croll 1995: 8).

Croll identifies three distinct moments in the emergence of the Chinese nation that shaped the identities of Chinese women: the early Republican period (1911–49), the years immediately following the Communist Revolution (1949–76) and the more recent period of Communist reform (1978 to the present). She characterizes women from the earliest era – who came of age amidst the radical social and political upheaval that accompanied the demise of the imperial government and the founding of the Republic of China in the early twentieth century – as "daughters of rebellion." For these women, rebellion often took the form of rejecting conventional patterns of behavior and pursuing new opportunities for education and employment, as well as the freedom to choose a marriage partner. The rise of Communism, however, marked renewed efforts by the nation to distinguish the future of the nation from the turmoil and chaos of the immediate past, prompting a new rhetoric of gender equality that inspired images of revolution rather than rebellion. Croll characterizes the young women of this era as "comrades of revolution," referring to the pervasive rhetoric that denied gender distinctions and welcomed women into the revolution as equals to men:

> What this meant in both rhetorical and practical terms was that women were invited to assume male qualities and enter male spaces, whether temporal or spatial, including work, the public sphere, the revolution or heaven as the male landscape of the future. They were invited to enter on terms that were the same as or equal to men's, with very few concessions to female-specific qualities.
>
> (Croll 1995: 7)

According to Croll, this rhetoric of gender equality resulted in a sense of estrangement from the revolution among women, who experienced the life cycle quite differently from their male counterparts. While the rhetoric of the first era depicts a world in which women are afforded the same opportunities as men, becoming what Croll describes as "boys of girls" or "honorary men," the rhetoric of the second era portrays women as "comrades," that is, androgynous model revolutionaries modeled, in fact, on the lives and experiences of men. Croll characterizes the third period, in turn, as an area of competing images and confusion, where open debate calls into question the relationship between gender and nationalism by focusing on whether women's identities should be defined as Chinese, modern or simply woman. While the lives and experiences of men offer a source of contrast and comparison, it seems worth noting that similar struggles did not take place in the construction of Chinese men's identities. Indeed, those struggles were (and continue to be) collapsed into the construction of a "generic" modern Chinese national identity. Although work such as Croll's makes a significant contribution to our understanding of the emphasis on gender in the construction of a modern Chinese nation, an exclusive focus on women remains complicit with a nationalist ideology that equates gender with women, thus failing to explore the importance of women as a social category for the construction of a modern nation. Efforts at gender equality focused exclusively on women, attributing the sole source of gender inequalities to the situation of Chinese women, without an investigation of the (often implicit) gendered qualities of men's lives and work.

The importance of gender in the construction of national identity can, perhaps, be seen more clearly through a comparison of two of the most important reforms of the early period of Chinese nationalism: the abolition of footbinding and the abolition of the imperial examination system. Footbinding abolitionists portrayed this practice as primitive and barbaric, calling for an end to the infliction of such tremendous physical pain and suffering on women. The abolition of footbinding marked a critical moment in the emancipation of Chinese women. By contrast, critics of the examination system portrayed the civil service bureaucracy as corrupt, inefficient and archaic, but the tremendous psychological pain and suffering inflicted on the countless men who prepared for (and often failed) these exams goes unmentioned. The end of China's "examination hell" was not about the emancipation of men from the rigid structures of the imperial bureaucracy, but about the emancipation of *China*, evidence of the conflation of the gendered sphere of men's activities with the nation.

Since the early years of Chinese nationalism, the emancipation of women from feudal oppression and the destruction of feudal gender hierarchies has been used as a powerful symbol of the eradication of the last vestiges of imperial Chinese society. However, social reforms carried out in the early years of the nation effectively did away with many of the "feudal" constraints on women, only to replace them with new constraints framed in the rhetoric of gender equality and outlined within the extremely specific and limited role assigned to women in the new nation. The Chinese nation eliminated *feudal* gender hierarchies, but gender hierarchies persisted, constructed as differences in the ways in which men and women might contribute to national culture and ushered into the nation under the cloak of the "traditional and distinctive culture of the Chinese people" – including the appropriation of one of the most powerful symbols of the moral authority of a class of ruling elite composed exclusively of men: the art of Chinese calligraphy.

The character(s) of Chinese calligraphy

The Chinese term *shufa* translates literally as "writing method" and refers both to the ritualized art of writing and to the visual images such writing produces. According to calligraphic tradition, one cannot, however, simply pick up a brush and begin to write. An elaborate set of highly technical rules governs the practice of calligraphy, regulating every detail, from the position of a writer's feet, to how a writer must hold the brush, to the shape and structure of the graphic forms inscribed. These technical rigors can be mastered only by adherence to a strict program of closely supervised, lengthy and intensive training comprised almost exclusively of imitation. Students began by copying classical masterpieces by rote in order to gain a firm grasp of basic techniques. Once students establish the necessary technical foundation, they move on to more abstract imitations of a selection of works by great masters that constitutes the classical canon of Chinese calligraphy. Technical mastery eventually paves the way for the development of a corollary system of aesthetic sensibilities. Thus mastery of the techniques and aesthetics of calligraphy requires complete submission to a set of rigid and narrowly defined conventions that will shape the terms of creative self-expression throughout a calligrapher's life. In other words, the physical and mental discipline that defines the art of calligraphy go well beyond a simple "writing method." Chinese celebrate the practice of calligraphy as a powerful system of moral cultivation and regard images of calligraphy as graphic expressions of moral virtue. The classic idiom *shu ru qi ren* expresses the belief that calligraphy makes visible the moral character of the calligrapher.

The bureaucratic structures of imperial China from as early as the second century B.C.E. drew directly upon the moral aesthetic of calligraphy. Calligraphy constituted a portion of the imperial civil service exams, the primary avenue of upward social mobility through which young men obtained official government positions and thus entered the highest echelons of Chinese society. Those who learned to read and write did so in accordance with a system of values and beliefs articulated in the technical and aesthetic principles of calligraphy. In his detailed account of the calligraphy of the eleventh-century master Mi Fu, art historian Lothar Ledderose (1979) describes the pervasiveness of calligraphy as a vehicle for the spread of these beliefs and values among the elite of imperial Chinese society:

> Everyone who aspired to government office had to be a proficient calligrapher; calligraphy therefore was the art form that was most widely practiced in China. Along with political, economical, ideological, and other factors, calligraphy was an important means of self-identification for the ruling class, and supported its social unity.
>
> (Ledderose 1979: 3)

One did not have to be a calligrapher, however, to appreciate the power of the brush. For those who could not read or write, calligraphic images emanated magic and mystery. Inscriptions decorated the doors and windows of the homes of both gentry and peasants. Temple committees, as well as the proprietors of small businesses such as tea houses or pharmacies, commissioned scholar-officials noted for their calligraphy to create inscriptions for places of worship and business. The moral force of calligraphy permeated life in imperial Chinese society, lending an even greater sense of exclusiveness to that small cohort of ruling elite who actually practiced this extraordinary art.

From the late eighteenth century until the early twentieth century the moral authority of calligraphy, particularly early calligraphic texts, acquired even greater significance as a link to an ancient past. An antiquarian program of "classical learning" promoted during the last centuries of imperial life included the mastery of a range of texts recovered from a distant past. In his classic work on early modern Chinese intellectual history, Joseph Levenson (1965) characterizes classical learning as a means of cultivating cultural rather than (useful) technical skills, resulting in a system of government based on leadership by "virtuous example," rather than "logical" professional qualifications:

> Artistic style and a cultivated knowledge of the approved canon of ancient works, the "sweetness and light" of a classical love of letters – these, not specialized, "useful" technical training, were the tools of intellectual expression and the keys to social power. . . . Amateurs in government because their training was in art, they had an amateur bias in art itself, for their profession was government.
>
> (Levenson 1965: 16)

Calligraphy served as a cornerstone of this system both in aesthetic and practical terms. Government edicts, coin inscriptions, medical prescriptions, even shop signs, as well as more artistic texts such as poetry and inscriptions on paintings, were judged as much by *how* they were written as by what they contained. The moral character exhibited in calligraphic texts served as a graphic depiction of Levenson's "virtuous example." The possibility for the cultivation of such virtue through the practice of calligraphy ostensibly existed in all human beings, yet a closer look at this complex bureaucratic system suggests that only a small portion of the general population possessed the potential for the kind of moral character necessary for government service. In fact, the ruling elite of imperial China consisted exclusively of men.

The gendered qualities of the imperial Chinese bureaucracy remain invisible largely because the system was never explicitly defined as gendered. No legal statutes dictated that only men could be appointed to public office, nor did any prohibitions exist that expressly prevented women from holding government posts, or even sitting for the examinations. Virtue and programs for the cultivation of moral character centered on a broadly based notion of *human* potential. Explicit references to gender in imperial Chinese society emerge only when we turn our attention to an extensive social code specifically governing the behavior of women and focusing almost exclusively on their roles and responsibilities within the Chinese kinship system.

An extensive set of restrictions and expectations regarding women made it virtually impossible for women to participate in the elaborate networks through which men climbed the social and political ladder of success. A legal code that regulated marriage and inheritance further supported a social structure that effectively eliminated the possibility for women to survive outside of kinship institutions. An ideology of gender conceived of exclusively in terms of sexuality (and, more specifically, reproduction) defined the place of women in society; first as daughters whose chastity must be preserved and protected so as to be bartered as an "untarnished" source of reproductive power to another lineage, then as wives and daughters-in-law who must reproduce in order to insure their place within the household and (peripherally) the lineage, then finally as mothers, mothers-in-law and, ultimately, grandmothers, whose sexual reproductive activities have earned them a position of power and authority over the relations of sexuality and reproduction among the younger members of their expanding household. Men, on the other hand, inhabited a world without distinctively gendered qualities – a world that appears to transcend gender categories and to encompass all of humanity. Practices such as calligraphy served as powerful symbols of the virtuous authority of the men who inhabited this world.

The growth of Chinese nationalism during the late nineteenth and early twentieth centuries, however, wrought dramatic changes in this world. Sweeping reforms gave rise to radical transformations in all aspects of Chinese society. The abolition of the imperial examination system in 1905 signaled the collapse of the imperial bureaucracy and effectively severed the link between government service and classical learning, including calligraphy. The structures that supported the practice of calligraphy as a symbol of the solidarity of the "ruling elite" rapidly disappeared. In their place, a set of structures and institutions emerged that would support a new understanding of calligraphy – as "the traditional and distinctive culture of the Chinese people, a symbol of their spirit" (Wang 1987: 3). Calligraphy no longer symbolized the moral authority of the ruling elite. Instead it came to symbolize the moral character of the Chinese nation – or, more precisely, a national ruling elite that sought some measure of legitimacy and authority by reference to "traditional" practices like calligraphy.

The national character of calligraphy

The appropriation of calligraphy as an emblem of the Chinese nation suggests more about the ideologies of the nation than about the practice of calligraphy. The rhetoric that portrays contemporary calligraphy as a deep and enduring connection to a glorious ancient past proves particularly instructive for understanding the construction of Chinese national heritage. A lengthy and well-documented historical record preserved in calligraphy supports this rhetoric and readily serves the aims of the nation, since it provides a material connection to the distant origins of Chinese people and culture. Elegant images dating back more than a thousand years testify to the technological and aesthetic sophistication of ancient China. Later works demonstrate both historical continuity and creative innovation. Yet the greatest significance of this history for the nation resides in the ways in which calligraphy signifies an imagined community spanning vast geographical spaces and centuries of historical time. The images of calligraphy embody a timelessness that transcends the particularities of any given historical moment, a sign of the persistence and endurance of the Chinese people. For a nation born out of nearly a century of humiliation and defeat at the hands of other nations, the historical transcendence symbolized in calligraphy holds tremendous appeal.

This transcendence rests in part in a uniform stylistic tradition – what the eminent historian of Chinese art, Max Loehr (1965), characterizes as a "closed system of forms" – exhibited in a well-defined canon of classical masterpieces. Detailed accounts of the technical, stylistic and aesthetic elements of this canon can be found in any introductory text to calligraphy. This canon is also the basis of most art historical studies.[3] For the purposes of the present discussion, however, I shall assume, rather than describe or explain, the artistic aspects of this tradition. Instead, I explore the emergence of this uniform stylistic tradition as a powerful social process that gives shape to the imagined community of the modern Chinese nation.

Over the course of more than a millennium, changes in the technology of writing and individual aesthetic experimentation gave rise to a range of *ti* or styles. The broadest distinctions in style can be seen in the systematization of gradual changes in the structure and shape of Chinese characters, which led to the development of five different scripts. Subtler distinctions can be found in the differences among various stylistic "schools" originating in the creative genius of particular historic individuals. While every calligrapher seeks his own *ti*, only a few emerge as great masters whose technical virtuosity and aesthetic imagination serve to inspire other calligraphers for centuries. Yet the reputations of these men were built slowly, at first through casual exchanges among peers and more formal relations between a master and his students. Exceptional works eventually circulated among collectors and other calligraphers, who often made copies and then passed the copies along. The personal papers and scholarly writings of scholars and officials throughout Chinese history, coupled with records that detail the movements of great masterpieces among collectors and scholars, reveal the paths along which calligraphic ideas spread and stylistic "lines of descent" grew. In other words, the historical record preserved in and through calligraphy provides a detailed map of social relations among China's ruling elite over the course of several centuries.

Contemporary calligraphers enter this web of social relations through intimate encounters with calligraphers from the distant past in the images of ancient masterpieces. Over the course of a lifetime of intensive study, a calligrapher will reproduce not only the mechanical processes but also the artistic spirit of countless classical masters. Every student of calligraphy must acquire a working knowledge of this large but clearly delimited canon through which a diversity of individual interpretation and historical experience is "telescoped into a uniform body of stylistic tradition" (Loehr 1965: 37). Departures from this tradition are either assimilated as transformations in style and technique, or discarded. In other words, the narrow conventions of style in classical calligraphy offers an excellent vehicle for constructing Chinese nationalism because it evokes a sense of immutability that readily symbolizes the impenetrability of Chinese tradition, achieved by condensing more than a thousand years of historical, regional and individual variation into a single, uniform tradition.

The appropriation of calligraphy by the nation as an instance of national heritage further contributed to representations of the nation as a singular, unified community. Since those who learned to write did so according to the technical and aesthetic principles of calligraphy, efforts to achieve mass literacy throughout the first half of the twentieth century essentially signaled mass instruction in calligraphy. Unlike the elite practices of the imperial past, all citizens of the modern nation could participate in the promotion and perpetuation of that distinctive culture that would give historical continuity and legitimacy to this emerging community of the Chinese nation. Moreover, calligraphy generally appears mysterious and exotic to Western viewers, thus furthering a sense of the unique Chinese nation *vis-à-vis* the West. In sum, the "traditional" practice of calligraphy embodied precisely those qualities that would give legitimacy and authority to the modern nation as an imagined community possessing, and sharing, a distinctive past.

The symbolic significance of this unchanging timeless Chinese tradition in the context of a modern nation, particularly the social construction of style discussed above, holds immediate and direct implications for China's women calligraphers. Among the countless classical masterpieces displayed in museums or reproduced in copybooks, few, if any, are by women.[4] Few women participate in either national or local calligraphy associations, nor do many women become calligraphy teachers. Contemporary calligraphers readily acknowledge, and offer short, simple and direct explanations for, the absence of women from the national heritage symbolized in and perpetuated by calligraphy. Explanations of women's calligraphic silence occasionally focus on the social structural position or essential nature of *women*. In other words, women were depicted either as too busy homemaking to devote the necessary time and energy or as incapable of the kinds of technical skills required to achieve calligraphic virtuosity on the same level as men calligraphers. Even more frequently, however, the issue of women calligraphers was framed in terms of the historical situation of women.

In my discussions with them, contemporary calligraphers maintained that, historically, women had not participated in those institutions that promoted and perpetuated the works of great calligraphers, namely, the imperial civil service. These calligraphers supported these historical explanations further with references to customs that had prohibited casual contact between men and women (particularly among the gentry), thus proscribing the participation of women in many of the informal gatherings of literati devoted to calligraphy where, as I have suggested above, stylistic "lines of descent" were established and maintained. Finally, they argued that since women could neither take the imperial civil service examination nor hold government office, and the value of young women rested solely in what they could bring to their natal households through marriage into a high-ranking family, few families invested in the education of their daughters. Instead, these girls received training in "less rigorous pursuits" such as needlework, as well as detailed instruction in the moral code that dictated proper behavior for women.

Ironically, since calligraphy is never explicitly defined as a gendered practice, nor is the development of stylistic lines of descent explicitly linked to sexuality, explanations for the absence of women from calligraphic communities do not focus on the art of calligraphy itself. Celebrations of China's national heritage in calligraphy thus privilege precisely those genres that have long served as an important means of self-identification for a ruling class of

Han Chinese men. Women must change, calligraphy cannot. I suggest, therefore, that it is not the immutability of tradition, but rather the appropriation of an ideology of immutability in the service of the nation, that renders Chinese women mute.

Constructing the sexualization of women in classical calligraphy

Despite the absence of women from the public record of calligraphy, during the course of my research I encountered many women who actively engaged in the study and practice of calligraphy in more private venues. Women enthusiastically attend calligraphy classes, workshops, exhibitions and other formal, scheduled events. However, in contrast to the men with whom I studied and worked, who invariably spoke with an air of expertise and confidence regardless of their actual training, knowledge or artistic abilities, most women refused to speak seriously about calligraphy, and dismissed their own work as merely a passing interest or casual pastime. In their interactions with men with regard to calligraphy, women generally assume the role of student rather than teacher and audience rather than artist. Despite both the numbers of women who now practice calligraphy and the intensity with which many pursue the practice, works by women rarely appear in exhibitions, galleries and competitions. When such works appear, they appear as the work of remarkable women whose extraordinary abilities have earned them a place alongside the ordinary accomplishments of men. Catalogues that document and promote contemporary calligraphers or commemorate a particular exhibition or competition do not often include women. Consequently, women's calligraphy does not achieve wide circulation, their works and names remain relatively unknown, and they rapidly disappear from the historical record.

The absence of women's calligraphy in representations of calligraphic communities intensifies in the social construction of calligraphic style that takes place in the casual, and often spontaneous, gatherings of friends and colleagues in tea houses, parlors and back rooms of storefront businesses after the other business of the day has been concluded. At these moments, stimulated by endless thimble-size cups of tea or wine, school teachers, local business people, government officials and bankers share their passion for calligraphy. Their debates surrounding the aesthetic qualities or technical difficulties encountered in the study of classical masterpieces often last for hours. While some men share the products of their own efforts to master the style of a particular work or artist, others offer praise or constructive criticism. Calligraphy and poetry also serve as vehicles for impromptu outbursts expressing the intense feelings of camaraderie inspired by an evening of good food and drink among old friends and colleagues. The casual spontaneity of these events belies their gendered quality and their significance as sites for the construction of a uniform stylistic tradition. The absence of women from these venues prohibits women from participating in the social construction of calligraphic style that takes place in these contexts and restricts them to the rote reproduction of the standard models assigned in calligraphy classes, outlined in public lectures and represented in museums and galleries. Men, on the other hand, engage in the creative interpretation of classical masterpieces that gives meaning to this traditional art as the heritage of the modern nation.

Among these men, a few great masters emerge and stylistic lines of descent continue to grow, much as they did in the imperial past, out of the gradual recognition of individual talents and skills. The works of these "model" contemporary calligraphers generally achieve such widespread circulation that even individuals who do not engage in the study of calligraphy can recognize the distinctive style of a particular individual. The continued cultivation of classical aesthetics and technical virtuosity links contemporary calligraphic communities to a "primordial" calligraphic past and demonstrates the viability of this "traditional" art in contemporary society. The promotion of the calligraphy of particular individuals as models echoes ideological elements from the imperial past, when calligraphy was regarded as a visual representation of the calligrapher's moral character. In other words, the calligraphy of these exemplars displays both technical and moral virtuosity. Through the images of their calligraphy, both ancient and contemporary, "great masters" thus serve as models of and models for proper moral conduct.

> The promotion of models that were both descriptive and prescriptive continued a long time Confucian tradition of journeying towards worthy or virtuous behavior in emulation of worthy or virtuous behavior. . . . The efficacy of the models as exemplars on revolutionary pedestals rested on the practice of emulation or on the willingness of the audience to replicate desirable and emancipatory gestures.
>
> (Croll 1995: 72–73)

Only rarely do representations of either women artists or their work appear among the representations of model calligraphers and model calligraphy found in exhibitions or catalogues. Some critics maintain that this absence of women in the contemporary context reflects the low quality of women's work at this historical moment, arguing that women artists have not yet achieved the level of virtuosity that would merit their inclusion in these publications. Others maintain that women lack the essential (moral) characteristics possessed by all great calligraphers. In the preface to the first catalogue devoted to the work of contemporary women calligraphers in post-liberation China, the vice-chairman of the National Calligraphers' Association, Shen Peng, suggests that both of these positions stem from the same source: an ideology of gender and sexuality that encompasses but is not limited to the art of calligraphy. He argues that while the recognition of women calligraphers a full fifty years after "liberation" is long overdue, it is also a clear sign of a more fundamental failure of the nation to eliminate the gender hierarchies of Chinese feudal society.

> In my reflections upon the collections of "great masters," I have discovered that women are as scarce as phoenix feathers and unicorn horns, and that the number of women artists [represented] is extremely unsuitable. Is this only because women have not reached the same level of achievement as men, or do [other] causes exist? I feel that this problem is worth investigating, and that the investigation should not be limited to the art of calligraphy itself.
>
> (Shen 1989: i)

Taking the practice of calligraphy as an example, he notes that in imperial China the classification of women calligraphers differed dramatically from the classification of men calligraphers. The marked category of *women* calligraphers (*nu shufajia*) not only served to exclude women from the community of (men) calligraphers, but also allowed for the evaluation of women's work according to a set of standards wholly different from the standards used to evaluate men's calligraphy. The technical, stylistic and aesthetic qualities of a man's calligraphy defined his place with regard to various stylistic schools. By contrast, the evaluation of women's calligraphy had little to do with elements of technique, style or aesthetics. Shen cites the nineteenth-century catalogue of women calligraphers, *The Jade Terrace History of Calligraphy*, which characterizes women calligraphers as "wives, mothers, daughters, sisters, concubines or courtesans," categories that locate a woman's calligraphy in relationship to a particular man, rather than the stylistic qualities of the visual images (see also Weidner *et al.* 1988: 88).

Shen (1989) maintains that, although these particular categories may no longer be used, the hierarchy which they represent persists. He argues that "the demand to seek a 'feminine spirit,' 'feminization,' or a 'feminine consciousness' in the works of women reflects a kind of male perversity" and serves to reinforce the subordination of women. While he acknowledges the present need to publish a catalogue devoted exclusively to calligraphy by women, readily admitting that works by women have been consistently judged inferior to works by men and thus have been less widely circulated, he also uses the publication of the catalogue as an opportunity to point out that the radical re-evaluation of gender promised by China's early revolutionaries in the 1940s has somehow failed to materialize.

The system of classification of women found in *The Jade Terrace* also points to the ways in which gender and sexuality in China are collapsed into a single category that defines women's morality, sexuality and worth almost solely in terms of their reproductive potentialities and abilities. The terms used to describe women ("wives, mothers, daughters, courtesans and concubines") emphasize women's sexual relationships to men, describing relationships of either reproduction or desire. A woman's calligraphy demonstrated moral character, but that character belonged to the man (or men) who had access to, responsibility for or control over her sexuality.

The relationship between women's artistic production and their sexuality can be seen in parallels between descriptions of Chinese women in the arts and of Chinese women in the traditional family. In a catalogue essay for an exhibition of the work of Chinese women artists from 1300 to 1912, Marsha Weidner characterizes women artists as "sustainers rather than innovators" (Weidner *et al.* 1988: 13). She argues that a combination of historical circumstances and social structural relations prevented women in imperial China from fully developing their creative artistic talents. Nonetheless, Weidner maintains, not only did these women make significant contributions to the perpetuation of certain stylistic traditions, but also they did so by constructing alternatives to the exclusive social processes through which artistic styles were established and maintained in late imperial China. According to Weidner, although women depended primarily on men for instruction and advancement in artistic endeavors, they also developed limited networks comprised of other women. Much like the social networks employed by men, this web of social relations gave women the opportunity to offer and receive instruction, to circulate their work to a larger audience and, ultimately, to gain a wider reputation. Unlike the formal political structures and institutions of imperial China, however, these networks were highly particularistic, developing out of individual rather than institutional relationships and enduring only as long as the women remained in contact. In her feminist critique of conventional studies of Chinese kinship and lineage organization, anthropologist Margery Wolf (1972) describes a similar pattern of relationships among women in response to the patriarchal structures of the traditional Chinese family:

> With a female focus . . . we see the Chinese family not as a continuous line stretching between the vague horizons of past and future, but as a contemporary group that comes into existence out of one woman's need and is held together insofar as she has the strength to do so, or for that matter, the need to do so.
>
> (Wolf 1972: 37)

Both Weidner and Wolf argue that the absence of women from art historical accounts and from descriptions of Chinese village lineage organization, respectively, stems largely from the fact that we have simply not looked in the proper places, that is, in the "domestic" sphere, a realm that is clearly distinct – both in structure and in practice – from the world inhabited by men. They maintain that once we have entered this alternate realm the lives and works of women become visible to us, and through these images we gain an even deeper understanding of the social construction of artistic style (and Chinese social life). Indeed, Weidner suggests that "As men have led us through the upper political and intellectual reaches of Confucian society, women can lead us into the heart of its fundamental institutions" (Weidner *et al.* 1988: 28). From Weidner's point of view, as we begin to understand

and appreciate artistic images by women on their own terms, we see them not as poor imitations of a loftier standard, but as a source of sustenance for the artistic traditions that gave rise to the great masters and great masterpieces of Chinese art history.

It is worth noting, however, that despite these elaborate alternatives to established networks of artistic production women artists did not use these networks, as men did, to advance their own technical and aesthetic innovations or to establish a distinctive and clearly identifiable "women's style." Instead women lent their energies and support to the perpetuation of those narrow conventions of style that symbolized the power and prestige of a class of imperial scholar-officials comprised exclusively of men. Although women did not participate directly in the political and social structures that defined the terms of calligraphic style, their calligraphy can (and should), nonetheless, be read as part of the single, unified and uniform art historical narrative appropriated by the nation as the tradition of Chinese calligraphy. Alternative networks for the circulation of their work allowed women to participate in existing communities of artists and calligraphers, but they effectively excluded them from the social production of calligraphic style that took place in the distribution of images and the recognition of individuals among the men who comprised the scholarly elite.

The images created by women in imperial China thus presented no direct challenges or disruptions to the cultural unity and historical continuity represented in calligraphy. Women employed the same calligraphic models, techniques and stylistic conventions that men used, in part because these models and techniques constituted the sole path by which women might gain some recognition for their artistic talents. In turn, by adhering to established stylistic conventions, women did not pose a direct challenge to men's goals of establishing a stylistic line of descent, "bulging to encompass all the members of a man's household and spreading out through his descendants" (Wolf 1972: 37). As I suggested earlier, this mode of participation in the male-dominated structures of the imperial ruling elite mimicked women's participation in the patriarchal structures of Chinese kinship. Women lived and participated in families, villages and lineages dominated by men because these structures insured their physical survival, and ultimately provided the only means for them to gain at least a modicum of power and influence.

Yet, despite women's efforts to conform as closely as possible to social expectations and aesthetic standards, the social system itself contained enormous tensions and contradictions. The intensely male focus of Chinese social organization rendered all women outsiders. Ironically, however, only these outsiders possessed the power to preserve a continuous line of descent. Chinese lineages depended so desperately on women's ability to reproduce their members that women's reproductive power was regarded as one of the greatest threats to the stability of the system. The power and danger of women's sexuality permeated all social constructions of women, including constructions of women artists. The sexualization of women's calligraphy mimics the sexualization of women's lives within the "traditional" Chinese family because both were part of a wider cultural order that shaped individual experience in imperial China. The rise of the modern Chinese nation in the early twentieth century, however, wrought dramatic changes in the Chinese cultural order, including definitions of gender and sexuality.

National exemplars: model calligraphers, model women

The nation's rhetoric of gender equality sought to insure that within the context of the modern nation Chinese women would no longer need, as they had in the imperial past, to rely on their relationships with men to enter that elaborate network of social relations that shapes the aesthetic and stylistic canons of calligraphy. The nation's social reforms further insured that these feudal avenues of social mobility would no longer be available to women. Yet, as I shall demonstrate in more detail below, without these relationships to the men who stand at the center of the social construction of calligraphic style to support and promote their work, even in the context of the modern nation, women calligraphers remain as "scarce as phoenix feathers and unicorn horns," despite the elimination of the "feudal" structures and institutions that had posed such tremendous obstacles to women's full participation in the calligraphic communities of the historic past. Much like the emergence of style in the imperial past, the social construction of calligraphic style in the modern nation continues to depend on an elaborate network of social, political and artistic connections. The model calligraphers at the center of this network in the modern Chinese nation bear a striking resemblance to the model calligraphers of the imperial past: they are all men. When women do appear, they appear as representations of model women, not model calligraphers.

Representations of model women in calligraphy begins with the construction of a distinct class of "women calligraphers." This class belies the nation's rhetoric of gender equality by encompassing women within national culture but restricting them to a space defined by gender and distinct from the sites of interpretation of calligraphic style as a meaningful part of the modern nation (through, for example, participation in informal gatherings or the circulation for their work). This construction of distinct gendered spheres of calligraphic production suggests strong parallels with the ideology of gender and women's sexuality that defined women's place within the traditional Chinese family solely in terms of reproduction and thus restricted women's range of movement to the "domestic sphere." Weidner's characterization of women artists in imperial China as "sustainers rather than innovators" echoes this ideology and points to the sexualization of gender that shaped the lives of all Chinese women, whether artists or not. The modern nation has appropriated this ideology in the construction of national heritage through the "tradition" of Chinese calligraphy, gendering women calligraphers and sexualizing their practices as a vehicle for the reproduction of the traditional and distinctive culture of the Chinese people.

The process of identifying a distinct group of "women calligraphers" imposes a singularity onto a diversity of lives, works and practices of those calligraphers defined as women and reiterates the contrast discussed earlier between the construction of Chinese national identity and the construction of Chinese women's identity. The term "woman calligrapher" points to the tensions between "woman" and "calligrapher," revealing the gender of this art by suggesting that women's calligraphic practices differ from the practices of other calligraphers. This tension resolves, however, in the *category* of women calligraphers which denies women membership in a broader category of calligraphers and delimits their practices to those appropriate to a woman. No matter the level of their skill or the degree of their talent, women calligraphers transgress the gendered boundaries of calligraphy and must be contained within a distinct realm of women calligraphers. While they become recognized as exceptional women, their calligraphy remains virtually invisible. A chosen few serve as exemplars of the proper relationship between women and national heritage, and their names and works become distinguished both from ordinary women and from ordinary calligraphers.

Representations of women in a catalogue of contemporary calligraphers (Yan Zheng 1988) published by the state-owned Fine Art and Design Press in Beijing provide an excellent example of the historical shifts in the rhetoric of gender equality and the construction of women's identity described above. The catalogue details the lives and works of 120 calligraphers, only 2 of whom are identified as women. The attribution of gendered qualities to these two calligraphers clearly marks their gender-differentiated relationship to the political, social, economic and ideological contexts through which calligraphy comes to symbolize national cultural identity. Yet it also establishes a similarity between what otherwise appear as radically different artistic practices and lives, thus giving substance to a category of "women calligraphers." This categorical similarity lays the groundwork for contrasts and comparisons between the two women, through which we can begin to see their significance as models of modern Chinese women, as gendered exemplars of the nation.

The catalogue describes Zhou Huijun as "the most outstanding among China's middle-aged women calligraphers." The biographical statement that accompanies the published images of Zhou's work focuses primarily on her lifelong commitment to the study of calligraphy and frames her successes largely in terms of her own determination and hard work. In contrast to those historical classifications that identified women as "wives, mothers, daughters, sisters, concubines or courtesans," the catalogue identifies Zhou as a calligrapher in her own right, albeit a *woman* calligrapher. In the context of the modern nation, Zhou Huijun's accomplishments as a calligrapher are no longer traced to the sponsorship of a particular man, but depend instead on her training and professional accomplishments – the same criteria purportedly used to locate men within the complex community of contemporary calligraphers. The image of Zhou as a fiercely determined and hard-working comrade unencumbered by relationships to a father, husband or children echoes images of the rhetoric of gender equality from the early years of the People's Republic of China, immediately following the Communist Revolution.

Ironically, this portrait of an "androgynous" calligrapher clearly points to the gendered qualities of Chinese calligraphic practice and the construction of calligraphic style. The catalogue briefly describes Zhou's training in the classical stylistic tradition of the great calligraphers of the past, listing the names of several noted contemporary calligraphers in Shanghai with whom she had studied, without specifying any of the details of her formal training (such as the dates, length of time or texts she studied) that would establish a definitive relationship to a particular stylistic line of descent. The use of vague but flattering phrases such as "easy and graceful, lucid and lively" to describe her calligraphy, coupled with the characterization of her work as exhibiting a "distinctive" appearance, further contribute to a general appreciation of Zhou's calligraphy, without assigning her work to a specific stylistic tradition. Unlike the catalogue entries for men of the same age, the entry for Zhou does not offer the kind of personal biographical information, such as occupation or educational background, that would connect her in specific ways to those social, political and artistic networks that define contemporary calligraphic style and make it meaningful within the modern nation.

Much of the brief biography focuses instead on Zhou Huijun's work as a calligrapher, including her participation in several national and international exhibitions and her expression of revolutionary zeal in an inscription of selected songs and poems by revolutionary leader Lu Xun, activities that demonstrate Zhou's commitment to the nation as the legitimate heir to the distinctive and traditional culture of the Chinese people. While this list of accomplishments contributes to an appreciation of Zhou's contributions to national heritage, it does not contribute to an understanding of her calligraphy in aesthetic or stylistic terms. The absence of stylistic commentary on Zhou's calligraphy in the catalogue restricts her to the margins of the social processes of production of calligraphic style, even as it makes her central to the perpetuation of traditional culture through the faithful reproduction of prescribed forms.

The most serious contradiction in this portrait of an androgynous model calligrapher comes in the opening sentence of the catalogue, which identifies Zhou as a *woman* calligrapher. The implications of this information for the interpretation of her calligraphic style never become explicit, but this early reference to Zhou as a woman frames a reading of her calligraphy in powerful ways, reminding viewers that the calligrapher is not "just" a calligrapher but also a woman. The vague and unenthusiastic discussion of the aesthetic and stylistic qualities of her calligraphy, as well as the lack of detail regarding her personal life and social identity, obscure the particular nature of Zhou Huijun's contributions to the art of calligraphy but do not challenge an ideology of women as "sustainers rather than innovators." The "feudal" system of classification of women in imperial China outlined in *The Jade Terrace* has been abandoned, but has not been replaced with another set of categories that might give insight into Zhou's social position and identity.

Clearly, gender remains the most important identifying characteristic in this portrait of Zhou Huijun. More than a model calligrapher, Zhou Huijun stands as a model woman, a living testimony to the ideology of androgyny of the Communist revolution articulated in slogans such as "women are the equal of men" or "anything a man can do a woman can do also" (Croll 1995: 71). What Zhou lacks in extraordinary calligraphic talent she makes up for in her contributions to the nation – as an extraordinary *woman* who happens to possess some talent for calligraphy.

In contrast to this catalogue portrait of Zhou Huijun as a *woman* calligrapher, the biography of contemporary calligrapher Shao Xian, the only other woman to be included in the *Catalogue of Contemporary Calligraphers*, clearly identifies Shao according to the classification system of *The Jade Terrace*, as daughter, wife and mother. This series of relationships – first to a father, then a husband, then children – marks Shao's path through the tumultuous events of twentieth-century China, documenting shifting definitions of gender and sexuality and their relationship to traditional culture.

Born in the final years of Qing imperial rule at the turn of the twentieth century, the catalogue describes how Shao spent her early years in an elite household that teetered between the imperial past and a nationalist future. While her father rejected the system of imperial bureaucracy and refused to hold government office despite his mastery of the necessary classical skills, he did not go so far as to provide a formal classical education for his daughter. Instead, as the catalogue documents, Shao Xian began her study of calligraphy simply by carefully observing (and secretly imitating) her father's practices. In the context of a radically changing China, the development of her artistic talents and skills, combined with her father's influence and instruction, eventually provided Shao with an alternative to the only path generally available to most young women: marriage. At the age of 23 she left the strict supervision of the domestic quarters of her father's household, but she did not immediately enter under the strict supervision of a husband's household. Rather she began the serious study of calligraphy under the tutelage of one of the most prominent Chinese social activists of the twentieth century, Kang Youwei. The catalogue notes that, at this point, "[her] style suddenly changed, escaping the 'women's quarters,' and sharing the male spirit. She therefore adopted the pen name '*Tuo Ge*' (escaping the quarters)." This sudden stylistic transformation marked an "escape from the women's quarters" that took Shao outside the physical confines of her father's home and into the studio of one of the greatest calligraphers of the early twentieth century. It also marks her movement beyond the classificatory schema of "wives, daughters, mothers, concubines and courtesans," terms that, as I have argued above, describe men's access to, control over and responsibility for women's sexuality.

As a student of Kang Youwei, Shao Xian began a relationship with the classical canon of Chinese calligraphy mediated through a master–student relationship, rather than through a relationship to a man responsible for maintaining control over her sexuality and reproductive powers. Although Shao had studied this canon under the direction of her father for more than ten years, it was not until she left her father's home that she began to approach these great masterpieces with what contemporaries characterized as a "male spirit," that is, through direct participation in the social processes of production of calligraphic style, rather than through the reproduction of a stylistic line of descent to which she could never belong. Shao's "escape from the women's quarters" transformed her identity *as a calligrapher* from woman to "honorary man," that is, one who shares the "male spirit" of producing a calligraphic canon.

Ironically, while Shao escaped the symbolic complex of the women's quarters in calligraphy, questions surrounding sexuality (framed in terms of an identity in relationship to a man) re-emerge in the catalogue in biographical details wholly unrelated to calligraphy. The catalogue describes how Shao eventually married an engineer, a man of modern productive skills and talents, and bore two children who became members of the People's Liberation Army. Like the classification of Zhou Huijun as a woman, the details of Shao's marriage and children do not contribute to a deeper understanding of her calligraphy (and go unmentioned in the biographies of all of the other calligraphers presented in this catalogue, including Zhou Huijun). Although Shao Xian escapes the women's quarters through her calligraphy, and appears to take on the gendered qualities of a man, she cannot escape the limits of her sex. Her mastery of the "male spirit" in calligraphy disrupts a cultural order defined in terms of lineages of men reproduced by women. Her calligraphic contribution to the nation takes the form of cultural production characteristic of men, rather than the cultural reproduction characteristic of women. But the catalogue reference to Shao's marriage and children compensates for the "gender-bending" qualities of her calligraphy by clearly demonstrating how her sexual reproductive powers have, in fact, been harnessed in the service of the nation. In contrast to Zhou Huijun, who is depicted as a woman who happens to possess a talent for calligraphy, Shao Xian is depicted as a calligrapher who happens to be a woman. In both instances, gender is used to identify the destination of the reproductive energies of these two calligraphers.

The catalogue's image of Shao Xian as a strong-willed rebel daughter closely parallels Croll's "daughters of rebellion," the rhetorical formula used to characterize young women in the earliest era of Chinese nationalism – those who came of age amidst the radical social and political upheaval that signaled the demise of the imperial government and the founding of the Republic of China in the early twentieth century. Croll writes that

> It is above all the notion of rebellion or rejection of prescription norms that is both the theme common to the published narratives of female

> lives during the first decades of the twentieth century and the predominant thread running through each individual life story. Each of the narrators not only primarily perceived herself as a rebel, but also denied having had a conventional Chinese girlhood. From the standpoint of grown women they recognized that from an early age they had a "lively curiosity," a "strong will" or an "independent spirit," all characteristics more likely to apply to sons than daughters. In categorizing themselves as girls of an unusual physical liveliness or curiosity of mind, they frequently thought of themselves as boys of girls.
>
> (Croll 1995: 43–44)

In the years immediately following the founding of the Chinese nation, girlhood rebellion took a variety of forms, framed in terms of girls' desires to move freely beyond the confines of the women's quarters, to receive an education and to participate in planning their own future. According to the catalogue, calligraphy served as a vehicle for Shao Xian to discover her own "lively curiosity" and "independent spirit," qualities that eventually earned her a position within calligraphy circles as an "honorary man." Much like the accounts of and by women from this era documented by Croll, Shao Xian openly rejected the singular path, or style, offered to women of the imperial past, that is, marriage. Instead, she cultivated her talents so that she could escape the confines of the women's quarters and "share the male spirit" of the early years of Chinese nationalism.

The gendered drama of Shao Xian's escape contrasts sharply with the near absence of gender in the catalogue entry for Zhou Huijun. I suggest, however, that the differences between the two constitute more than the idiosyncratic characteristics of "exceptional women." The life stories of Shao and Zhou constitute iconic representations of shifting constructions of gender in the modern Chinese nation – representations that correspond closely to the rhetoric of gender equality described by Croll and embedded in the rhetoric of national heritage embodied in calligraphy.[5] The absence of representative numbers of women calligraphers, as well as the images of those women calligraphers that do appear in catalogues, exhibitions and other public spaces must be read as political strategies rather than social realities, constructed in order to insure both the "purity" of the Chinese nation's traditions and the nation's control over women's (re)productive activity.

Dissent, resistance and subversion: the agency of Chinese women

In the opening paragraphs of this chapter I suggested that images of women in constructions of a modern Chinese nation must be read as a political strategy that seeks, on the one hand, to insure the "purity and beauty" of Chinese tradition, and, on the other hand, to define and control the locus of (re)productive activity in the nation. Images of women calligraphers and their practices serve as a particularly powerful instance of this strategy in large part because calligraphy inspires a sense of mystery and awe among most Western and many Chinese viewers, thus readily deflecting challenges to its gendered qualities. Indeed, there are those who argue that calligraphy represents the unique and distinctive characteristics of Chinese history and culture and must therefore be preserved exactly as "tradition" dictates, regardless of its gendered qualities. Others promote calligraphy as a culturally appropriate vehicle for the democratization of both language and art required by a modern nation, encouraging the practice of calligraphy by women as a means of incorporating them into the nation, without an investigation of the gendered qualities of calligraphy. Women are, thus, invited to assume male qualities, enter male spaces and engage in male practices as citizens of the modern nation. In both instances, a reverence for tradition obscures its profoundly gendered qualities.[6] Rey Chow makes a similar argument, linking the successful appropriation of Chinese tradition by the Chinese nation directly to Western scholarly practices:

> In the long run . . . the problems *embodied* by Chinese women with regard to Chinese tradition and China scholarship in the West would serve as focal points through which the reverence for authority must be attacked. In a field where such reverence is, in the foreseeable future at least, clearly immovable, and where investments in heritage are made with strong patriarchal emphases, the emergence of Chinese women as feminist subjects (rather than as objects of study) is difficult.
>
> (Chow 1991: 95, original emphasis)

In other words, as long as the appropriation and representation of Chinese heritage in practices such as calligraphy are regarded – and revered – by researchers as an exotic and ancient art form, the overwhelming silence of women can be dismissed under the weighty authority of "tradition," without an investigation of the ideologies, institutions, processes and practices that render invisible the experiences, work and lives of Chinese women.

Scholarly research on Chinese calligraphy generally focuses on constructing an historical narrative based on the visual imagery found in classical masterpieces. This work has contributed a great deal to our understanding of the historical significance of calligraphy for the political and social structures of imperial Chinese society, and the ways in which shifts in technique and style were shaped by the interests of the political and scholarly elite. These studies, however, concentrate almost exclusively on those works that have survived and which today form the core of those popular representations of calligraphy that are constructed in the public spaces of museum collections, art galleries and auction houses. Scholarly investigations rely upon, and lend scholarly legitimacy to, these popular representations of ancient Chinese tradition. Given the limited range of visual images that it has represented, it is not surprising that scholarly research on Chinese calligraphy "reveals" a remarkable degree of homogeneity, in both practices and people.

Women violate that homogeneity, and are consequently confined to a separate category that not only erases the distinctiveness of individual lives but also denies the agency possessed by all human actors and realized by many in the practice of calligraphy. Even those women who seek to stay within the borders of a distinctively "feminine" style violate the gendered boundaries of calligraphy. Few, however, are as articulate as the Yuan dynasty artist Guan Daosheng (1262–1319), who expressed the irony of women's artistic practices in an inscription that appears on one of her own paintings: "To play with brush and ink is a masculine sort of thing to do, yet I made this painting. Wouldn't someone say that I have transgressed [propriety]? How despicable; how despicable" (Weidner *et al.* 1988: 67). Guan's inscription adroitly captures the ambiguities and contradictions of the practice of calligraphy by women by cleverly informing us that all women who engage in this practice subvert those narrow conventions that make calligraphy such a powerful symbol – and implicitly exclude women – even as women reproduce these images in their work.

Silence constitutes the only alternative to this kind of subversion (or other forms of resistance or dissent). Women could refrain from such "despicable" practices and preserve their virtue through silence, much as they could preserve their virtue through chastity. But by doing so they would fail to fulfill their primary obligation to the social order: to perpetuate a line of descent. In other words, all of women's calligraphic practices violate the socio-cultural order and, while most women are not as clever as Guan Daosheng, nonetheless they "make visible a crack, a fault line in the dominant male discourse of gender and power, revealing it to be not monolithic but contradictory and thus vulnerable" (Gal 1991: 197). These "cracks" grow out of the contradictions inherent in an ideology of reproduction that constructs a category – "women" – of individuals who are responsible for perpetuating a system to which they will never belong.

Women become visible when we recognize their calligraphy as cultural agency. This understanding of women's calligraphic practices goes well beyond the images of "daughters of rebellion" or "comrades of revolution" constructed within the rhetoric of the Chinese nation, maintaining that women's calligraphic practices must be recognized as forms of dissent, rebellion and subversion of those genres that serve to perpetuate the dominance of Han Chinese men. The varieties of women's calligraphic practices, including the maintenance of silence in the public realm, can be read as "strategic *responses* – more or less successful – to positions of relative powerlessness" (Gal 1991: 182). Clearly, this is not a distinct realm of isolated experience. Rather, such responses require active engagement with and disruption of the synthesis of empire/*guo* and patriline/*jia* that constitutes the modern Chinese nation.

Acknowledgments

Portions of this research were carried out in Taiwan and the People's Republic of China from 1987 to 1990 supported by grants from the Center for Chinese Studies and the Pacific Cultural Foundation. An earlier version of this chapter was presented at the 1996 annual meeting of the American Ethnological Society. I am grateful to Amber Ault, KarenMary Davalos, Nina Dorrance and Linda Sturtz for their lively commentary and insightful critiques on various versions of this chapter. Tamar Mayer provided much-needed patience, persistence and editorial guidance in preparing the chapter for publication. Any errors or omissions remain my own.

Notes

1 My translations of these terms was inspired by Tani Barlow's (1994) essay on the term *funü*. She suggests the significance of the construction of the term *guojia* in a footnote, where she remarks that the term *guojia* "partakes of an older social formation, *guo* (empire), and *jiating*, meaning a contemporary domestic unit that formed in part as a reaction to a *jia* (patriline)" (Barlow 1994: 278).

2 This occurs in classical calligraphy where differences are mediated through a singular version of the past in which the diversity of historical experience is "telescoped into a uniform body of stylistic tradition" (Loehr 1965: 37).

3 See, for example, Chiang Yee's *Chinese Calligraphy*, Kwo Da-Wei's *Chinese Brushwork*, L. Ledderose's *Mi Fu and the Classical Tradition of Chinese Calligraphy* or Shen Fu's *Traces of the Brush*.

4 Several possible exceptions where women might appear in the art historical record of calligraphy include the Tang Dynasty empress, Wu Zi-tian, the Qing Dynasty Empress Dowager Ci Xi and the Yuan Dynasty artist Guan Dao-sheng, wife of the well-known artist Zhao Meng-fu.

5 Women of the third period, characterized as "women of reform," came of age during the early years of Communist reform following the death of Mao. Those among them who began the study of calligraphy, even as children, would just now be reaching the conventional age of artistic maturity (around 40 years old). It is not surprising, then, that the *Zhongguo Dangdai Shufa Daguan* (*Catalogue of Contemporary Calligraphers*) has no "model women" from this period. Published in 1988, the catalogue includes only a very small percentage of calligraphers under 40 (among them, Zhou Huijun, born in 1949) and no instances of calligraphers under the age of 30.

6 Others, particularly artists who produce forms of modern art, maintain that calligraphy has no relevance for the modern nation and they dismiss women's artistic endeavors in this field, along with the efforts of all calligraphers, as useless antiquarianism. They express disdain, rather than reverence, for "tradition," and ultimately fail to recognize how the social production of style in calligraphy embodies, parallels and mimics social production throughout the modern nation.

References cited

Barlow, T. E. (1994) "Theorizing woman: *funü*, *guojia*, *jiating* (Chinese woman, Chinese state, Chinese family)," in A. Zito and T. E. Barlow (eds.) *Body, Subject and Power in China*, Chicago: University of Chicago Press, pp. 253–289.

Chiang Yee (1938) *Chinese Calligraphy: An Introduction to its Aesthetic and Technique*, Cambridge, MA: Harvard University Press.

Chow, R. (1991) "Violence in the other country: China as crisis, spectacle, and woman," in C. Mohanty, A. Russo and L. Torres (eds.) *Third World Women and the Politics of Feminism*, Bloomington: University of Indiana Press, pp. 81–100.

Croll, E. (1995) *Changing Identities of Chinese Women*, London: Zed.

Gal, S. (1991) "Between speech and silence: the problematics of rcsearch on language and gender," in M. di Leonardo (ed.) *Gender at the Crossroads of Knowledge*, Berkeley: University of California Press, pp. 175–203.

Jayawardena, K. (1994) *Feminism and Nationalism in the Third World*, London: Zed.

Kwo, Da-Wei (1981) *Chinese Brushwork: Its History, Aesthetics, and Techniques*, Montclair, N.J.: Allanheld & Schram.

Ledderose, L. (1979) *Mi Fu and the Classical Tradition of Chinese Calligraphy*, Princeton, N.J.: Princeton University Press.

Levenson, J. R. (1965) *Confucian China and its Modern Fate: A Trilogy*, Berkeley: University of California Press.

Loehr, M. (1965) "Art-historical art: one aspect of Ch'ing painting," *Oriental Art* 35–7.

Shen Fu (1980) *Traces of the Brush*, New Haven: Yale University Press.

Shen Peng (1989) "Huhuan Qingxu," in *Zhongguo Dangdai Nu Shufa Jia Zuopin Huicui*, Beijing: Beijing Tiyu Xueyuan Publishing House.

Waltner, A. (1996) "Recent scholarship on Chinese women," *Signs* 21, 21: 410–428.

Wang, P. Y. (1987) *Shufa Quan Neng Dacai Zuopin Zhan* (*Catalogue of Calligraphy Competition Exhibition*), Taipei: Fine Arts Museum.

Weidner, M. *et al.* (1988) *Views from the Jade Terrace: Chinese Women Artists 1300–1912*, Indianapolis: Indianapolis Museum of Art.

Wolf, M. (1972) *Women and the Family in Rural Taiwan*, Palo Alto, CA: Stanford University Press.

Yan Zheng (1988) *Zhongguo Dangdai Shufa Daguan* (*Catalogue of Contemporary Calligraphers*), Beijing: Wenhua Yishu Publishing House.

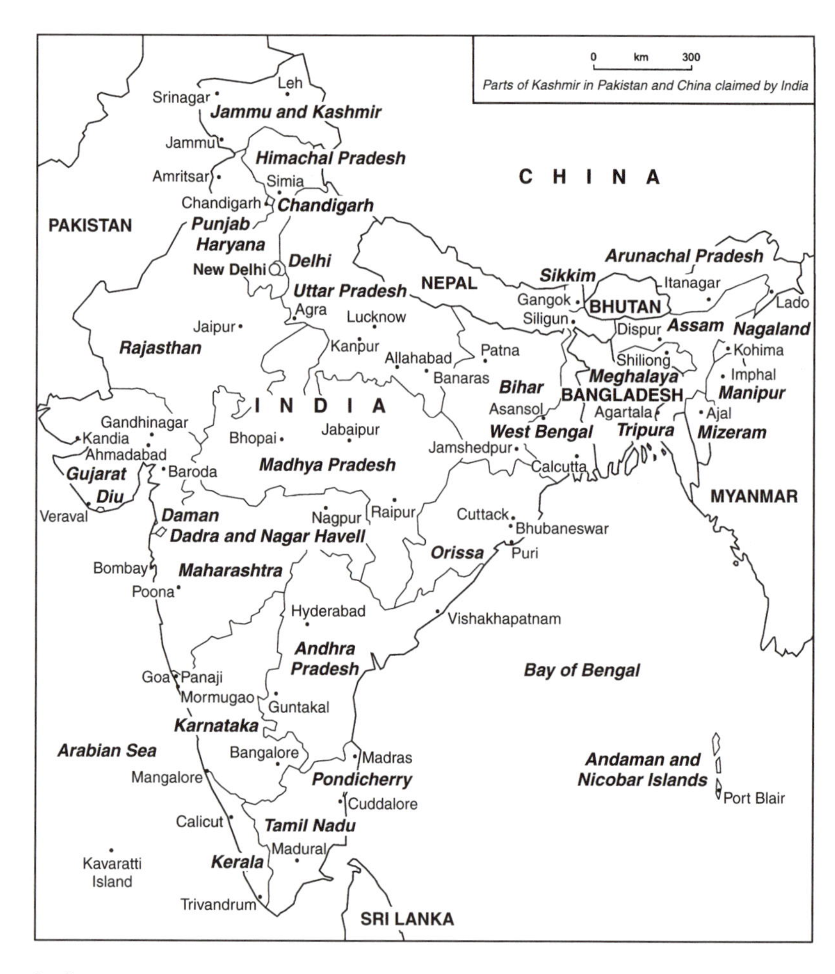
0 km 300
Parts of Kashmir in Pakistan and China claimed by India
Leh
Srinagar
Jammu and Kashmir
Jammu
Himachal Pradesh
Amritsar
Simia
CHINA
Chandigarh
Chandigarh
PAKISTAN
Punjab
Haryana
Delhi
New Delhi
Arunachal Pradesh
Sikkim
Itanagar
Uttar Pradesh
NEPAL
Gangok
BHUTAN
Lado
Agra
Lucknow
Siligun
Jaipur
Dispur
Assam
Nagaland
Rajasthan
Kanpur
Allahabad
Patna
Kohima
Shiliong
Banaras
Meghalaya
Imphal
Bihar
BANGLADESH
Manipur
INDIA
Asansol
Agartala
Ajal
Gandhinagar
Kandia
Bhopai
Jabaipur
West Bengal
Tripura
Mizeram
Ahmadabad
Jamshedpur
Gujarat
Baroda
Madhya Pradesh
Calcutta
Diu
MYANMAR
Veraval
Daman
Nagpur
Raipur
Cuttack
Bhubaneswar
Dadra and Nagar Havell
Orissa
Puri
Bombay
Maharashtra
Poona
Hyderabad
Vishakhapatnam
Andhra
Pradesh
Bay of Bengal
Goa
Panaji
Mormugao
Guntakal
Karnataka
Arabian Sea
Bangalore
Madras
Andaman and
Nicobar Islands
Mangalore
Pondicherry
Port Blair
Cuddalore
Calicut
Tamil Nadu
Madural
Kavaratti
Island
Kerala
Trivandrum
SRI LANKA

India

10 Men's sexuality and women's subordination in Indian nationalisms

Steve Derné

Asserting control over one's own body as a way of rejecting the alien forces of colonialism, secularism and modernity has been an important component of men's nationalism in India. The best-known example is Mahatma Gandhi's celibacy – a self-control that he believed translated directly into both public and private power (Alter 1994: 45). In an influential article, Joseph Alter argues that *brahmacharya* (celibacy) "developed as a strategic concept opposed to Westernization" (1994: 49). Alter's research on contemporary male wrestlers in Banaras and Dehra Dun, India, suggests that for many Indian men, practicing *brahmacharya* is a form of resistance against alien seductions. Yet purifying the body through celibate self-control is only one politics of Indian nationalism which works on the level of the body. Emphasizing sexual potency is another way that Indian nationalists have responded to colonialist discourses that have characterized Indian men as effete. And – perhaps even more important – despite their differences, a politics of celibacy and a politics of sexual potency have both provided nationalist Indian men with a feeling of power by emphasizing sexual control of Indian women.[1]

This chapter grew out of my skepticism – based on my own research with Indian men – about Alter's argument that *brahmacharya* is the fundamental component of Indian nationalism. The wrestlers Alter interviewed in Banaras and Dehra Dun in 1987 emphasized celibacy as the main means of reinvigorating both Hinduism and Indian masculinity. Yet few – only 6 percent – of the Hindu men whom I interviewed in Banaras in 1987 accepted wrestlers' notion that sex weakens the male body (Derné 1994a: 254). Despite having a strong Hindu and Indian identity, most of the men I interviewed in Banaras in 1987 and Dehra Dun in 1991 embraced aspects of modernity like cinema and tea-drinking that wrestlers criticized as intoxicating weaknesses.[2] Like the wrestlers, the men whom I interviewed felt that modernity keeps them from controlling their own lives. But rather than focusing on invigorating their sense of themselves as men and as Hindus through self-control and body-building, they focused instead on enforcing the practices of Hinduism among their wives and daughters. Rather than seeking compensation for their sense of national powerlessness by controlling their own bodies, they pursued a feeling of power by controlling women. Considering both the wrestling culture which has directly influenced the recent resurgence of Hindu nationalism (Kakar 1996, van der Veer 1994) and the contradictory filmgoing culture of other men highlights how a masculine identity that emphasizes dominance over women has become a fundamental component of many men's national identity as Indians. Hence, male power has become central to most forms of contemporary nationalism in India.

By focusing on the concerns of common men (including wrestlers and filmgoers), I try to highlight some of the reasons why nationalisms which center on controlling female sexuality appeal to Hindu men. The chapter draws on interviews about family life that I conducted with Hindu men in Banaras in 1987 (Derné 1995a), as well as on interviews and fieldwork that I conducted with male filmgoers in Dehra Dun in 1991 (Derné 1995b). More than 90 percent of these men embraced cinemagoing and romantic connections with women that wrestlers regard as alien to Indian traditions, but most of these men also complained that city life, Western education and the corrupting influences of modern culture have disrupted Indian families – and Indian national values – by leading women to desire more independence. These concerns show how much men's national identity is embedded within a particular gender regime.[3]

Nationalist revitalizations of Hinduism and Indian masculinity

As Uma Chakravarti (1990 [1989]:35–36) shows, cultural nationalism emerged in India in the 1830s in response to attacks on Hindu civilization made by Christian missionaries and colonial administrators. When the colonial state abolished *sati* (the immolation of a woman on her husband's funeral pyre), Hindu nationalists perceived this as an intrusion into the most sacred sphere of Hindu society – the Hindu family (see also Kosambi 1995). Constructing Hindus as morally inferior by selectively highlighting the ways in which Hinduism oppressed women, colonialists prompted a nationalist response that celebrated Hinduism. A well-known example of this form of nationalism is Gandhi's use of religious language in describing independence as *Ram Rajya* – the rule of the mythological Hindu king, Ram (Chhachhi 1989: 569). As Peter van der Veer points out, the independence movement was ostensibly secular, but the "most important imaginings of the nation" in India have always been religious (1994: 22).

In recent years, Hindu nationalist organizations like the Bharatiya Janta Party (B.J.P.) have flourished because many Hindus believe that Hinduism itself is under assault. Many Hindus were discomfited by 1981 news reports that untouchables in a South Indian village had converted to Islam (van der Veer 1994: 28, Mujahid 1989). Many Hindus were also upset when, in 1985, the Indian government preserved the principle that the Muslim community should have its own personal law for regulating family matters (see Pathak and Rajan 1989).[4] And the refusal of the ruling Congress (I) government to bow to Hindu fundamentalist demands that a disputed mosque in Ayodhya be destroyed and rebuilt as a Hindu temple angered so many Indians that it swept a new coalition into power at the national level in 1990. Capitalizing on the call to rebuild the ancient Hindu temple, the B.J.P. gained power in Uttar Pradesh, India's most populous state, and in the 1996 elections the B.J.P. emerged as the single largest party in Parliament.

Many of the Hindu men whom I interviewed in Banaras were uneasy with the changes that they feel are being brought about by a "Westernized" elite that has embraced "modern" principles (Andersen and Damle 1987: 72). They complained, in particular, that the greed and corruption that they associate with Westernization have poisoned Indian families: brothers are no longer willing to live together in joint families; fathers of sons have become more concerned with dowry than with the purity of potential daughters-in-law; influenced by the West, women want to move about outside the house (Derné in preparation). One 26-year-old husband complained that traditional Hindu arranged marriages are being challenged by *filmwalle* (filmmakers and filmgoers) and "government" officials, like the then prime minister Rajiv Gandhi, whom he described as "trying to mix Indian culture with dirt." A 76-year-old man similarly complained that

> India is becoming more and more like Western places. There is an increasing attraction to the artificial and materialistic [*bhautikavadi*]. No one moves according to the heart [*hradaya*, *dil*] and soul [*atma*] any more. Instead, people move according to what they see in the outer world, whether it is eating, clothing, or lifestyle [*rahansahan*].

Although the tensions that men describe are probably an inherent part of joint-family living, arranged marriages and restrictions on women's movements outside the home (Derné 1995a),[5] many men nonetheless imagine in the past a happier era in which brothers lived harmoniously together, parents found suitable spouses for their children and women happily accepted the restrictions placed on them. In short, many Hindu men continue to feel that the Hindu family is under assault by modern practices and principles.[6] As Ashis Nandy (1980: 113) argues, Hindu fundamentalist groups have been most successful with people in the "semi-modern classes" who are "anxious about their rootlessness and constantly doubting their own authenticity as Indians and Hindus" (see also Chhachhi 1989: 569).

Yet nineteenth- and twentieth-century Indian nationalists were also responding to colonialist challenges to Hindu men's sense of masculinity. Alter (1994: 54–55) describes how colonialist writings constructed Indian men as impotent sexually, politically and militarily. European colonialist writings emphasized Indians' defeat by a dominant power, and portrayed "Indian men, in general, and upper-caste Hindu men in particular [as] effete" (see also Pandian 1995). Chakravarti (1990 [1989]: 47) reports that these

European writings led some Hindu men to feel that they had failed to show sufficient resistance to Muslim, and, later, British invaders, and that this failure indicated the limitations of their manhood (see also Nandy 1983: 10, 25). One of the first responses of Indian nationalists was to celebrate the manly, martial spirit. As Chakravarti reports (1990 [1989]: 48), some nineteenth-century nationalist writers traced the martial spirit back to traditional Kshatriya values,[7] suggesting that "authentic Indianness in the regenerated Hindu . . . lay in" a martial "Kshatriyahood" (see also Nandy 1983: 52, van der Veer 1994: 72, Andersen and Damle 1987: 29). By associating such masculinity with Hinduism's indigenous martial tradition, nationalists aimed at reinvigorating both masculinity and Hinduism.

Today, many men still regard Indian masculinity as threatened by modern influences. Alter reports that wrestlers see the seductions of modernity – music, romance, hairstyling salons and cinema – as undermining the strength and vitality of the masculine Indian body. Wrestlers believe that the scented oils, hair dyes, facial creams and aftershaves provided in the hairstyling salons that surround Indian cinema halls inhibit the "natural, healthy glow" of the wrestler's body (Alter 1992: 240). Seeing "modern Indian cinema" as the most "hedonistically debauched" aspect of "modern life," wrestlers complain that the dancing of scantily clad heroines in Indian film seduces young men away from the self-control that they value, producing an "erotic mood" that makes eyes "lose their brightness and become hollow" (Alter 1992: 242; see also Alter 1994: 49). To some of these wrestlers, male filmgoers even appear feminized because they use hair dyes and facial creams and embrace the seductive appeal of sex that many Indians associate with women (Carstairs 1967 [1957], Raheja and Gold 1994, Bennett 1983).

The heroically "masculine" physique that Indian wrestlers build through exercise, diet, celibacy and other ascetic practices is meant, according to Alter, to combat modern influences by preventing modern seductions from affecting the body (Alter 1994: 46). Wrestlers see the way of life that produces bodies of which they are proud as a "form of protest against self-indulgence and public morality," and they believe that these strong bodies can transform modernity's corruptions: that if more police officers wrestled, the police would not put personal gain before public justice; that a wrestling dairy farmer would never adulterate milk (Alter 1992: 251). They critique the "dystopia of modernity" by imagining a world in which "*akharas* [gymnasiums] replace tea shops and cinema halls, milk replaces tea . . . , earth replaces facial cream, [and] mustard oil replaces scented hair tonic" (Alter 1992: 250). Renouncing the intoxicating temptations of the modern world would "produce a citizen who embodies the essence of national integrity and strength" (Alter 1994: 46). While Alter argues that wrestlers' masculinity is counterposed against a masculinity that emphasizes domination, self-gratification and control of others, even wrestlers try to build strong bodies which could be used to control the world through violence – and they do, after all, hope to defeat other wrestlers in the *akhara*.

This emphasis on reinvigorating the masculine body has also been important in some contemporary forms of religious nationalism in India. Van der Veer (1994: 72–73) connects wrestling practices with the physical training of militant Hindu nationalist groups, like the Rashtriya Swayamsevak Sangh (R.S.S.), a youth organization that has spread throughout North India since its founding in Maharashtra in 1925. The wrestling pit and the R.S.S. training grounds both aim at producing spiritual purity as well as physical strength (van der Veer 1994: 72–73, Andersen and Damle 1987: 35–36). The R.S.S.'s celebration of warriors' weapons aims at reinvigorating Indian masculinity in order to fight off alien intrusions. In contrast to Gandhi's model of non-violence, R.S.S. leaders see independence as requiring a Kshatriya model based on "militancy, vigor and domination." Paramilitary training appears to be an important part of the R.S.S.'s continuing appeal (van der Veer 1994: 71–72, Chhachhi 1989: 569, Andersen and Damle 1987: 29).

While celebrating Hinduism and Indian masculinity, Indian nationalists have sometimes actually transformed traditional notions. Challenged by the colonialist assertion that Hindus were unfit to rule themselves because of the abject status of Indian women, some nationalists emphasized reforming Hinduism to improve the condition of women. One nineteenth-century strategy of cultural nationalism, for example, was to argue that the classical religious texts describe the ideal society as one in which women participate fully in all areas of public life (Chakravarti 1990 [1989]). These arguments blamed Muslim threats as the source of contemporary restrictions on women, and urged that some restrictions be modified.[8] And in the twentieth century, Gandhi urged the rejection of those aspects of classical Hindu texts which could not command the respect of "men who cherish the liberty of woman as [much as] their own and who regard her as the mother of the race" (Gandhi 1970: 85, as cited by Chatterjee 1989: 627). While Gandhi was an *aficionado* of Tulsi Das's medieval Hindi rendering of the Hindu epic *Ramayana*, he nonetheless rejected the famous couplet which justifies violence against women (and lower-caste people): "the drum, the fool, the s[h]udra [lower-caste], the animal and the woman – all these need beating" (Patel 1988: 380), and he criticized existing Hindu customs, like child marriage, *sati*, dowry and *purdah* (seclusion) as preventing women from developing themselves (Patel 1988). Gandhi included women in the nationalist struggle, suggesting that by wearing and producing Indian-made clothes women could help make the *swadeshi* (self-reliance) movement a success (Patel 1988: 380).[9]

Still, India's cultural independence movement offered a very limited liberation for women. Describing the nineteenth-century nationalist movement, Chakravarti (1990 [1989]: 53) argues that only the crisis of colonialism "justified the delinking of wifehood from the 'enclosed' space of domesticity." While nineteenth-century nationalists praised women for work in the independence struggle, they still criticized women's wage work (Chakravarti 1990 [1989]: 64). And in the twentieth-century independence movement, Gandhi offered women a role in politics, but still embraced the dominant

Hindu focus restricting women to the home (Patel 1988: 380). By urging women to take a *swadeshi* vow, Gandhi "introduced nationalist politics into the household without breaking the domestic space which defines women" (Patel 1988: 380). In fact, as I show below, Indian nationalisms aimed at reforming Indian patriarchy usually ended up reinventing it.

Just as the colonialist argument that Hinduism worked against women prompted reform in Hinduism, so colonial discourse portraying Hindu men as effete prompted a reformed masculinity. Seeing the might of the British Empire, Gandhi recognized that independence could not be attained by trying to beat the British at their own game through masculine violence (Nandy 1980: 87). Gandhi insisted instead on a militant *non*-violence as the route to independence. By urging men to be non-violent, Gandhi "was trying to fight colonialism by fighting the psychological equation which a patriarchy makes between masculinity and aggressive social dominance and between femininity and subjugation" (Nandy 1980: 74).

Part of Gandhi's efforts at overcoming masculine violence centered on celibacy. Gandhi believed that supernatural powers could be achieved through disciplining the body. Following his wife's death Gandhi himself "tried to increase his powers by taking naked young women to bed with him in order to test his detachment" (van der Veer 1994: 97). Gandhi believed that women, as well as men, could achieve a higher moral and spiritual role by rejecting sexuality, reproduction, and family life (Patel 1988).

Gandhi's emphasis on "denying the sex instinct" explains his fascination with Hindu widows' renunciation of sex (Patel 1988: 384). God, Gandhi said, "created nothing finer than the Hindu widow" (Gandhi 1927: 141, cited by Patel 1988: 384). Yet, if Gandhi's praise elevated their status, it nonetheless affirmed the limitations Hinduism places on Hindu widows' opportunities to participate fully in family life.

Alter's description of the popular literature on *brahmacharya* suggests that ascetic control of sexual desires continues to be an important way for some Hindu men to reclaim masculinity that many feel has been threatened by colonial domination.[10] To advocates of *brahmacharya*, the proliferation of sex manuals, the overflowing sexuality of film heroes and the alarming rate of population growth shows that "modern life" is excessively "preoccupied with intercourse and erotic stimulation" (Alter 1994: 58). *Brahmacharis* critique modern "libertines" as seeing masculinity in terms of "domination, self-gratification, and the control of others" (Alter 1994: 58). The celibate male instead defines himself by the "sex he does not have, and by the semen that he does not allow to flow" – and thus resists the "legacy of colonial sexuality" which defined virility in terms of the ability to spend semen (Alter 1994: 56, 58).

Nationalism and women's subordination

The nationalism and masculinity that emphasizes men's celibacy and martial strength in India is not purely self-referential. While men may gain a feeling of power by controlling their sexual desires or building strong bodies capable of defeating outsiders, this nationalism has nonetheless also focused on women's place in the Indian nation. Many Indian men have, for example, framed nationalism in terms of protecting women (and especially women's sexuality) from assaults by foreigners. As Partha Chatterjee (1989: 630) shows, many male nationalists have identified feminine self-sacrifice, devotion and religiosity as a "sign for nation," and have justified their own departures from Hindu tradition by highlighting women's continued adherence to it (see also Whitehead 1995). Nationalist Hindu men's sexual self-control, moreover, has sometimes aimed at producing strong male bodies that would later produce healthy Hindu children. These linked notions suggest that many nationalists have celebrated the Hindu nation by emphasizing women's adherence to a Hindu tradition that has kept women in the home – despite nationalist reforms. Indeed, their emphasis on men's control of women in the home may be an important source of the appeal of Indian nationalisms for male followers.

While Gandhi and others saw celibacy and sexual self-control as an important part of Indian nationalism, the emphasis on controlling women's sexuality by limiting their contact with outsiders has been just as important a part of nationalist discourses about sexuality. As van der Veer argues, "nationalist discourse connects the control over the female body with the honor of the nation" (1994: 113). Activists in the independence movement used the India-as-mother simile to emphasize the importance of protecting the Indian nation from rapacious outsiders (Bose 1996, Vasudevan 1995: 97, Whitehead 1995). Beginning in the nineteenth century, cultural nationalists emphasized that young men needed to defend women from molestation by British soldiers, and nationalist opposition to British efforts to outlaw child marriages was driven by men's fear that the British were trying to control Indian women's sexuality (Chakravarti 1990 [1989]).

Today's religious nationalists continue to invest the protection of women from lecherous outsiders with a simultaneously nationalistic and masculine charge. Amrita Chhachhi shows how Hindu nationalist "propaganda is full of the fall from greatness in the past, challenge of foreign domination today, the need to prove strength, courage and manliness. What better way to prove manliness than by showing that women are under your control?" (1989: 575). As Chhachhi points out, the speeches of Hindu communalists repeatedly refer to the "violations of our sisters and mothers" and exhort men to "take revenge and prove that the men of that particular community are still men" (Chhachhi 1989: 575; see also van der Veer 1994: 104).

In India, the nationalist emphasis on punishing outsiders who threaten women has been coupled with a focus on protecting women by restricting them to the home. Thus, at the same time as nineteenth-century nationalists valorized women's participation in the independence movement they remained critical of women who work in the labor market (Chakravarti 1990 [1989]: 64). In the twentieth-century independence movement, Gandhi criticized traditional Hinduism for treating women as "playthings" with no creativity; but he himself nonetheless limited women's "creativity" to their roles as wives and mothers (Patel 1988: 378). Gandhi saw a role for women in the independence movement, but at the same time insisted that women had no role in the workplace:

> There ought to be no need for women . . . to go to work . . . for the sake of an extra income of [a few] rupees. . . . [Women] have plenty of work in their own homes. They should attend to bringing up their children. They may give peace to the husband when he returns home tiredly, minister to him, soothe him if angry and do any other work they can staying at home.
>
> (Gandhi 1920: 47–51, cited by Patel 1988: 380)

Gandhi warned members of a trade-union movement against sending "your women out to work. Protect their honor, if you have any manliness in you. It is for you to see that no one casts an evil eye on them" (Gandhi 1920: 47–51, cited by Patel 1988: 380). He saw the *swadeshi* (self-reliance) movement not only as "a means of protecting India's wealth" but also as a way of protecting women's honor by encouraging them to produce goods within the confines of the home (Gandhi 1920: 326, cited by Patel 1988: 381).

Chhachhi (1989: 572) notes how today's Hindu fundamentalists also advocate that "women return to the home and give up their jobs in favor of unemployed men." As Mira Chatterjee similarly puts it, "these communalists' prescription for us is that we stay at home and lose ourselves in our husbands, children and religious texts. To do otherwise is to be a 'witch', a fallen woman and an insult to manhood."[11]

The formulation that the nation's identity rests with Indian women's domesticity continued to restrict women to traditional (subservient) roles, while simultaneously freeing men to embrace modern values (Chatterjee 1989, 1990 [1989], Chakravarti 1990 [1989], Patel 1988). By emphasizing the home as the "principal site for expressing the spiritual quality of the national culture," nationalists constructed women in the home as the primary carriers of the nation's identity (Chatterjee 1989: 626–627, 1990 [1989]: 234). For Gandhi, women's non-violence and self-sufficiency made them "the repository of the spiritual and moral goodness of society," and protecting women's honor became for him a fundamental part of the independence movement (Patel 1988: 383). Nationalists were preoccupied with assuring that women maintain family cohesion and observe traditional food taboos and religious rituals – even as Indian men who participated in the modern economy and modern politics embraced new dress and food habits and modified religious observances (Chatterjee 1989, 1990 [1989]). Thus, as Partha Chatterjee (ibid.) shows, the nationalist focus on the Hindu home as the defining characteristic of Indian identity made room for nationalist men to embrace Western science, technology and economic organization.

Today, India's religious nationalism continues to construct women as "symbols of culture and tradition" (Chhachhi 1989: 575). Religious fundamentalist calls to return to Indian culture and tradition are addressed to women, rather than men. Fundamentalists urge women to reject modern garb in favor of traditional *saris* and *salwar kamizes* – but do not demand that men stop wearing suits and ties (Chhachhi 1989: 575). Thus, shifting the site of national identity to the family links national identity to women's subordination in their homes. As van der Veer argues, the honor of the nation has come to depend on "modesty, and submissiveness of the female body" (1994: 85).

My research on male filmgoers in Dehra Dun, India, suggests that many men continue to emphasize the importance of women's adherence to traditional cultural norms as a way of bolstering their own sense of Indian identity – even as men themselves embrace important aspects of Western culture. The most popular Indian films since the late 1970s contrast traditionally modest Indian heroines with "Westernized" Indian women who are represented as immodest and overly forward. In the big hit *Maine Pyar Kiya* (*I Fell in Love*),[12] for example, the heroine is Suman, a village girl who wears *saris* and *salwar kamizes* and is devoted to her father. Suman's modesty, sincere feelings, religiosity, devotion and non-violence attract the hero, and are highlighted by the contrast that the film draws with an antiheroine who smokes, wears Western clothes and boldly tries to win the hero's affections by wearing a short red dress, sporting a short hair cut and spreading out on his bed. Yet while *Maine Pyar Kiya* and other films denigrate women like this antiheroine who seems to embrace too many Western values, the heroes of these films are often young men who are quite Westernized. Prem, the hero of *Maine Pyar Kiya*, studies abroad, smokes cigarettes, wears jeans and leather jackets, and decorates his room with posters of Western icons like Madonna, Sylvester Stallone and Michael Jackson. Prem falls in love with the traditionally modest Suman, but shows his own modernity by marrying her against the wishes of his father, who hoped to instead marry Prem to the daughter of a business associate. Like men in the independence movement, Prem's movement toward modernity does not threaten his Indian identity because of the continued adherence of women like Suman to the traditional gender regime.

Indian film journalists describe the appeal of *Maine Pyar Kiya* and other recent Indian hits by emphasizing the modesty of female characters like Suman. One writer for a political mainstream magazine, for example, believes that "what seems to have really got to people is [*Maine Pyar Kiya*'s] advocacy of a particularly Indian virtue moth-balled for a long time: sharam (coyness)" (Jain 1990: 65).[13] This writer describes the "key scene" as the one in which Prem gives Suman a skimpy dress to wear. "At this point, the camera goes behind the girl as she draws open the cloth. The camera sees nothing except her covered back and [Prem] in a state of rapture. But he quickly gathers himself and drapes the cloth around her. Modesty restored"

(Jain 1990: 65). *Maine Pyar Kiya* celebrates traditional *salwar kamiz*es and *saris* which protect the body from men's gaze, and denigrates skimpy miniskirts and bikinis which do not.

The celebration of modest Indian women is equally apparent in Indian actresses' preoccupation with defining themselves in opposition to modern women who expose their bodies. In Indian film fan-magazines, actresses like Bhagyashree (who played *Maine Pyar Kiya*'s Suman) repeatedly emphasize their own modest images:

> Today, I am in a different position because I have a different image. When producers sign me, they want something different. Bhagyashree is completely different from those who wear miniskirts.
>
> (Tandan 1991: 19)

In nearly all of her film-magazine interviews, Bhagyashree reiterates this same point:

> My image is very honorable [*sammanapurn*] and dignified [*garimaya*]. I would never think of becoming a sex symbol. Therefore, any producer who wants me to wear a miniskirt is wasting his time [*bekar*].
>
> (*Chitralekha* 1991: 36)

Moreover, Bhagyashree claims that her on-screen modesty is real, and emphasizes her adherence to traditional Indian forms of greeting:

> I am a simple girl. Will you believe I had never gone to a discotheque until I got married and the first time I visited one was with [my husband]? To me "namaste" comes more naturally than shaking hands. In fact, my two younger sisters are more modern than me.
>
> (Venkatesh 1991: 24)

Bhagyashree's husband, too, draws a judgmental contrast between tradition and modernity, praising women who do not need to be instructed in the traditional respect for elders:

> [Bhagyashree] has always been a traditional person. She still is. For example, if *mausi* [mother's sister] comes here, Bhagyashree and I touch her feet. She does not need any kind of prompting.
>
> (Venkatesh 1991: 25)

Other Indian actresses who are vying to move up the ladder of success are similarly anxious to portray themselves as modest. Manisha Koirala quit a film in order to protect her "royal image [English phrase, transliterated into Hindi]," because she was "not prepared to expose herself to the public's view [*ang paradarshan wale drshya*]" (*Filmi Kaliyan* 1991a: 32). Chandni, similarly, emphasizes that she is "not prepared to expose for any price" (*Filmi Kaliyan* 1991b: 14). Her refusal to show "leg or cleavage," she says, is motivated by her desire to spare her family embarrassment: "My grandfather, my father, and the rest of the family should be able to sit together in a theater and watch my film without cringing. I am very possessive about my family and will not do anything to upset them" (Meena 1991: 69).

Fan-magazine writers in India praise heroines who act appropriately Indian by protecting their modesty, and they question actresses about any racy scenes they have appeared in, and columnists in film-magazines rail against actresses who expose their bodies for the camera. The following criticism of one actress who exposes "too much" is typical:

> What Shilpa Shirodkar did in . . . *Trinetra* was shocking to say the least. One enlightened collegian was on the brink of losing her cool and said: "What does she think she is Madonna or Samantha Fox . . . ? Is she come here to show her talent or to shamelessly display her body?"
>
> (Ali 1991: 10)

Significantly, in making the comparison with Madonna and Samantha Fox, the columnist identifies the heroine who "crosses all limits of decency" (Ali 1991: 10) with the shameless West.

The Indian film world's valorization of modest Indian women is shared by the male filmgoers I interviewed in Dehra Dun in 1991, all of whom praised Suman's traditionally Indian modesty – even as male filmgoers themselves embraced aspects of modernity, like romance, Western dress and cinemagoing. One unmarried 22-year-old college student who was wearing jeans and a t-shirt with an English slogan complained nevertheless of Western influences on Indian women:

> According to the Indian culture, [women] should wear *salwar kamizes*. Wearing jeans and smoking cigarettes are things of a different country and are separate.

Some men praised Suman by contrasting her willingness to do the family's housework with the independence of modern women like the antiheroine. An 18-year-old high-school student said, for instance, that he "like[s] the role taken by Suman because she serves her family in a good way. She is one who understands [*samajhdar*]." A 20-year-old man who likes "women who wear *salwar kamizes*" said that "a modern woman would not be able to do the good work of serving the family [*ghar*]."

Men's taste in film heroines is similarly consistent with their preference for modest, traditional Indian women. All of the men I interviewed who identified heroines they disliked named those, like Kimi Katkar, who are famous for "exposing." An unmarried 19-year-old man complained that some heroines "wear bikinis in one film after another"; an unmarried 24-year-old criticized actresses who are "openly used to attract the audience for a reason other than their acting." Of the heroines whom men mention as their favorites – Madhuri, Sridevi, Rekha, Smita Patel, Dimple, Bhagyashree and Juhi Chawla – none has a reputation for exposing her body.

While the male filmgoers I interviewed praised women who embrace traditional dress and family roles, these men often embraced for themselves practices that male wrestlers would criticize as based on alien modern influences. Filmgoers enjoy the very movies that wrestlers regard as the most debauched aspect of modern life. They patronize the hairstyling salons that wrestlers see as purveyors of facial creams and scented lotions that weaken the body. They enjoy "promenading in public with no other purpose than to show off – a practice that wrestlers also criticize" (Alter 1992: 242). While emphasizing that women should wear traditional Indian clothes, male filmgoers themselves wear "modern" Western clothing. While wrestlers believe that romance undermines the strength and vitality of masculine Indian bodies, more than 80 percent of the filmgoing men I interviewed like romantic films.

In sum, male Indian filmgoers continue to equate women's religiosity, traditional attire and maintenance of the home with Indianness. Even as Indian men have bureaucratic jobs in the modern economy, enjoy the romance of Indian films and wear Western clothes, they also feel a strong connection to Indian tradition because of subservient, *sari*-wearing women in the home.

Moreover, rather than embracing celibacy, many Indian men emphasize the importance of reproducing another pure generation of Hindus. Their commitment to restricting women to the home, to assure that they fulfill household duties and to protect women from sexual relationships with outsiders, is combined with a commitment to harness women's sexuality in the service of Hindu men. As Chakravarti (1990 [1989]: 60) argues, "one form of nationalist resolution of women's sexuality" has been to "use her biological potential for child bearing in the service of the physical regeneration of what was seen as a now weakened" race. Indeed, the nineteenth-century nationalist reformer Dayananda suggested that the practice of *brahmacharya* perfected "reproductive elements" so that the children born of a man who had practiced *brahmacharya* are of "a very superior order" (cited by Chakravarti 1990 [1989]: 58). Alter (1992: 246–247) similarly reports that wrestlers believe that by practicing *brahmacharya* they will build strong bodies that can later "produce strong, healthy children." Even the emphasis on building strong bodies through celibacy can, then, be linked to reinvigorating the nation through sexual intercourse with Hindu women. In some recent forms of Hindu nationalism, the appeal to have more children is explicit. Many Hindus believe that Muslims' fertility rate is higher than that of Hindus, and Hindu nationalist organizations like the V.H.P. fallaciously assert that Muslims will outnumber Hindus in less than a quarter century (Patwardhan 1995, van der Veer 1994: 27, Chhachhi 1989:

575–576). These myths work to pressure women to produce more children. Patwardhan's (1995) documentary film of the Hindu fundamentalist movement *Father, Son and Holy War* shows leaders praying that each Hindu woman will have eight sons and that each of these sons will in turn have eight sons. As a result of such pressures, "women increasingly lose control over their bodies" (Chhachhi 1989: 576).[14]

Indeed, while wrestlers embrace celibacy, most of the male filmgoers whom I have interviewed embrace sexual relations as part of their Indianness (Derné 1995a: 90–93). One 47-year-old Kshatriya who migrated to Banaras from Sind (Pakistan) as a child during partition called the sort of sexual limitations advocated by wrestlers "unfathomable":

> I take enjoyment with my wife. In the morning, we kiss and love [*prem karna*; *mohabbat karna*] [each other] until I go to [work]. Her best quality is making love [English phrase.] She never refuses. Whenever I want, I just say it and she makes herself ready to please me. If I don't receive love [*pyar*] the whole day, then life is useless, isn't it?

This husband connects these sexual energies to his Indianness by linking his sexual appetite to his Kshatriyaness (while defining himself in opposition to Brahmans whose vegetarianism, he says, limits their sexual desires). By pointing to Kshatriya traditions, this man rejects the notion that unrestrained sexuality is foreign to Hindu culture. He, moreover, continues to embrace traditional restrictions on women's movements outside the home. He says that when "a woman takes even one step outside the home, her mind will go outside, too. But if she remains in the house, then she remains controlled." This husband wants sexual connections with his wife and control over her – and this is the promise that contemporary nationalism offers.

Nationalist emphasis on women's religiosity, service in the home and sexual service to the nationalist cause is clearly appealing to men who want to maintain patriarchal privileges against the assaults of modernity. The nationalist focus on women's reproductive contributions to national strength legitimizes the sexual relations most men want, just as the nationalist call to protect women's honor insures women's restrictions to the home. Such appeals may be particularly important because many men fear that modern influences are weakening male control in the home.

The Hindu men I have interviewed often emphasize that an Indian wife's willingness to "obey" her husband is both her most ideal and her most "Indian" quality (Derné 1995a: 22–23, 1994b, 1994c). One married 30-year-old man whom I interviewed in Banaras in 1987 described shyness [*sharm*] and shame [*lajja*] as the "most important ornaments of Indian women." A married 27-year-old man whom I interviewed in 1987 said that India's "most wonderful tradition" is that a woman's shyness [*sharm*] will prevent her from complaining about anything in her husband's house – even if she has real difficulties. "In Hindustan," he says, "shyness [*sharm*] and deference [*lihaj*] are women's status [*man*] and prestige [*maryada*]." Like nationalist activists, the men I have interviewed associate Indianness with women's adherence to traditional gender roles.

The colonial state's emphasis on the need to improve women's status appeared to nineteenth-century nationalists to be an intrusion into the sacred sphere of the home. Contemporary men, similarly, believe that alien influences of modernity are a threat to women's willingness to obey their husbands (Calman 1992: 12, Mankekar 1993: 485, Derné in preparation). Phoolchand, a married 28-year-old man whom I interviewed in Banaras in 1987, described how his wife's labors in the home provide him with many of the pleasures of life:

> My wife remains at home doing all kinds of service, adorning the house, the door, and herself. She does household works so completely that when I come home and see the beauty I become very happy.

Phoolchand contrasted his happy situation with that of a man whose wife "has gone outside to earn":

> There is daily quarrel. The husband comes into the house and knows that his wife has gone outside to earn. The tired husband makes his own tea. He does everything by his own hands. He even has to make his own tea! From this, his mental condition starts deteriorating and every form of corruption is born in the mind.

Phoolchand fears that modern influences may put more and more husbands in this situation. He believes that modern influences make women want to move around (*ghumna*), attend parties and become independent. He laments that "progressive [*pragatishil*] ways of thinking [*vichardhara*]" have even led some women to work outside the home. According to Phoolchand, such work is useless (*vyarth*) because it spoils (*bigar*) and corrupts (*vikrit*) her family. Phoolchand emphasizes how women's wage work jeopardizes their children's upbringing by describing difficulties of finding a maid servant who is full of motherliness (*matrtav*). He indicts the woman who "wants to move ahead standing shoulder to shoulder with men" as forgetting her obligation to her son who "needs his mother 24 hours a day." For men like Phoolchand, nationalism may be appealing because it contests the forces that he believes are a threat to the gender regime that benefits men.

Fathers-in-law similarly complain that education which has been influenced by foreign principles is making women unwilling to do the domestic work needed in the family (Derné 1994c: 92–96). A 76-year-old man said, for instance, that "if [the daughter-in-law] is educated then it is difficult for her to get along [*patna*] in the family." Another 45-year-old grandfather similarly complained of the difficulties that an educated daughter-in-law would pose: "I get up early in the morning and recite the *Ramayana*. How would it be if the girl got up and started singing English poems?" This man fears that education will lead women to embrace foreign cultural traditions like singing English poems, neglecting religious observances like chanting the *Ramayana*. He believes that "modesty [*shilta*]" and the "feeling [*bhavna*] of adjusting in the family" are the most important qualities of a daughter-in-law and fears that an educated woman might neglect housework by failing to "rise before her mother-in-law and successfully complete all of the arrangements for the house." For these fathers-in-law, nationalism may be appealing because it reasserts the gender order in the home that they value.

A fundamental appeal of nationalism may thus be its focus on women's adherence to restrictive household roles as a defining feature of Indian identity and the Indian nation. Men who value women's domestic service and feel that such service is being disrupted by modern forces like education may be attracted to nationalist discourses that make women's subservience in the home a distinctive feature of Indianness. A central "foreign" threat that makes nationalism appealing is the perceived challenge to male privilege.

Nineteenth- and twentieth-century nationalists who felt threatened by a colonialist discourse that constructed Indian men as effete emphasized their masculinity. Contemporary men's attraction to strong male bodies seems to similarly work to bolster a sense of Indianness, masculinity and male dominance. Like militant religious nationalists (like those in the Rashtriya Swayamsevak Sangh), wrestlers emphasize the discipline that builds strong bodies. While wrestlers criticize filmgoing men's tea-drinking and romantic desires as weakening their bodies, filmgoing men nonetheless also celebrate male strength. Films show Indian heroes' strong bodies as providing the power to beat up the villains. Film stars, fan-magazines and filmgoers all comment on male actors' strong bodies. Fan-magazines repeatedly reprint stills that show the broad chest of the hero of *Maine Pyar Kiya*. While heroines' reputations are ruined if they display too much of their bodies, heroes are often willing – even obligated – to show male power. As one film hero told a fan-magazine, "if you have a great body which can be displayed, then I don't think there's anything wrong in flaunting it" (Jain 1991: 29). Nearly half of the young men whom I interviewed in 1991 said that the ability to fight and dance is what makes a hero their favorite – and that scenes of heroes fighting and dancing allow actors to display their strong bodies. Several of the filmgoers I interviewed commented on how heroes who fight and dance well display excellent bodies. One 22-year-old man said, for instance, that Sunny Deol is his favorite actor "because his acting is the best, and his body is also the best. When he dances, his body comes out. His body is the best." A 17-year-old described the actor Anil Kapoor as his favorite hero "because he is manly [English phrase]." Despite their different stances toward film, tea-drinking and men's hairstyling salons, both wrestlers and filmgoers are attracted to cultural practices that celebrate strong male bodies – and the masculinity that such strength represents.

As it did for earlier nationalists, the image of Indian men as powerful provides today's men with a feeling of Indianness and masculinity. But the male power that such strength provides may be appealing as well. As feminists like Michael Messner argue, men often maintain power over women by elevating the "male-body-as-superior through the use (or threat) of violence" (Messner 1992: 15; see also Nelson 1994, Bryson 1987). Indian films' displays of male bodies are often associated with male violence. The rape scenes that are a standard part of most Indian films (Khanzada 1991) portray controlling male bodies and controlled female bodies. In *Maine Pyar Kiya*, for instance, the hero has to protect the heroine against a rape attempt by the antiheroine's brother. More disturbing, films often depict heroes' violence against the heroines they love. In *Maine Pyar Kiya*, the hero's body is celebrated in a punching-bag workout, stills of which are often reprinted on posters and in fan-magazines. Emphasizing how the violence could be directed against the modest heroine, the filmmaker shows the heroine's head snapping back (and wincing with pain) each time the hero punches the bag – even as the heroine is simultaneously telling the hero that she loves him.[15] While Alter (1992: 132) suggests that wrestlers avoid the questions of male dominance by assiduously avoiding the company of women, some of the men I have interviewed do recognize male force as an important component of male control. One 28-year-old husband told me, for instance, that the reason a woman cannot compete with a man in business is that "the whole society has the desire to rape the woman." For at least some men, moreover, a willingness to use violence against women is a distinctively Indian characteristic. Beatrix Pfleiderer interviewed a man from a village outside Banaras who emphasized male violence as the source of Indian marriages' superiority to Western marriages: "There are two kinds of culture in India": the Western type "where people get 24 divorces in 24 hours" and the Indian type, which is so different that it "cannot be compared with" the "foreign" culture. In England, he claimed, the dispute between a husband and wife would be solved with "a divorce, but here the work [is] accomplished with a slap" (Pfleiderer 1985: 67–68).

Like the focus on women's adherence to restrictive household roles, the focus of nationalist activists, wrestlers and filmgoers on building strong bodies invigorates a sense not just of Indianness and masculinity, but of male dominance of women as well.

Conclusion

While nineteenth-century Indian nationalists focused on reforming Hinduism to improve women's position, the independence movement ended up inventing new ideologies of male dominance that identify women with the traditional and men with the modern, women with the home and men with the outside world. As Chatterjee argues, nationalism advocated a "new patriarchy" that "conferred upon women the honor" of new social responsibilities, while binding "them to a new, and yet . . . legitimate, subordination" (1989: 629). Nationalists focused on women's household labor as vital to the *swadeshi* movement. They identified women's reproductive and childrearing roles as vital to regenerating the Hindu nation. And they identified women in the domestic sphere as the fundamental basis of men's Hindu identity. These identifications continue to be salient in India today, bolstered by diverse cultural forms that include popular Indian films.

Of course, limiting women to the home has long been a cornerstone of male dominance in India. Restrictions on women make it difficult for women to participate in public, political and economic life. Even when women have jobs outside the home, restrictions on their movements limit their opportunities: government officers refuse to send women to district jobs, preventing them from getting experience which is vital for promotion (Liddle and Joshi 1986: 134). Male bosses will not allow female engineers to do work on site (Liddle and Joshi 1986: 136–137) and refuse female accountants permission to work late, do out-of-town audits or work in dangerous areas of the city, thus also cutting short the routes by which they might earn job promotions.

Even worse, while Gandhi and others saw the household as the arena in which women are "protected" (Patel 1988: 380), contemporary Indian feminists recognize the domestic sphere as central to women's oppression, especially insofar as the home constitutes the site of husbands' violence against their wives (Gandhi and Shah 1992, Calman 1992). By making the domestic the realm of authentic Hinduism, nationalists have contributed to the taboo against intervening in the family sphere – the very arena in which women are most oppressed.

This chapter identifies important differences among various forms of Indian nationalism. While men's control of their own sexual urges through *brahmacharya* and other ascetic practices is one way Indian men have sought to reinvigorate masculinity, other men see sexual relations with modest, home-loving women as the main indicator of manliness (see Nandy 1980: 86). While R.S.S. members embrace the martial spirit in order to counter their sense of emasculation by modern forces, followers of Gandhi advocate non-violence. While wrestlers bolster their sense of Indianness and masculinity by rejecting modern seductions like cinema halls, male filmgoers find a sense of Indianness and masculinity in modest film heroines who protect Indianness by maintaining family life, wearing traditional clothes and observing traditional religious practices – even as these men embrace for themselves many modern practices.

Despite these differences, diverse forms of Indian nationalism share a few common themes. First, virtually all Indian nationalists focus on controlling bodies, both male and female. Male wrestlers and followers of Gandhi focus on controlling their own sexual desires. Rashtriya Swayamsevak Sangh members focus on building strong bodies that can militarily defeat alien intruders. Both sorts of nationalists emphasize protecting women's bodies from these outsiders by controlling women's sexuality. Nationalism is a response to alien intrusions that appear to threaten the social body (Andersen and Damle 1987: 72). But as a discourse grows which emphasizes these threats, many Indians come to feel that their individual bodies are threatened as well. Thus, control of bodies (and especially, sexuality) is one important form that nationalism takes.

Second, most male nationalists in India focus on protecting their identity as men, usually emphasizing an oppressive Hinduism that reinforces women's subordinate position. Both wrestlers and filmgoers seem concerned with being rendered helpless by modern bureaucracies (Alter 1992: 244–245; Derné in preparation: ch. 2). But neither filmgoers nor wrestlers confront bureaucracy directly. Rather, they bolster their own sense of Indianness and freedom by emphasizing their adherence to particular standards of masculinity. Wrestlers feel Indian through their participation in physical disciplines that build the masculine bodies they desire. Male filmgoers strengthen their sense of Indianness by defining themselves in opposition to men who are seduced by "modern" women, like the antiheroine of *Maine Pyar Kiya*.

Third, male nationalists handle anxieties about modernity by emphasizing a gender identity which reconstitutes traditional differences between men and women. Male nationalists emphasize the maintenance of women's restriction to the household as fundamental to protecting Hindu, Indian and male identity. Filmgoers' focus on women's traditional household roles and wrestlers' focus on the strong male bodies similarly construct differences between men and women that bolster male dominance in India today. Men's identity as men and Indian, then, is based on the continued subordination of Indian women.

Acknowledgments

The published work of Joseph Alter, Uma Chakravarti, Partha Chatterjee, Amrita Chaachhi, Ashis Nandy, Sujata Patel and Peter van der Veer provides the descriptions of nationalism that I rely on in this chapter. My 1987 interviews with Hindu men in Banaras were funded by a Fulbright-Hays grant from the U.S. Department of Education. My 1991 fieldwork with male filmgoers in Dehra Dun was funded by a senior fellowship from the American Institute of Indian Studies. Nagendra Gandhi, Parvez Khan, A. Ramchandra Pandit and Awadesh Kumar Mishra provided research assistance in Banaras. Narender Sethi and Vimal Thakur provided research assistance in Dehra Dun. Portions of this chapter were presented at the meetings of the American Anthropological Association (Atlanta, GA, 1994) and the Eastern Sociological Society (Boston, MA, 1993). Joseph Alter, Tamar Mayer and Allen Shelton made useful comments that improved the chapter.

Notes

1 My focus is on men's nationalism. Women were involved in India's independence struggle, but their influence was limited. As Partha Chatterjee (1989: 632) argues, nationalist discourse has been men's "discourse about women: women do not speak." More recent forms of religious nationalism have been overwhelmingly male dominated (Chhachhi 1989: 569, van der Veer 1994: 71). While women participated in nationalist discourse (Chakravarti 1990 [1989]), the concerns of the male leadership have most strongly shaped nationalism in India.

2 About 90 percent of the upper-caste, upper-middle-class Hindu men whom I interviewed in Banaras in 1987 at least occasionally watched Hindi films in cinema halls.

3 In describing men's concern with what they perceive to be "modern" influences, I am not asserting that their descriptions of change are accurate, but am simply arguing that "modernity" has become an important indigenous concept men use in their imagination of an era in which women happily accepted restricted roles in the family.

4 India lacks a uniform civil code. Rather, each religious community has its own personal law governing family matters. In 1985, a divorced Muslim woman successfully petitioned the court for maintenance from her husband under the criminal code. When Muslim groups rallied around the perceived threat to Islam, the ruling Congress (I) party bowed to their demands and passed an act reinstating the supremacy of Muslim personal law in family matters, like divorce. In response Hindu fundamentalist groups, including the Vishva Hindu Parishad (V.H.P.), Rashtriya Swayamsevak Sangh (R.S.S.) and Shiv Sena, complained that Muslim personal law threatened India's Hindu identity.

5 Older ethnographies (Luschinsky 1962) and recent analyses of nineteenth-century folk theater (Hansen 1992) and women's long-standing oral traditions (Raheja and Gold 1994) indicate that such tensions are, in fact, an inherent part of Indian family practices.

6 The men I interviewed blamed modern influences for family tensions that are, in fact, an inherent part of Indian family structures. Many authors note that Indian men describe modern forces as the source of family tensions (Bhattacharjee 1992: 30–31, Calman 1992: 12, Chakravarti 1990 [1989], Chatterjee 1989: 630, Mankekar 1993: 484–485, Gandhi and Shah 1992: 59, Chhachhi 1989: 568).

7 This represents a decision to emphasize the values of the Kshatriya (warrior) caste, rather than the Brahmanical (priestly) caste; this also emphasizes the martial tradition as well as the long history of Hindu civilization.

8 Such an explanation continues to be salient: several of the Hindu men whom I interviewed in Banaras in 1987 asserted that protecting Hindu women from Muslim invaders was the reason for restrictions on women's movements outside the home.

9 Gandhi emphasized using Indian-made goods to end reliance on colonial suppliers.

10 Alter (1994: 49) suggests that this popular literature is "widely read" by teenage boys and young men "who would be the primary consumers of films" and pornography.

11 Mira Chatterjee, "Religion, secularism and organising women workers," presented at the 1986 Indian Association of Women's Studies National Conference in Chandigarh (India), cited by Chhachhi (1989: 572).

12 Between 1975 and 1993, *Maine Pyar Kiya* was one of two films to gross 30 million rupees. By July 1991, the December 1989 release had grossed five times that amount (Venkatesh 1991: 24). More than 40 percent of the men I interviewed in 1991 named *Maine Pyar Kiya* as their favorite film 18 months after its release, and one young man told me he had seen the film 27 times. In

the summer of 1991, it was still fashionable to wear belts and shirts bearing the words "Maine Pyar Kiya." Posters of Bhagyashree, the traditionally modest heroine of *Maine Pyar Kiya*, were more available in Dehra Dun markets than the also common posters of Samantha Fox, a Western sex symbol who has become popular with some Indian men.

13 Coyness is the writer's translation. *Sharm* also connotes shyness and modesty.

14 Consistent with Foucault (1978: 25–26, 139), a concern with population is linked to discourses about sexuality.

15 In films like *Karma* (1986) and *Trinetra* (1991), the heroine's love is prompted by the hero's forcible kisses.

References cited

Ali (1991) "Ali's notes," *Screen* [Bombay] August 9: 10.

Alter, J. S. (1992) *The Wrestler's Body: Identity and Ideology in North India*, Berkeley: University of California Press.

—— (1994) "Celibacy, sexuality, and the transformation of gender into nationalism in India," *Journal of Asian Studies* 53, 1: 45–66.

Andersen, W. K. and S. D. Damle (1987) *The Brotherhood in Saffron: The Rashtriya Swayamsevak Sangh and Hindu Revivalism*, Boulder, CO: Westview.

Bennett, L. (1983) *Dangerous Wives, Sacred Sisters: Social and Symbolic Roles of High-Caste Women in Nepal*, New York: Columbia University Press.

Bhattacharjee, A. (1992) "The habit of ex-nomination: nation, woman, and the Indian immigrant bourgeoisie," *Public Culture* 5, 1: 19–44.

Bose, P. (1996) "Engendering the armed struggle: women, writing, and the Bengali 'terrorist' movement," *Genders* 23: 145–183.

Bryson, L. (1987) "Sport and the maintenance of masculine hegemony," *Women's Studies International Forum* 10, 4: 349–60.

Calman, L. (1992) *Toward Empowerment: Women and Movement Politics in India*, Boulder, CO: Westview.

Carstairs, G. M. (1967 [1957]) *The Twice Born: A Study of a Community of High-Caste Hindus*, Bloomington: Indiana University Press.

Chakravarti, U. (1990 [1989]) "Whatever happened to the Vedic *Dasi*? Orientalism, nationalism and a script for the past," in K. Sangari and S. Vaid (eds.) *Recasting Women: Essays in Indian Colonial History*, New Brunswick, N.J.: Rutgers University Press [originally published by Kali for Women, New Delhi, 1989], pp. 27–87.

Chatterjee, P. (1989) "Colonialism, nationalism, and colonized women: the contest in India," *American Ethnologist* 16, 4: 622–633.

—— (1990 [1989]) "The nationalist resolution of the women's question," in K. Sangari and S. Vaid (ed.) *Recasting Women: Essays in Indian Colonial History*, New Brunswick, N.J.: Rutgers University Press [originally published by Kali for Women, New Delhi, 1989], pp. 233–253.

Chhachhi, A. (1989) "The state, religious fundamentalism and women: trends in South Asia," *Economic and Political Weekly* [Bombay] March 18: 567–578.

Chitralekha (1991) "Bhagyashri Ab Age Kahin Nahin?" August: 35–36.

Derné, S. (1994a) "Violating the Hindu norm of husband–wife avoidance," *Journal of Comparative Family Studies* 25, 2: 249–267.

—— (1994b) "Hindu men talk about controlling women: cultural ideas as a tool of the powerful," *Sociological Perspectives* 37, 2: 203–227.

—— (1994c) "Arranging marriages: how fathers' concerns limit women's educational achievements," in C. Mukhopadhyay and S. Seymour (eds.) *Women, Education, and Family Structure in India*, Boulder, CO: Westview, pp. 83–101.

—— (1995a) *Culture in Action: Family Life, Emotion, and Male Dominance in Banaras, India*, Albany, N.Y.: SUNY Press.

—— (1995b) "Popular culture and emotional experiences: rituals of filmgoing and the reception of emotion culture," in C. Ellis and M. Flaherty (eds.) *Social Perspectives on Emotion*, vol. 3, Greenwich, CT: JAI Press, pp. 171–197.

—— (in preparation) *Movies, Masculinity, and "Modernity": An Ethnography of Men's Filmgoing in India.*

Filmi Kaliyan (1991a) "Manisha Koirala," August: 31–34.

—— (1991b) "Filmi Samachar," August: 12–14.

Foucault, M. (1978) *The History of Sexuality, Volume 1: An Introduction*, New York: Random House.

Gandhi, M. K. (1920) *Collected Works*, vol. 17, New Delhi: Publications Division.

—— (1927) *Collected Works*, vol. 67, New Delhi: Publications Division.

—— (1970) *Collected Works*, vol. 64, New Delhi: Publications Division.

Gandhi, N. and N. Shah (1992) *The Issues at Stake: Theory and Practice in the Contemporary Women's Movement in India*, New Delhi: Kali for Women.

Hansen, K. (1992) *Grounds for Play: The Nautanki Theatre of North India*, Berkeley: University of California Press.

Jain, M. (1990) "Cinema: return to romance," *India Today* May 15: 62–69.

—— (1991) "Cinema turns sexy: films become increasingly raunchy, ribald and explicit," *India Today* November 15: 28–34

Kakar, S. (1996) *The Colors of Violence: Cultural Identities, Religion and Conflict*, Chicago: University of Chicago Press.

Khanzada, F. (1991) "Roopesh Kumar: a long night's journey . . . ," *Screen* [Bombay] July 5.

Kosambi, M. (1995) "Gender reform and competing state controls over women: the Rakhmabai case (1884–1888)," *Contributions to Indian Sociology* (n.s.) 29: 265–290.

Liddle, J. and R. Joshi (1986) *Daughters of Independence: Gender, Caste and Class in India*, New Brunswick, N.J.: Rutgers University Press.

Luschinsky, M. S. (1962) "The life of women in a village of north India: a study of role and status," Ph.D. dissertation, Cornell University. Ann Arbor, MI: University Microfilms.

Mankekar, P. (1993) "Television tales and a woman's rage: a nationalist recasting of Draupadi's 'disrobing'," *Public Culture* 5: 469–492.

Meena (1991) "CB spotlight," *Cineblitz International* 3, 6: 67–69.

Messner, M. A. (1992) *Power at Play: Sports and the Problem of Masculinity*, Boston, MA: Beacon Press.

Mujahid, A. M. (1989) *Conversion to Islam: Untouchables' Strategy for Protest in India*, Chambersburg, PA: Anima Books.

Nandy, A. (1980) *At the Edge of Psychology: Essays in Politics and Culture*, Delhi: Oxford University Press.

—— (1983) *The Intimate Enemy: Loss and Recovery of Self Under Colonialism*, Delhi: Oxford University Press.

Nelson, M. B. (1994) *The Stronger Women Get, the More Men Love Football: Sexism and the American Culture of Sports*, New York: Harcourt Brace.

Pandian, M. S. S. (1995) "Gendered negotiations: hunting and colonialism in the late nineteenth century Nilgiris," *Contributions to Indian Sociology* (n.s.) 29: 239–264.

Patel, S. (1988) "Construction and reconstruction of women in Gandhi," *Economic and Political Weekly* [Bombay] February 20: 377–387.

Pathak, Z. and R. S. Rajan (1989) "Shahbano," *Signs: Journal of Women in Culture and Society* 14: 558–582.

Patwardhan, A. (director) (1995) *Father, Son and Holy War*, First Run/Icarus Films.

Pfleiderer, B. (1985) "Rural audience reactions," in B. Pfleiderer and L. Lutze (eds.) *The Hindi Film: Agent and Re-Agent of Cultural Change*, New Delhi: Manohar, pp. 58–78.

Raheja, G. G. and A. G. Gold (1994) *Listen to the Heron's Words: Reimagining Gender and Kinship in North India*, Berkeley: University of California Press.

Tandan, C. (1991) "Bhagyashri," *Mayapuri* August 11: 19.

van der Veer, P. (1994) *Religious Nationalism: Hindus and Muslims in India*, Berkeley: University of California Press.

Vasudevan, R. S. (1995) "'You cannot live in society – and ignore it': nationhood and female modernity in *Andaz*," *Contributions to Indian Sociology* (n.s.) 29: 83–108.

Venkatesh, J. (1991) "Himalay-Bhagyashree: partners in a plot," July 9: 22–26.

Whitehead, J. (1995) "Modernising the motherhood archetype: public health models and the child marriage restraint act of 1929," *Contributions to Indian Sociology* (n.s.) 29: 187–210.

The Caribbean

11 Nationalism and Caribbean masculinity

Linden Lewis

The concept of nationalism has generated renewed interest in recent years, particularly because a number of issues which were previously assumed to be settled firmly within the national imagination are being contested in various ways. Many countries continue to struggle over how best to address these issues. The ideological underpinnings of nationalism are being challenged by subjects who were long thought to have been interpellated into the nationalist project. In short, as civil society becomes more crisis-ridden, and as the pace of global production and technological advancement quickens, old forms of hegemony based on certain national mythologies give way, under the pressure of new searches for collective and individual identities.

Anthony Smith (1991: 14) has defined the nation as "a named human population sharing an historic territory, common myths and historical memories, a mass, public culture, a common economy and common legal rights and duties for all members." Nationalism is an important complement to the nation, facilitating colonial conquest and subjugation. It also serves as the ideological vehicle through which the nation attempts to homogenize dissimilar social elements into the nationalist project. Indeed, there is no singular conception of the nation. Each nation is constructed in accordance with the priorities which it deems important in the process of unifying disparate social forces. As used in this chapter, "nationalism" refers to the ideological practices articulated by the state and reinforced by other social institutions. These practices are designed to reproduce the many and diverse human groups as a common community of individuals. Nationalism is also a phenomenon which manifests itself, at both the political and cultural levels, as a movement for the creation of an autonomous, sovereign space and unique identity.

Historical forces have left their mark on the nations of the region at the economic, political, cultural and even psychological levels. Caribbean conceptions of nation remain bound up with the historical processes of slavery, colonialism and imperialism which shaped them. Hence, the nationalism of the English-speaking Caribbean is different from that of the Dutch and Spanish Caribbean, and the French Caribbean, namely Martinique, Guadeloupe and French Guyana. This chapter will focus mainly on the Anglophone Caribbean or what was formerly known as the British West Indies.

The terrain of nation formation is quite familiar to most students of nationalism and of the struggles for self-determination in the Caribbean and in other parts of the so-called third world. An increasingly important dimension of this discourse on the nation is the relationship between gender and the struggle for self-determination which has been so largely masculinist in orientation. Though considerable work has begun on this issue at the global level (see Walby 1996, McClintock 1995, Anthias and Yuval-Davis 1993, Smith 1991, Mosse 1985, *inter alia*), the specifically gendered terms of the nationalist project remain largely underdeveloped in the literature on the Caribbean.

One may ask the question why – even though the nationalist project is so blatantly framed in masculinist terms – this feature has remained for the most part ignored and untheorized in the Caribbean. The answer to this query is perhaps to be found in the hegemony of the masculinist articulation of the project of self-determination. In the first place, men were largely in control of framing the nationalist project; and even though some of their language clearly indicated the gender orientation of the project, they did not perceive the issue of self-determination as an essentially male project but instead as intended to benefit everyone who would later constitute the nation. In short, the masculinist orientation was normalized as a general process of struggle rather than a specific struggle, by one group of men to wrestle control from another, more powerful group of men. In the case of the Anglophone Caribbean, the assumption of leadership by men in the nationalist project appeared to both men and women as a natural evolution, given the relation of men to power, access to resources and privilege.

Second, given the hegemonic ideology of masculinity, it was not necessary for men to articulate their project as a male-centered one, even though from time to time they may have referred to their own suitability for leadership at this level. Privileged ideologies are seldom obliged to articulate the basis of their power; it is only in the discourse of resistance that the need arises to delineate the anguish of marginalization associated with lack of access to apparatuses of power. Mosse (1985) has certainly demonstrated that there is generally a convergence in the relationship between nationalism and the interests of men. Moreover, this convergence manifests itself as a particular type of hegemonic masculinity associated with conquest, control and the consolidation of power by privileged men.

Feminist discourse has also been able to provide some clarity about the gendered nature of the national imaginary. Feminism has unmasked ways in which masculine aspirations, vulnerabilities, imaginings, contradictions and fears have been inscribed on the nationalist project. Anne McClintock (1995), for example, notes that

> all nationalisms are gendered, all are invented and all are dangerous. . . . Not only are the needs of the nation here identified with the frustra-

> tions and aspirations of men, but the representation of male national power depends on the prior construction of gender difference.
>
> (McClintock 1995: 261–262)

McClintock also argues that irrespective of the ideological appeal to people for popular unity, the nation has historically been constructed as a site for the institutionalization of gender differences. According to this argument, men and women have differential access to the protective rights and to the resources of the state (see McClintock 1995: 353). One should be careful to indicate, however, that though gender always seems to underlie nationalism, there are other social forces such as religion, ethnicity, race and class, operating individually or simultaneously, which may preponderate in determining nationalism's form. In turn, these social forces may themselves be mediated by considerations of gender.

What is quite evident is that while women in many parts of the world have been present in these struggles for self-determination, they have not been equally incorporated into the national project. This is as true for Puerto Rico as it is in the Anglophone Caribbean. As Roy-Fequiere (1997: 122) argues, Puerto Rican women intellectuals made important contributions to the nationalist debate but were never fully recognized. What is also clear from the experience of the Caribbean is that irrespective of their contributions to the nationalist struggle, women have not reaped rewards commensurate with their efforts. The argument that, given their occupations, some women could have supported the nationalist cause only in an indirect way, or could not have taken some of the key decision-making positions, clearly could not have applied to all women who actively participated in the nationalist project. One has therefore to concur with Sylvia Walby when she opines:

> Sometimes women may support a different national project from that of men. There is a struggle to define what constitutes the national project, and women are, typically, heard from less than men in this. Thus gender relations are important in determining what is constituted as the national project. Where the national project includes women's interests, then women are more likely to support it.
>
> (Walby 1996: 245)

Framing the nationalist project

For men and women in the Anglophone Caribbean, the nationalist project involved a dynamic process of mass mobilization in support of the ideals of political autonomy for the territories of the British West Indies. Nationalism became a signifier of the political maturity and the readiness of the peoples of the Caribbean to administer their own affairs.

The bulk of this leadership was made up of men from the emerging indigenous elite, as well as working-class men who had established popular bases in the fledgling trade union movement. The leadership of this nationalist project articulated a vision of the Caribbean based on a notion of self-determination, first in the form of a federation. This federation was intended to unite the territories of the British West Indies into a single political unit, administered by a regional government with a federal parliament. It was generally agreed, by both the British colonial administrators as well as by Caribbean nationalists, that these territories were too small and not sufficiently viable economically to survive as separate autonomous units at the time. Second, the nationalist project had as its goal the political break, if not complete separation, from the colonial authorities as the territories became more politically and economically developed. Colonialism was seen as a fetter on the forces of development and progress, and therefore had to be abandoned. Colonialism was also seen as a hindrance to the development of a West Indian identity and to the fostering of notions of self-worth and self-affirmation. In this regard, the nationalist project of the people of the former British West Indies was quite typical of such movements in other parts of the world. What was unique was of course the specific historical forces which coalesced around this project. This nationalist project essentially crystallized between the closing years of the decade of the 1920s to the decade of the 1960s, by which time many of these islands had become independent. One can, however, begin to see some early indications of the development of a nationalist consciousness in the polemic of J. J. Thomas (1889), of whom more will be said subsequently.

The struggle for self-determination therefore extends from the return to the Caribbean of the men of the British West Indies Regiment, to the first set of reverberations of the labor revolt in Belize in 1934. Along the way, this struggle encompasses the philosophy associated with the teachings of Marcus Garvey and Black Nationalism, as well as the emergence of an indigenous elite and the spread of the labor riots from St. Kitts to Guyana. All these events combined to politicize and to foster the development of a sense of national consciousness among the middle-class and working-class strata of the region. In short, a number of social forces coalesced in the making of the nationalist project in the Caribbean.

The economic conditions of the region

By the late 1920s and early 1930s, the Caribbean had a faster growing population than the economy was able to service adequately. The impact of the Great Economic Depression on the Caribbean was significant. It was during this period that the sugar industry collapsed. The political system at the time was characterized by authoritarian colonial rule with a deeply ensconced powerful white community which embraced pre-emancipation world views and attitudes. At another level, poor working and living conditions for the majority of the people, along with an inadequate educational system, exerted some political pressure on the emerging national bourgeoisie for some improvement of their conditions of existence. In short, a combination of the increasing contradictions of colonial rule and the deteriorating conditions under which the masses lived, led to the emergence of a nationalist movement which represented a challenge to the colonial order. In summarizing the conditions obtaining in Trinidad, Daniel Guérin employed sexist language typical of the period to describe the situation which applied to both men and women: "The Trinidad worker became aware of his dignity as a man long before his comrades living elsewhere in the Caribbean" (Guérin 1961: 129).

In the interstices of these economic conditions of the English-speaking Caribbean there emerged an indigenous middle class, with aspirations for political leadership and motivated to take over from the colonial administrators. Though this middle class included some important women, it was dominated by educated, professional men. Most of these men viewed themselves as the natural heirs of the colonial mantle of leadership. This indigenous middle class was not without its contradictions, containing both a radical and a conservative faction. Moreover in contesting the colonial order, members of this class tended to stop short of problematizing the gender order. This point will be further addressed in a subsequent section.

National consciousness and Black Nationalism

An important development in the evolution of the nationalist project in the Anglophone Caribbean was the convergence of Black Nationalist ideology and nationalism in its broader sense. Careful examination of such important early contributions as J. J. Thomas's *Froudacity*, which was originally published in 1889 and which represented a systematic rebuttal to James Anthony Froude's *The English in the West Indies* (1888), revealed a strong link between a growing black consciousness and an emerging national pride, for Froude's diatribe against the Caribbean people revolved around the notion that they were intellectually incapable of self-rule.

> The real offence of Froude's book, in the light of the record, was not that it was full of English Negrophobia in extremis, but that it gave open and unashamed expression to prejudices that the English official mind, with typical hypocrisy, preferred to keep sub rosa.
>
> (G. Lewis 1968: 108)

Denis Benn (1987: 60) argues that: "Froudacity remains an important intellectual contribution as one of the earliest and clearest challenges to the ideological validations of colonial rule." Indeed, Thomas's rebuttal not only disputes Froude's Eurocentric assumptions about Caribbean people but also boldly affirms both a racial and national consciousness throughout the text. A similar conflation of race and national consciousness can also be seen in the work of Theopilus Scholes. According to Howe (1994: 30), the Jamaican

Theophilus Scholes also engaged in a major rebuttal to E. A. Freeman, who maintained that Blacks were incapable of becoming fully civilized.

Garvey, Black Nationalism and trade unionism

In the English-speaking Caribbean around the turn of the century, this early link between Black Nationalism and national resistance was solidified through the preaching, teachings and activism of Marcus Garvey, the internationally known leader of the Black Nationalist and "Back to Africa" movement. Globally Garvey's movement articulated the merger of racial pride and self-determination for people of African descent. Though the Garvey movement seemed to have had a fairly decent record with respect to the status of women in the United Negro Improvement Association (U.N.I.A.), it still fundamentally concerned itself with validating Black masculinity, in defiance of attempts by the White establishment to infantilize and marginalize Black men.

Garvey was also an important link between the nationalist movement and the labor movement in the Caribbean. Both movements sought to reclaim the psyche of the "Black man," one through racial uplift and the other through the struggle for better working and living conditions in the 1930s. Both movements also helped to provide the impetus for nationalist resistance. Given the movement's emphasis on racial pride and self-determination, it is no surprise that followers of Garvey in the Caribbean would play a significant role in the mobilization of workers and would establish important contacts with the major labor leaders of the day.

In some countries in the Caribbean, labor organizing was facilitated through the assistance of Garveyites. One of the leading Garvey scholars, Tony Martin (1986: 53), notes that Garvey was either in contact with or had influenced such important Caribbean labor leaders as Captain A. A. Cipriani, A. Bain Alves, D. Hamilton Jackson and Hubert Harrison. These men were well known in the region for their views on racial uplift for people of African descent, on labor organizing and on the struggle for self-determination. Martin (1986: 54) has demonstrated, further, that the leadership ranks of the Trinidad Workingmen's Association, one of the earliest forms of union organizations in that country, dovetailed with the leadership of U.N.I.A.

One is also able to discern the convergence of interests between the two social movements – Garveyism and labor – in Phillip Sherlock's comments, made in reference to the Caribbean labor leader A. A. Cipriani:

> he [Cipriani] pressed home three demands, protection for the working man through a strong trade union movement, a claim to self-government based on the capacity of the West Indians to run their own affairs, and racial harmony.
>
> (Sherlock 1973: 285)

Considering that Cipriani was a White Trinidadian, this remark meant that racial unity would have meant some expectation on his part of improved status for Blacks. In fact, Cipriani's record in this regard was quite remarkable for his time. He was commonly hailed as working in defense of the "barefoot man." His exposure of the treatment of Black West Indian soldiers in World War I is also worthy of note and won him much respect (see Martin 1986, Joseph 1971) and Cipriani himself served in the British West Indies Regiment and was stationed at Taranto, Italy, with the Fifth Reserve (Joseph 1971: 118). Cipriani embodied this ideological troika of racial uplift, nationalism and trade union organizing. His support for these causes therefore endeared him to the Garveyites operating in Trinidad at the time.

Trade unionism and nationalism

The history of the trade union movement in the Caribbean is very well documented (see Hart 1988, Post 1981, Henry 1979, Manley 1975, A. Lewis 1977, G. Lewis 1968, Mark n.d.). Most important for this chapter, however, is the relationship between trade unionism and colonial resistance in the Caribbean.

Men seized the opportunity provided by the trade union movement in the Caribbean for social advancement and political leadership. In Jamaica, for example, the first categories of work to be unionized were those among the skilled trades – trades entirely dominated by men (see Hart 1988). In other parts of the Caribbean, similar patterns emerged. It is therefore no surprise that men became the leaders of the trade union movement despite the presence of women, and despite women's contribution to these struggles. Men were as a result well placed to seize the opportunity for leadership roles made possible through union mobilization. The trade union movement served as an important base of power: a number of Caribbean political parties emerged from trade unions, while some trade unions were established by political parties in the region.

The trade union struggle was therefore at the heart of the nationalist project, a project which had already established a decidedly gendered bias, as evident from the leadership patterns, hierarchical arrangements and influence of the movement. It was union men and staunch nationalists such as T. A. Marryshow of Grenada, Arthur Andrew Cipriani, Uriah Tubal Butler and Albert Gomes of Trinidad, Norman Manley and Alexander Bustamante of Jamaica, Grantley Adams of Barbados, Robert Bradshaw of St. Kitts-Nevis-Anguilla and Hubert Nathaniel Critchlow of Guyana, whose names would later become closely associated with the process of decolonization, federation and independence. Early trade union mobilization was the integument within which masculinity resided, shaping and profoundly influencing the character, direction and scope of the nationalist imagination.

The men of the British West Indies Regiment

Another factor which should be considered as an important strand in the nationalist project in the Anglophone Caribbean is the treatment, experiences and return of the men of the British West Indies Regiment (B.W.I.R.). Approximately 15,204 West Indians were deployed for military duties in Europe, Africa and the Middle East (Howe 1994: 27). What is remarkable to note here is that despite the British War Office's initial rejection of the idea of establishing a contingent of West Indian men, some of these men arrived in England at their own expense, ready to do battle on behalf of the Crown. The fact that these colonial subjects were not conscripted but actually volunteered to do battle suggests some misguided allegiance. The action of these West Indians is consistent with a universal mythology of masculinity which valorizes gallantry and warriorhood.

Many of these veterans of World War I were not involved in any combat action but were assigned to non-combatant roles in labor battalions. They were assigned to cleaning dirty linen and toilets, they worked on the quays in construction, they were ammunition carriers, they unloaded stores and loaded lighters for ships (Howe 1994, Joseph 1971: 113). Their physical stamina, their ability to fight, their intelligence – all these features of the readiness of West Indian soldiers for combat were seriously questioned by the British military establishment (Howe 1994, Joseph 1971). This treatment was clearly discriminatory, for British soldiers were never subject to such acts of devalorization. West Indian soldiers had been promised equal treatment upon enlistment, but Howe notes that West Indian soldiers, who were mostly Black, were subject "to racial abuse and neglect in the areas of housing, food and health-care which contributed to much sickness and death among them" (1994: 55).

> When Major Thursfield protested to Brigadier-General Carey Bernard, the South African camp commandant, he was told that he (Bernard) had no intention of treating West Indians like British troops, that they were only niggers and were better fed and treated than any nigger had a right to expect.
>
> (Joseph 1971: 119)

Thursfield's comment goes to the heart of the issue of race and British mistreatment of the West Indian soldier, and is dramatically echoed by the following incident cited by Bolland:

> Samuel Haynes, who had been a corporal in the BWIR, wrote to the *Belize Independent* in 1919, complaining of their treatment by British soldiers in Egypt in 1916. Arriving hungry and tired, the Belizean soldiers were singing "Rule Britannia" when British soldiers demanded, "who gave you niggers authority to sing that? Clear out of this building, only British troops admitted here."
>
> (Bolland 1995: 29)

The above incident not only demonstrates blatant racism but also raises a number of issues related to national identity. First, the Belizean soldiers above seemed so preoccupied with their colonial identity that they expressed their allegiance to England by singing the most patriotic of British songs. Second, the incident illustrated the extent to which race may mediate the formation of national identity, as Blackness was clearly disavowed in this context. Third, the British soldiers demonstrated that there existed a significant difference between the status of colonial subjects and that of full citizenship. The account above revealed all the contradictions of incorporating colonial subjects into the British nationalist project while denying them full rights and protections of citizenship.

The practices of the British military establishment basically denied West Indian men the opportunity to engage in one of the more enduring rituals of masculinity, the displaying of symbols of power and military conquest, the drama and the destruction of war. British hegemonic masculinity sought instead to subordinate Caribbean Black men, restricting their membership in the symbolic universe of the "tried, tested and true" notion of masculinity and clearly demonstrating the extent to which race mediated the construction of masculinity in the Caribbean. In short, the British exploited the colonial status of Caribbean men, to invent new vulnerabilities and to create doubts over their claims to manhood.

In addition, some West Indian men must have internalized colonial ideas about bourgeois democracy without interrogating them or situating themselves culturally, politically or racially in relation to those ideas. Had they been more critical of colonial bourgeois notions of democracy, perhaps they could have spared themselves the humiliation and indignity of these experiences. Note for example the contradiction and political naïvety of a protagonist of nationalism, T. Albert Marryshow, as he describes the benefits of fighting for the "Motherland":

> the grand spectacle has been left for us, West Indians, most of whom are the descendants of slaves, fighting for human liberty with the immediate sons of the Motherland in Europe's classic fields of war.
>
> . . . it is a question of dutiful service to the King and Empire; a question of establishing for all time the good name of the West Indies and West Indians.
>
> (cited in Sheppard 1987: 10)

Instead of validating the ideals of masculine honor and reputation, the war experience of members of the British West Indies Regiment resulted in disillusionment and alienation.

Yet the largely negative experience of these Caribbean men also led to a secretly formed cell within the B.W.I.R. called the Caribbean League (see Bolland 1995: 28), made up of men whose racial consciousness had been heightened by such developments. Bolland notes:

> They demanded that "the black man should have freedom to govern himself" and declared that they were prepared to use force to that end, and to strike for higher wages after demobilisation.
>
> (Bolland 1995: 28)

Here, the resonance with the Black Nationalist philosophy of Marcus Garvey is evident. Like Garvey before them, these men had concluded that to improve their social and economic conditions they needed to have some degree of political control over their destiny, and that one way of achieving this control was through labor organization. In fact, Martin argues that some of the returning members of the B.W.I.R. became members of the Garvey movement and were believed to have brought back with them and to have distributed copies of Garvey's newspaper *The Negro World* (Martin 1986: 56).

In summary, therefore, the nationalist project in the Anglophone Caribbean was largely formulated in the crucible of projects of colonial resistance, be they the result of trade union mobilization, intellectual rebuttals of English condescension toward the abilities of Caribbean people, experiences in the British West Indies Regiment or involvement in radical race conscious movements. The masculinist character of the nationalist movement in the Anglophone Caribbean was not, however, merely the result of the predominance of men, even though this was an important condition for the movement's success. More important, was the extent to which the nationalist project manifested and reproduced male forms of domination – institutional, political and social – which essentially formed a continuum from the colonial times to the construction of a post-colonial society. Put differently: having seized the opportunities for leadership provided through nationalist struggles, men then proceeded to consolidate their positions in power at every level of society. That women in the Caribbean did not occupy such positions until relatively recently suggests their systematic exclusion from participation, despite their willingness to do so.

Nationalism, masculinity and ambivalence

Masculinity and nationalism merged in the Caribbean, at an important intersection, with race as well. Not only was there a struggle against colonial conditions but also one against metropolitan racial stereotyping. The Caribbean individual was seen by the colonial establishment as unfit intellectually to handle the challenges of self-determination. Late-nineteenth- and early-twentieth-century British opinions of Caribbean individuals are characterized by racist condescension (see James 1992). One gets a sense of these metropolitan attitudes from C. L. R. James's summary of prevalent views of the English during this period. He notes that the colonial authorities at various times found Caribbean people to be savage, uncivilized, criminal, unintelligent, child-like and woefully lacking in the requisite number of men "of calibre necessary for administering his own affairs" (James 1992: 50).

James argues that the Englishman who came to the Caribbean was frequently surprised by the level of sophistication, education, dress and manner of West Indians, and had to resort to other ways of affirming his superiority:

> Men have to justify themselves, and he falls heavily back on the "ability of the Anglo-Saxon to govern," "the trusteeship of the mother country until such time" (always in the distant future) "as these colonies can stand by themselves."
>
> (James 1992: 52)

James's perception of the English in the Caribbean is corroborated by the work of Gordon Lewis (1968). Lewis argued further, that the English rationalized their contempt for Caribbean individuals in the following terms:

> The "blacks" were utterly incapable of self-government; if they came to rule the island [Jamaica] not one white man would remain; it would become a "lawless republic" like Haiti. What the island needs is a solid population of "solid white men," not the migratory English adventurer who looks for a fortune in ten years and then a suburban home near London. Labour difficulties would disappear under the rule of such a new settler type, for the "black man" is very much what his white employer makes him.
>
> (G. Lewis 1968: 170)

Given these racist perceptions, the protagonists of the Caribbean nationalist movement felt the need to affirm their status as adult men, as mature and responsible enough to conduct their own affairs (see Sheppard 1987, Gomes 1974, Caribbean Labour Congress 1945). In welcoming delegates to the Caribbean Labour Congress (C.L.C.) conference in 1945, Grantley Adams, the head of the Barbados Workers' Union, adopted a tone of approval-seeking that also reflects that aspect of colonial socialization which stressed respect for imperial authority. He was moved to say:

> We in the West Indies are asking the British Government to feel that if, in the same way, they extend to us the fullest measure of self government we will prove ourselves to be worthy of it. (Cheers)
>
> (Official Report of Caribbean Labour Congress Conference 1945: 2)

Referring to Grantley Adams, T. A. Marryshow similarly reminded delegates of the C.L.C. conference in Jamaica in 1947, in a presidential address, that Adams had expressed the desire to tell the Colonial Office: "Thanks for your nursing; but we are now of age" (C.L.C. 1945: 7). This discourse – with its metaphors for nurturing and maturity – underscores the way in which

Anglophone Caribbean nationalist leaders continued to define themselves against the stereotypical colonial representation of their manhood.

Thus an important factor in understanding the Anglophone Caribbean nationalist project was the power, force and legacy of colonial socialization on the men who were in the vanguard of colonial resistance. Understanding the internalization of this colonial socialization helps to explain some of the contradictions within the process of resistance and the nationalist project that emerged. J. M. Lee has argued that the men who were recruited from Britain for service in the colonies operated on the assumption "that they themselves could only be replaced by an indigenous elite which had enjoyed a political education similar to their own" (Lee 1967: 1). Since, as Cynthia Enloe (1989) argues, British leaders worked to ensure that appropriate forms of masculinity were internalized to reproduce the empire, whether through sport or the Boys' Scout movement. The efficacy of the colonial socialization of Anglo-Caribbean men is evident in the pervasiveness of European gender ideologies throughout the social institutions. The main contradiction at this level revolved around the desire to be politically autonomous while at the same time maintaining fidelity to the colonial establishment. In this regard, Charles Moskos (1967) mused:

> The West Indies had been under British rule for so long, and exposed to Britain's cultural penetration to such an extent, that even such militant early nationalists as Cipriani frequently coupled their demands for political reform with sincere expressions of fealty to the Crown. There was little hatred of the British in the West Indies; indeed, many of the educated colored [*sic*] colonials acquired the air, manner, and accents of cultivated British gentlemen – and so they were.
>
> (Moskos 1967: 50–51)

In analyzing A. A. Cipriani's contribution to the nationalist struggles, Gordon Lewis emphasized Cipriani's ultimate belief in the justice of the British Empire:

> It seemed at times indeed that he was not so much against colonialism as such but merely unenlightened colonialism as his remark in 1933 to a visitor that "if the Mother Country would send us her best brains to govern us we should have nothing to say; but she does not" would appear to indicate.
>
> (G. Lewis 1968: 207)

Cipriani was not alone in his devotion to the Crown. Indeed, in a remarkable moment of self-reflection, if not self-denigration, Albert Gomes, the vice-president of the Federated Workers' Trade Union and member of the Caribbean Labour Congress, opined:

> There's a Briton lurking somewhere in every West Indian and his influence is hardly diminished even when the tendency is to disavow him.
> (Gomes 1974: 229)

The fact that these Caribbean nationalists did not recognize contradiction within the positions they held may indeed provide some insight into the nature of the colonial challenge itself. The apparently divided loyalties inherent in the Caribbean nationalist project may have been a strategy of assuring the colonial authorities that the nationalist challenge was not threatening or particularly disruptive. In short, the contradictory positions of some nationalists may have sent an early signal that the nationalist project was essentially reformist and not revolutionary.

T. A. Marryshow, the Grenadian trade unionist and journalist, himself a great admirer of the British family, also expressed these conflicting views about the British Empire. At the 1945 C.L.C. conference in Barbados he proclaimed, embracing all the values of responsible citizenship:

> We are loyalists. We may have our own method of approach to things. But we do feel that we want to make all the West Indian people a respected and self-respecting people, making conditions such that we in the West Indies will hold and control our own Government and run our own way of life. We believe that in doing so we will be making a solid contribution to the stability of the British Commonwealth, (Loud Cheers), and for that particular reason we are animated the more by the desire to make the West Indian people West Indian.
> (C.L.C. 1945: 7)

Here Marryshow seems less concerned with nation-formation than with securing British accord. A bit later in the same speech, however, Marryshow links responsible citizenship to men of goodwill, thereby gendering the nation:

> Those of you who feel that we are wild men who have come to the Island to turn things upside down, disturbers of the peace, and things like that, I will tell you you are making a Himalayan mistake. We have come here as men who desire not only to be politicians but to take a broad view of things.
> (C.L.C. 1945: 7)

The examples above demonstrate some of the contradictions of colonial resistance. Gordon Lewis in fact captured the essence of these contradictions when he surmised that Cipriani's basic flaw was that he was conflicted over the ambivalence associated with "trying to be Empire loyalist and West Indian patriot at one and the same time" (G. Lewis 1968: 207).

Given the colonial socialization of the leadership of the nationalist movement, one is forced to conclude that there was considerable collusion of Caribbean men with the colonial authorities to reproduce colonial hierarchies. Insofar as Caribbean nationalists did not contest the colonial gender order, they participated in its hierarchical arrangement and benefited from the patriarchal rule of the colonizers. These early nationalists also benefited from a system of patriarchal privilege which assured them leadership roles, provided them with access to power, and provided them with access to valued resources. The contradiction inherent in this form of colonial resistance lies in its failure to address gender as an integral part of the struggle against injustice. The privileges that men had opened to them as part of the nationalist project were not available to women in the Caribbean. Moreover, anti-colonial leaders were not always vocal or particularly insistent that such privileges be extended to women. As a result, the emerging indigenous elite was able to manipulate the apparatuses of state in order to consolidate itself in post-colonial Caribbean society from a base rooted in patriarchal power.

The extent of male domination of the Anglo-Caribbean nationalist project can be gleaned from the dramatis personae of the nationalist struggle at every level. Some of the most influential committees, involved in the most important discussions and decisions about self-determination for the Caribbean, did not even have female representation. The list of names of delegates at the Caribbean Labour Congress included no female names, even though women were in attendance at the major conferences. There were no women, for example, among the members of the Standing Federal Committee or the Standing Closer Association Committee. Women were also not among the delegations invited to the Lancaster House Conferences. Indeed, only two women served in the Federal Parliament during the lifetime of the federation: Florence Daysh, an independent parliamentarian from Barbados, and Phyllis Allfrey, a Dominican, who was the only woman to become a minister in the Federal Parliament. Allfrey was Federal Minister of Social Affairs for the duration of the West Indian Federation, 1958–62. Richard Hart – President of the Jamaican Government Railway Employees' Union, member of the Executive Committee of the Peoples' National Party, Secretary of the Caribbean Labour Congress and one of the few surviving figures from this period – stated – in a recent conversation with this author – that he could come to no other explanation for the exclusion of women than that "It was pretty much a man's world." He noted:

> Unlike the case of women in England, women in the Caribbean had always played a decisive economic and important social role. Yet they were excluded from political responsibility.
>
> (Richard Hart, conversation with author, 1997)

In summary, the exclusion of women from political responsibility represents a major failing of the nationalist project in the English-speaking Caribbean.

Leaders of the nationalist movement in the former British West Indies remained ambivalent about how to situate women in the nationalist project. Their vision of that project remained male centered, as did the institutional base of power established by the colonial authorities and subsequently appropriated by male nationalists.

Men, women and the nationalist project

Though Caribbean men were unclear about how to reward the contributions of women to the nationalist struggle, they bore no illusions about women's value to that movement. Men recognized that a struggle for self-determination could not be conducted exclusively by themselves. Women were too large a constituency in all of the islands to be ignored. The support of women in the project of decolonization was critical. Caribbean nationalists, however, weighed the contribution of women against the benefits of patriarchal privilege inherent in the status quo and in effect decided that the latter best suited their interests. The extent to which men benefited from the established gender order can be seen in the apparent unanimity they shared about the status of women. Both progressive and conservative forces in the nationalist movement participated in the marginalization of women. Linnette Vassell's (1995) initial work on what she describes as "an anti-feminist position" of these political tendencies during the 1950s is worth citing here.

> Colonial ideology stressed women's "service" and warned against the promotion of women's "rights"; the anti-Communist tendencies which had developed in Jamaica seemed to also feel that feminism would split the working-class movement; hence women's pursuit of "rights" was restricted to the elaboration of trade union demands and did not embrace calls that questioned or sought to change power relations between men and women in society.
>
> (Vassell 1995: 318–319)

In short, patriarchal privilege and power transcended political ideology at the level of gender in the struggle for self-determination in the Caribbean.

One factor which should not be overlooked in this discussion is the relative underdevelopment in the region of a feminist consciousness and a feminist movement. There might have been more of a challenge to the masculinist orientation to self-determination had such a developed feminist consciousness existed in the English-speaking Caribbean at the time of the nationalist struggle. Active involvement of women in mobilization around the issue of self-determination did not necessarily or automatically suggest a feminist consciousness. For example, Rose Leon of Jamaica, member of the Jamaica Labour Party, urged St. Lucian women to take a more active part in the affairs of their country. However, she was not making a point for the empowerment of women. She went on in that address to argue that there were many ways in which St. Lucian women could help the men in their political life (*The Advocate* 1957: 1).

Women's ambivalence over the question of gender can also be seen in responses from Daphne Campbell, one of the more prominent women in the leftist movement in Jamaica. When asked about how male leadership of the left forces in Jamaica affected the kinds of issue addressed, or the demands made on behalf of women, Campbell reflected:

> we knew that our needs wouldn't be attended to under the system we were under, you understand. The first thing we wanted was freedom from the shackles there so that we could demand these things.
>
> (Vassell 1995: 329)

Rhoda Reddock (1994) also noted that some Trinidadian women voted against the move for enfranchisement in that country. In short, in keeping with the domination of men over the nationalist project, it was not only that men placed other issues that they considered more pressing on the national agenda but also that women did not have the same political cache in the construction of the nation. They were not really encouraged to express or articulate their own political visions and concerns. In addition, at some level in the process of resisting colonial oppression, women wittingly or unwittingly participated in this patriarchal practice and relegated their own needs to a lower priority than those of men. Perhaps they had some understanding that issues affecting them as women would be addressed once colonial rule had ended. Enloe (1989: 57) has argued that it is pointless attempting to negotiate gender change ex post facto. The issue of gender has to be integrated into the process of change from the beginning (ibid.). Moreover, Sharp (1996) argues that such negotiations of gender concerns have implications for the way the state becomes organized:

> the postponement of a consideration of gender issues in the name of the construction of the nation-state will irrevocably alter the direction the emerging nation-state will take. The very nature of national identity will be different depending upon whether or not it deals with gender issues at the outset. National identity, therefore, is not something which can be retrieved intact from the past and slotted back into the heart of contemporary culture unchanged, it is constituted in particular times and places through relations of power already existent in society.
>
> (Sharp 1996: 103)

Though Sharp's point above is an important contribution to the discursive strategy of the nationalist project, it nevertheless forecloses the possibility of future gendered change. National identity is neither unalterable nor static. The possibility of change must remain a viable option irrespective of the historical circumstances. What was therefore problematic in the Caribbean was the gendered priorities of the process of colonial resistance.

Many socially and politically active women in various parts of the Caribbean supported the efforts for self-determination. Women such as Daphne Campbell (Jamaica), Florence Daysh (Barbados), Phyllis Allfrey (Dominica), Audrey Jeffers (Trinidad), Una Marson (Jamaica), Rose Leon (Jamaica) and Janet Jagan (Guyana) were all in their various ways active participants in the struggle for self-determination. However, none of these women gained the kind of prominence and place of importance accorded the men who negotiated a change of political status for the region. Gordon Lewis similarly concludes about Caribbean women activists:

> But whether due to the general social environment or inherent tradition the scope of influence of such individuals was generally restricted to the narrow sphere of their own circles. To the general public they were little more than legendary figures, seldom reported, and usually denied widespread recognition by means of opportunities of public service at the top levels.
> (G. Lewis 1968: 82)

In the end, the fact remains that a specific political space was not made available for Caribbean women. The place they occupied restricted their participation to certain types of activities. The classic public–private dichotomy seemed to have prevailed for men monopolized the public sphere of politics, leaving women marginal to that arena. Some have argued that in terms of the politics of operation, the state is male (Brown 1995) while the nation is female (Anthias and Yuval-Davis 1993). However, in the case of the nationalist movement in the Caribbean, whatever notions of femaleness may have existed about the nation, women were not really given any opportunity to participate in its construction, to delineate its parameters or to mediate its apparatuses of power. These determinations remained the prerogatives of indigenous men who had been schooled in the "best" tradition of colonial patriarchy and its deployment of power.

Nationalism and the failure to conceive of alternative visions

The male-centered orientation of the Caribbean nationalist project raises a number of questions about the failure of men of the region to fashion and articulate a more broadly based, inclusive, democratic vision of post-colonial society. Much of the explanation for this missed opportunity is rooted in the power of colonial socialization alluded to earlier. Caribbean men seemed to have been so caught up with defining their project in accordance with the examples set down by European men, that they left themselves little room for creative alternatives. Here, the wisdom of Gramsci (1980) remains compelling:

> Subaltern groups are always subject to the activity of ruling groups, even when they rebel and rise up: only "permanent" victory breaks their subordination, and that not immediately.
>
> (Gramsci 1980: 55)

Fashioning an alternative vision of society would have meant that Caribbean men would have had to engage in new constructions of masculinity which would move them away from tradition and from inherited hegemonic European models. Given the character of gender relations during the struggle for self-determination, it is perhaps unrealistic to expect that the vanguard of the nationalist movement would have been more sensitive to matters of gender equality.

One may also ask why such myopia existed? Part of the answer to this question has to do with the normalization of patriarchy. Undoubtedly, many of those involved did not even see inherent limitations in the way they approached the nationalist project. Some, like Hart (1988), have had the vision to revisit the period and to ponder the way the nationalist drama unfolded. Others remain loyal to the structures and practices of patriarchy. In a rather insightful and historiographically engaging recent interview, Caribbean economist and political activist Lloyd Best reflected on the intellectual landscape of the discourse on self-determination in the Caribbean. In many ways he reaffirmed the male-centered orientation of this period, he mentioned few women in his reflection on the leading intellectual figures in the Caribbean at the time, and he did not provide any analysis of the kind of post-colonial Caribbean envisioned by women such as Elsa Goveia, Esther Unger and Lucille Mair. These women were simply referred to by Best as facilitators assigned by the University of the West Indies to provide post-lecture discussions for the Open Lecture series held at the Mona Campus of the university. Kari Levitt, with whom Lloyd Best has collaborated in his work, a distinguished Canadian scholar with a specialty in third world political economy, is one of few other women discussed in this interview. All of the associates, colleagues and intellectuals whose ideas were analyzed by Best are men, all of whom, interestingly enough, have gone on to become major technocrats, politicians and well-known academics (for details see interview by Scott 1997: 119–139).

What is quite clear is that the discourse of colonial resistance in the Anglophone Caribbean did not in many ways embrace a radical philosophy, except perhaps among the left wing of the Caribbean Labour Congress. Even within this organization, the radical ideas for a strong federal state, including British Guiana and British Honduras, with full self-government and Dominion status, were modified by the time the federation was established. In short, the nationalist project was essentially a reformist project. Insofar as it was reformist, it would seem to be somewhat naive to expect that gender issues, gendered hierarchies and the whole idea of change would be interrogated seriously. In the Anglophone Caribbean the struggle for self-determination remained separate from the struggle for improvement in the conditions of existence of women. Men's goals defined the project of nationalism and were not only given priority but also constructed as being beneficial to all members of the nation. Hence, women's issues were expected to wait to be addressed until after the nation was consolidated. Indeed, women's goals did not coincide with the building of the nation, except insofar as women were needed to support the efforts at decolonization – by mobilizing popular support.

Finally, what has been the legacy of this male-centered nationalist project and vision for the English-speaking Caribbean society? Based on past political practice in the Anglophone Caribbean, men have continued to monopolize the positions of power in the state and the nation. While there has been much improvement in the status of women, with a few notable exceptions women still remain largely outside of the realm of political power. Women in the Caribbean also remain largely outside of the important decision-making processes in business and at the political level. What the masculinist orientation of the nationalist struggle has actually done is to reinforce and build upon the patriarchal hierarchies of the colonial era.

The Caribbean region is at a crucial historical juncture in which the process of global restructuring has begun to dislocate certain types of social relationships, including gender relationships. This dislocation, though bringing conflict and controversy in its wake, also opens up a space for social transformation. A mature, confident and increasingly energetic feminist practice, which has been building in a systematic way since the 1970s in the Caribbean, seems well poised, at this stage, to seize the opportunities made possible by these social changes to push for greater demands for women and to expand the horizon of possibility for gender equality in the region. It is only when women have equal access to the apparatuses of power and the mechanisms of distribution of resources in the broadest sense, that these islands can begin to build truly democratic nations.

Acknowledgments

The author would like to thank Nigel Bolland, Eddy Souffrant and Tamar Mayer for comments and suggestion on an earlier version of this chapter.

References cited

The Advocate (1957) "Mrs. Leon explains socialism and new federal party aim," *The Advocate*, May 30: 1.

Anthias, F. and N. Yuval-Davis (1993) *Racialized Boundaries: Race, Nation, Gender, Colour and Class and the Anti-Racist Struggle*. London and New York: Routledge.

Benn, D. (1987) *The Growth and Development of Political Idealism in the Caribbean 1774–1983*. Jamaica: Institute of Social and Economic Research, University of the West Indies.

Bolland, N. (1995) *On the March: Labour Rebellions in the British Caribbean, 1934–39*. Jamaica: Ian Randal.

Brown, W. (1995) *States of Injury: Power and Freedom in Late Modernity*. Princeton, N.J.: Princeton University Press.

Caribbean Labour Congress (C.L.C.) (1945) Official Report of Conference held in Barbados, September 17–27. Barbados: Advocate.

Enloe, C. (1989) *Bananas, Beaches and Bases: Making Feminist Sense of International Politics*. London: Pandora.

Froude, J. A. (1888) *The English in the West Indies: or, The Bow of Ulysses*. London: Longmans, Green.

Gomes, A. (1974) *Through a Maze of Colour*. Trinidad: Key Caribbean.

Gramsci, A. (1980) *Selections from the Prison Notebooks*. New York: International Publishers.

Guérin, D. (1961) *The West Indies and their Future*. London: Dennis Dobson.

Hart, R. (1988) "Origin and development of the working class in the English-speaking Caribbean area: 1897–1937," in M. Cross and G. Heumann (eds.) *Labour in the Caribbean: From Emancipation to Independence*. London: Macmillan, pp. 43–79.

Henry, Z. (1979) *Labour Relations and Industrial Conflict in Commonwealth Caribbean Countries*. Trinidad: Columbus.

Howe, G. (1994) "West Indian Blacks and the struggle for participation in the First World War," *Journal of Caribbean History* 28, 1: 27–62.

James, C. L. R. (1992) "The case for West Indian self-government," in A. Grimshaw (ed.) *The C. L. R. James Reader*. Oxford: Blackwell, pp. 49–62.

Joseph, C. L. (1971) "The British West Indies Regiment 1914–1918," *Journal of Caribbean History* 2: 94–124.

Lee, J. M. (1967) *Colonial Development and Good Government: A Study of the Ideas Expressed by the British Official Classes in Planning Decolonization 1939–1964*. London: Oxford University Press.

Lewis, A. (1977) *Labour in the West Indies: The Birth of a Workers' Movement*. London and Port-of-Spain: New Beacon.

Lewis, G. (1968) *The Growth of the Modern West Indies*. New York: Monthly Review Press.

McClintock, A. (1995) *Imperial Leather: Race, Gender and Sexuality in the Colonial Contest*. London and New York: Routledge.

Manley, M. (1975) *A Voice at the Workplace*. London: André Deutsch.

Mark, F. (n.d.) *The History of the Barbados Workers' Union*. Barbados: Barbados Workers' Union.

Martin, T. (1986) "Marcus Garvey and Trinidad, 1912–1947," in R. Lewis and M. Warner Lewis (eds.) *Garvey: Africa, Europe, the Americas*. Jamaica: Institute of Social and Economic Relations, University of the West Indies, pp. 52–88.

Moskos, Jr., C. (1967) "Attitudes toward political independence," in W. Bell (ed.) *The Democratic Revolution in the West Indies*. Cambridge, MA: Schenkman, pp. 49–67.

Mosse, G. (1985) *Nationalism and Sexuality: Respectability and Abnormal Sexuality in Modern Europe*. New York: Howard Fertig.

Post, K. (1981) *Strike the Iron, a Colony at War: Jamaica 1939–1945*, vol. 1. Atlantic Highlands, N.J. and The Hague: Humanities Press and the Institute of Social Studies.

Reddock, R. (1994) *Women, Labour and Politics in Trinidad and Tobago: A History*. London: Zed.

Roy-Fequiere, M. (1997) "The nation as male fantasy: discourse of race and gender in Emilio Belaval's Los Cuentos de la Universidad," in J. M. Carrión (ed.) *Ethnicity,*

Race and Nationality in the Caribbean. Puerto Rico: Institute of Caribbean Studies, pp. 122–158.

Scott, D. (1997) "The vocation of a Caribbean intellectual: an interview with Lloyd Best," *Small Axe* 1, March: 119–139.

Sharp, J. (1996) "Gendering nationhood: a feminist engagement with national identity," in N. Duncan (ed.) *Body Space: Destabilizing Geographies of Gender and Sexuality*. London and New York: Routledge.

Sheppard, J. (1987) *Marryshow of Grenada: An Introduction*. Barbados: Jill Sheppard.

Sherlock, P. (1973) *West Indian Nations: A New History*. Jamaica and London: Jamaica Publishing and Macmillan.

Smith, A. (1991) *National Identity*. London and New York: Penguin.

Thomas, J. J. (1889) *Froudacity*. London: New Beacon.

Vassell, L. (1995) "Women of the masses: Daphne Campbell and 'left' politics in Jamaica in the 1950s," in V. Shepherd, B. Brereton and B. Bailey (eds.) *Engendering History: Caribbean Women in Historical Perspective*. New York: St. Martin's Press, pp. 318–333.

Walby, S. (1996) "Woman and nation," in G. Balakrishnan (ed.) *Mapping the Nation*. London and New York: Verso, pp. 235–254.

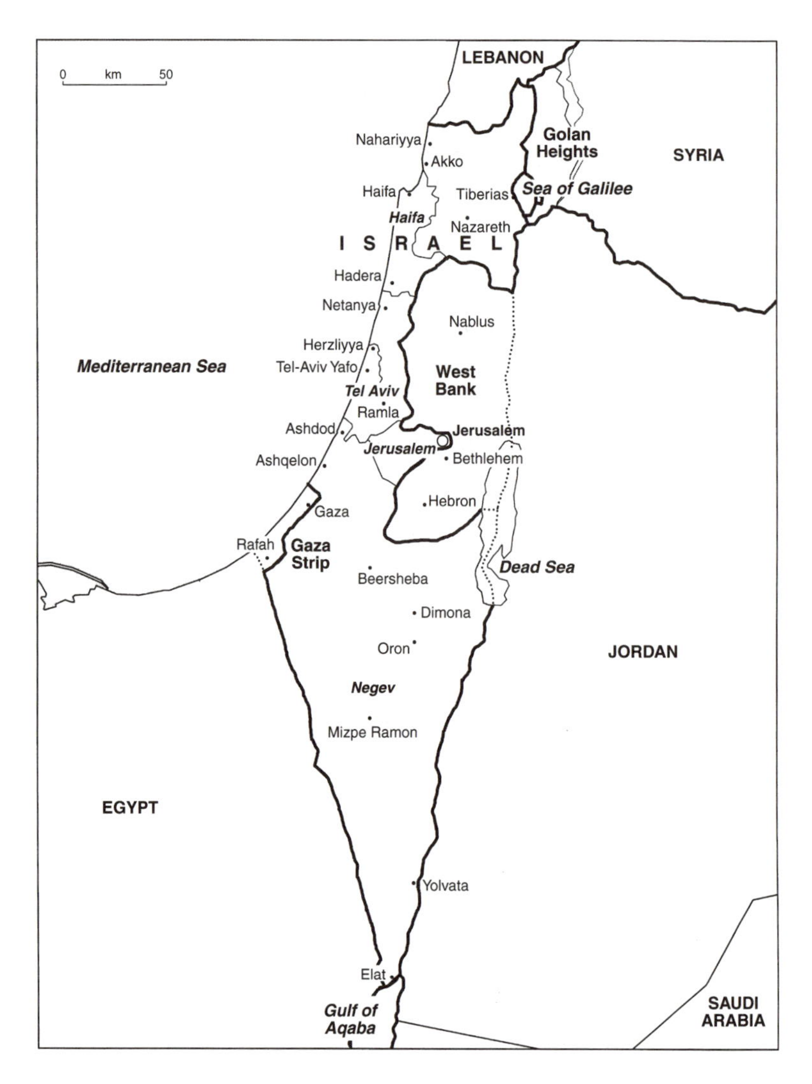
0 km 50
LEBANON
Nahariyya
Akko
Golan Heights
SYRIA
Haifa
Tiberias
Sea of Galilee
Haifa
Nazareth
I S R A E L
Hadera
Netanya
Nablus
Herzliyya
Mediterranean Sea
Tel-Aviv Yafo
West Bank
Tel Aviv
Ramla
Ashdod
Jerusalem
Jerusalem
Ashqelon
Bethlehem
Hebron
Gaza
Rafah
Gaza Strip
Dead Sea
Beersheba
Dimona
Oron
JORDAN
Negev
Mizpe Ramon
EGYPT
Yolvata
Elat
Gulf of Aqaba
SAUDI ARABIA

Israel

12 From zero to hero

Masculinity in Jewish nationalism

Tamar Mayer

In late 1994, a year after the signing of the Oslo Accords, one of Israel's major newspapers, *Yediot Achronot*, carried a lead story entitled "[W]e used to be men, now we are zero" (November 11, 1994). This story concerned members of an elite military unit who had deserted their post because they were so disappointed by the turn that their military service had taken, once Israel began training soldiers for peacekeeping missions after pulling out of the Gaza Strip and beginning its withdrawal from the West Bank. No longer were these young men able to perform the tasks which had motivated them to join this elite unit, during the days of the *Intifada*, and for which they had trained to be, in their own words, "killers" who enforced Israeli military rule among Palestinians (Shachor 1994: 6).[1] Now instead, as a result of the Oslo Accords, the soldiers said, they were being assigned to guard daycare centers in Jewish settlements. "What started as an attempt 'to be a man,' turned into an addiction for 'action'" (Shachor 1994: 6), said one of the deserting soldiers in an attempt to justify his unit's act. These soldiers ran from duty, as they said in the article, because in peace missions there is no "action," no glory, no "rush."

In short, several of them said in their testimony, they had wanted to join the elite unit because they believed that in the Israeli Defense Forces (I.D.F.) they would get a chance to become "real men," they would be transformed "from nerds, from zero" into units of a "mighty machine" that would enable them to give everything, even their blood, "for the country, for the flag" (ibid.). In 1994 – just as peace seemed a real possibility – those young men actually claimed to feel cheated by the military: instead of becoming real "men," they now felt they were zero.

While this story of desertion may be uncommon among Israeli soldiers, it provides nevertheless a rare inside look at the way many young Israeli men have come to feel about volunteer service in elite military units. The strong relationship between masculinity, militarism and Jewish nationalism[2] articulated by these men has its origins in the early days of Zionism when Jews, first in Europe and later in Palestine, felt forced to defend themselves against the Other (first European nationalists whose anti-Semitism was already well established by the early days of Zionism, and then the indigenous Arab population of Palestine who resented attempts by Zionist immigrants to take over and "Judaize" Arab lands). Over the course of the twentieth century the constant threat (real or imagined) of annihilation has made Israeli Jews rely heavily on military and physical strength; in turn, militarism has become intimately connected to the construction of both Jewish nationalism and Israeli Jewish masculinity. Although both Jewish men and women in Israel are conscripted into the Israeli Defense Forces, it is men who "do the *real* defense work" (Izraeli 1994): they are the ones who actually serve in combat and eventually risk their lives for Israel's survival. Especially in the most decorated units, as a result, a cult of heroism has developed among young Jewish men. And the continuous sense of threat from the Arab world to Israel's existence has further sanctified the I.D.F. as a major institution in Israeli society – justifying, reinforcing and even sharpening the image of the Jewish warrior, whose masculine identity has become intertwined with Israel's security. For many Israeli Jewish men, in fact, the military has become the only rite of passage into manhood. As long as Israelis believe that they have enemies who remain committed to Israel's annihilation, the priority of defense will continue to shape Jewish nationalism in Israel, and the military will continue to prevail as one of the major institutions in the lives of Israeli Jews and an important arena of masculinity.

In this chapter I shall examine the close relationship between nationalism and masculinity in Zionism during the first several decades of the twentieth century. I focus here on the foreign, specifically German, influence on the construction of the masculine *New Jew*, and on the subsequent development of the Jewish warrior ideal in Palestine. The central role that Zionist youth movements, the new Hebrew education and paramilitary activities played in this construction are also discussed here, because they too have been instrumental in the construction of the *New Jew*, the mythological symbol of Jewish nationalism.

Although it is well established by now that gender identities are generally constructed in opposition to one another and that we cannot understand the construction of the "masculine" without understanding the construction of the "feminine," in Israel and in other societies that have perceived themselves to be "under siege," militarism continues to be instrumental in the construction of gender identities. I focus here, therefore, on the interrelationship in Israel between militarism and masculinity, and leave a thorough analysis of masculinity in opposition to femininity to others.[3] The relationship between masculinity and militarism has had a profound effect on the kind of nationalism that was born with Zionism and which later has been refined in Jewish Israel. The almost intimate relationship between masculinity and Jewish nationalism is, I believe, a product of the initial Zionist project, which assigned men and women different positions in society, and an outcome of modern Jewish history's survivalist orientation, which has led to a prioritizing of national security needs that has constructed men as superior to women. This notion of male superiority is further anchored in

Jewish religious tradition, especially as there is no constitutional separation in modern-day Israel between "church" and "state."

Jewish nationalism and masculinity

Almost from its inception, Jewish nationalism has been closely intertwined with masculinity. The transformation of "the political status, the socio-economic profile, and the psychological self-image of the Jews" in Europe (Shimoni 1995: 3) – central to Zionism – was based (in large part) on the construction of the *New Jew*, the *Muscle Jew*. Most of the gender references to the *Muscle Jew* are to men, illuminating the connection between masculinity and Zionism and the invisibility of women in Zionism.

Although Zionist writings appear as early as the mid-nineteenth century, Zionism became an organized movement only in the late 1890s, as Theodor Herzl – the father of modern Zionism – organized the first Zionist Conference in Basel (in 1897). At a time when many people in Europe perceived *nation* as the legitimate foundation of the state, Herzl appreciated Bismarck's success in mobilizing the German masses around the nationalist banner and dreamed of a similar future for the Jewish nation.[4] He dreamed that nationalism would free the Jews from problems caused by 2,000 years of living in exile. Significantly, as also in the German case, Herzl's quest for freedom was associated with a complete transformation of the national as well as the individual character – both of which, in his view, involved notions of manliness. Herzl's explicitly gendered contempt for European Jewry is captured well in his diary when on June 8, 1895, after visiting with some well-to-do and educated friends, he wrote: "they are Ghetto creatures, quiet, decent, timorous. Most of our people are like that. Will they understand the call to freedom and manliness?" (Herzl 1956: 39). Clearly for Herzl manliness and freedom were closely tied together, and both directly connected to militarism and patriotism. Herzl planned to call up historical events of mythical proportions, such as the legendary Maccabee fighters (almost all men), as a way to set the stage for and to mobilize the nation and the actors of Zionism. On June 7, 1895 he wrote:

> I must train the youth to be soldiers. But only a professional army. Strength: one tenth of the male population; less would not suffice internally. However, I educate one and all to be *free and strong men*, ready to serve as volunteers if necessary. Education by means of patriotic songs, the Maccabean tradition, religion, heroic stage-plays, honor, etc.
>
> (Herzl 1956: 37, added emphasis)

For Herzl the most important idea of Zionism was to teach Jewish men – the principal figures of Zionism – to be free and to reclaim the masculine past of the nation. This was necessary, he believed, because years of life in the Diaspora had given Jews many feminine characteristics and made them, as a result, easy targets for anti-Semitism. As Zionism would free the Jews of Europe from their constant battles with anti-Semitism, create a Jewish national culture, "normalize" Jewish national life and offer Jews the tools with which to negotiate with both modernity and anti-Semitism, Herzl hoped it would also free them from their "feminine" nature.

Although women, too, were ghetto dwellers and were integral to the makeover of the Jewish people that Herzl hoped, women were clearly not essential to Herzl's program for national change. In Herzl's vision women did not need to be transformed in the way men did because there was no real dissonance between their behavior and their gender identity. As a result, women were not as central to Zionism as men as a diary entry (June 11, 1895) makes clear: "[N]o women or children shall work in our factories. We want a vigorous race. The state takes care of needy women and children. 'Old maids' will be employed in kindergartens and in mothering the orphans of the working class. I will organize these spinsters into a corps of governesses for the poor" (Herzl 1956: 41). As Herzl saw Zionism, women's importance seems limited to their role in reproducing and sustaining the Jewish nation. Women did not feel in the early days of Jewish nationalism that they occupied secondary roles. They rather believed that Zionism offered them a new kind of equality: the right to vote and to be elected to office. But once they actually emigrated to Palestine to create the new society women began to realize that their dream of equality between the sexes was not likely to be achieved; that in fact their position in the "new" society and in the "new" land of Israel was not much different than it had been in Europe (Bernstein 1987).

Zionism and the *New Jew*

Thus the national project of Zionism was to transform Jewish life in the Diaspora – and this was possible, many believed, only by creating a new person, the *New Jew*, the *New Hebrew*. The *New Jew* was to be the antithesis of the "ghetto Jew" whom Herzl and other Zionist thinkers saw as helpless, passive and feminine and thus in need of major transformation. Ironically, early Zionists' notion of the Jews' passivity and femininity was actually in many ways an internalized version of the prevailing anti-Semitic view of the time. The impulse to feminize the Other was not new in the late nineteenth century:[5] it had enabled modern society to build cohesiveness (Mosse 1996) and influenced much of the anti-Semitic rhetoric of the time. The Jewish man's passivity was often caricatured and ridiculed on the streets of Europe and in European newspapers.[6] Even more, the Jewish man's body was seen as "aged, weak or effeminate," calling up yet another countertype to modern masculinity: homosexuality (Mosse 1996: 70). The Jewish male's stereotyped body was "given specific bodily features and measurements to demonstrate [his] difference from the norm" (ibid.). Like the homosexual, the Jewish man was seen as limp and slim, and both the Jewish man and the homosexual were condemned as transgressors of a masculine standard of beauty (ibid.).

Given the prevalent anti-Semitic discourse in Germany during the late 1890s, it is no surprise that the fathers of Zionism dreamed that national liberation would bring "freedom and manliness" to the Jews of Europe. And because most of the Zionist writings in the late nineteenth and early twentieth centuries focus on the Jewish man's body, it seems that the Jewish woman and her body remained invisible: "In the collapse of Jewish masculinity into an abject femininity, the Jewish female seems to disappear" (Pellegrini 1997: 109). Therefore, even as Jewish women escaped the anti-Semitic ridicule to which Jewish men were subjected, they also were left out of the body and character reform advocated by Zionist thinkers such as Herzl and his close associate and second in command Dr. Max Nordau.

A major element of the Zionist reform agenda involved social engineering that intended to create a dignified, masculine *Muscle Jew*. Nordau called the Jews to reconnect with their Jewish past and with ancient Jewish heroes like Bar Kochba, "to again become deep chested, strong limbed, and fierce looking man" (Nordau 1900: 10). The *New Jew*'s characteristics were to mimic those of the gentiles: tall, virile, close to nature and physically productive. The *New Jew* was to become in some sense an *Übermensch*,[7] a super-human, whose fit body would help his Jewish mind to excel and who would thus be able to stand up to anti-Semites.

The transformation of the Jewish man's body would best be accomplished, Dr. Nordau believed, through involvement in gymnastics. As a psychiatrist, Nordau believed that he saw many physical and mental similarities between Jews and "degenerates," who in his view were "not [just] criminals, prostitutes, anarchists, and pronounced lunatics [but also] authors, and artists" (Nordau 1895: vii).[8] Because of physical and mental similarities that the two groups exhibited – physical frailty and a tendency toward nervousness – Nordau prescribed gymnastics as a healing regimen for both Jews and "degenerates." Gymnastics, Nordau believed, would be the most effective way for Jews to develop their bodies – "[S]olid stomachs and hard muscles would allow Jews to overcome their stereotype . . . to compete in the world . . . and to recapture [their] dignity" (quotes from Nordau's essays in Mosse 1993: 164) – and to calm what he observed as both groups' nervousness. Specifically, a physically fit body, according to Nordau, would lead to creating a masculine identity (Mosse 1993: 165) and to what later would be referred to as the *Muscle Jew*.[9]

Given the historical events of the twentieth century, it seems ironic as well that much of the Zionist ideology of nation and masculinity was derived from the German experience. Nordau's and Herzl's commitment to gymnastics as a way to achieve the desired transformed body was in fact greatly influenced by the German philosophy and practice of the day. Gymnastics had been essential both to the construction of masculinity and to national ideology throughout Europe but most specifically in Germany since the early nineteenth century (Krüger 1996, Mosse 1996, 1993, 1985, Hoberman 1984). In Germany gymnasts became the “national stereotype in the making” (Mosse 1985: 50) and gymnastics festivals became such an important way to organize the crowds and mobilize the masses that they became “a part of the national liturgy” (Mosse 1975: 132). For gymnastics was uniquely suited both to enabling the individual to develop his own body and, at the same time, to building the group solidarity which nationalism requires.

The nineteenth-century gymnastics society in Germany, better known as the German Turner Movement, became crucial to the development of the German nation-state, the *Deutsche Reich*.[10] Practiced in schools, clubs and in the army, and as an expression of order and discipline (Krüger 1996), *Turnen* (gymnastics) became “a system of rationalized and formalized exercises” (Krüger 1996: 413) which helped develop a specific culture of the male body connected to a specific *Deutsche Kultur* (German culture). The development of gymnastics in clubs and of regional and national gymnastics festivals which used flags, ribbons, uniforms and songs helped to create and sustain the us–them distinction so essential to nationalism (Krüger 1996). Over the years, as they drew more and more people, these festivals became a way to pioneer the ideals of German national self-representation (Mosse 1975), by demonstrating the benefits of controlling mind and body as well as loyalty to the Reich (Krüger 1996). Significantly, the Turner Movement of Germany was exclusively male and thus helped to build a connection between the development of male culture and bonding and the development of German nationalism (Reulecke 1990).

Herzl, Nordau and other Zionist leaders who were influenced by German culture and impressed by German nation-building achievements advocated similar programs for Jewish nation-building.[11] Nordau called on the delegates to the second Zionist Congress of 1898 to establish and join gymnastics clubs. The establishment of new Jewish gymnastics clubs and their spread throughout Western and Central Europe testifies to the importance of Nordau’s plan for the *Muskeljudentum* (Muscle Jewry).

Although some of these clubs predated Nordau, their numbers were small and no national ideology was associated with them. After Nordau, almost all of the pre-existing Jewish sports clubs joined the Zionist movement, and many of the newly established ones were given the names of ancient Jewish male heroes like Bar Kochba and the Maccabees. The *New Jew*, the *Muscle Jew*, was to take as role models Jewish heroes of the past, especially those whose battles with the Romans and the Greeks dramatized their willingness to fight for the land and to sacrifice their lives, if necessary, for their belief in the national cause (Shapira 1992). Nineteenth-century Jewish athletes – exclusively men – trained in the explicit spirit of nation-building: mimicking the German gymnastic model of order and discipline; performing as a group with banners and ribbons; using Hebrew in their drill exercises and singing Hebrew songs (Berkowitz 1993: 108). The motto of these Jewish athletes reflected the us–them requirement of nationalism, the fighting spirit and the idealized masculinity of the gymnastic clubs: "We fight for Judah's honor/Full strength in youth/So when we reach manhood/Still fighting ten times better" (quoted in Berkowitz 1993: 109).[12]

Jewish youth movements, especially the ones that developed in Germany and in Central Europe, also borrowed their format – and their emphasis on masculinity – from their German counterparts. Like the Turner Movement, nineteenth-century German fraternities and youth movements advocated a return to pre-industrial nature through hiking – enabling them at once to spread their ideology throughout Germany and, at the same time, to connect the German landscape with the spirit of the nation (Mosse 1975) – and through songs, dances and plays, which helped reinforce the love of both physical movement and the nation.[13] As in the German case, Jewish youth movements rejected family traditions, revolted against bourgeois values, and emphasized instead a return to nature, simplicity and male comradeship (Reinharz 1996: 279). Emerging Zionist youth movements were influenced as well by Lord Baden-Powell's Scouting movement in Britain;[14] some of the Slavic youth movements also emphasized a return to nature and a rejection of the values associated with modern industrialism (Naor 1989). At the same time as Zionist youth movements, first in Europe and later in Palestine, adopted the rhetoric and structure of these other youth movements, they also added their own unique pioneering mission. Because Jewish nationalism could truly develop only through a return to a historical homeland of *Eretz Yisrael*, the goal of all Zionist youth movements became, ultimately, emigration to Palestine.[15] It was only there, according to Zionism, that Jews could become rooted, control their own economics and politics, create a Jewish majority, revive the Hebrew language – and thus achieve national liberation (Berkowitz 1993, Luz 1988, Vital 1982, 1975).

Youth movements, education and nationalism in Palestine

The Zionist youth movements which were formed in the early twentieth century in Europe – initially the *Maccabees* and, later, *Hechalutz* (pioneer), the *Scouts*, *Hashomer Hatsair*, *Hanoar Hatsioni*, *Beitar* and *Gordonia* – had a profound impact on the construction of Jewish nationalism and masculinity, as both developed together in Jewish Palestine.[16] Although these movements were ideologically different from each other, they all shared the Zionist commitment to a national revival in Palestine (Lamm 1991). As they spread throughout Europe and in Palestine (between the two world wars) all of them trained Jewish youth for life on the farm and instilled in both boys and girls the commitment to personal fulfillment of the national goal, including the sacrifice of the material comforts of home for the sake of rebuilding the homeland, and the Zionist ideals of heroism, love of the land and physical labor (S. Almog 1984).

The ideal *New Jew* – the youth movement graduate turned pioneer settler (*chalutz*), colonizer and defender – became the emblem of Zionism. Although women were *chalutzot*, too, and their contributions were crucial to the success of the Zionist project, they did not come to symbolize Zionism's achievements. While both men and women opened up the frontier, built *kibbutzim* and created a new Hebrew culture in Palestine, it was mostly men who were involved in fighting the indigenous population of Palestine who took the more publicly visible agricultural jobs, which called for greater physical endurance. In his memoirs about life in *Sejera*, one of the first Jewish colonies in Palestine, David Ben Gurion (who later became Israel's first Prime Minister) wrote in 1907 about gender differences: "All members of the *moshava* [colony] work. The men plow and plant their land. The women work in their garden and milk the cows. The children herd the geese on the farm and ride horses towards their fathers in the fields" (Ben Gurion 1971: 35). It appears that despite the Zionist ideal of gender equity, which some of the women had hoped for before they came to Palestine, men and women were assigned different tasks in the Jewish colonies which reproduced an all too familiar set of inequities.

Members of the youth movements identified so wholly with the revolutionary message of Zionism that we can say they were "the soldiers of the Zionist revolution" (Guri 1989). In turn, and in tandem, as the message of Zionism spread throughout the many Jewish communities of Europe and the number of young emigrants to Palestine increased, the *Yishuv* (Jewish community in Palestine) found that it had to design an educational plan which would accommodate and reinforce the message of Zionism. As in other national movements, like the German and the Slavic ones, both formal and informal education were crucial to the spread of Jewish nationalism, and teachers and youth movement leaders became the agents of such change. Education in the *Yishuv*, especially after the late 1890s, was in Hebrew (a language that most new immigrants did not know), and emphasized the tie to the historical homeland and to the Jewish heroic past as well as physical activities and the development of a close tie with nature. Furthermore, in the spirit of Zionism, students in the *Yishuv*, during the 1930s, 1940s and after statehood, studied agriculture as a topic in school, and several agricultural boarding schools were established so that students could learn how to be productive farmers. And students received training in self-defense and paramilitary activity, so that they would be able to defend themselves.

Just as it had in Zionist youth movements, Hebrew education in the *Yishuv* had an anti-intellectual orientation which emphasized applied subjects and activities that developed physical fitness and strength rather than the life of the mind. At the center of these curricula were activities which developed physical endurance, bravery and heroism among youth (Reichel 1997, Alboim-Dror 1996) and which trained young settlers, especially the men, for their future defense duties. Long, exhausting hikes in the rugged terrain of Palestine became an important tool for merging the Zionist message of love of the land and the building of physical strength. These hikes were the climactic events in each year in school and in the youth movement, increasing in difficulty as each youth moved up the movement hierarchy. Eventually they would also become an important rite of passage in the Israeli Defense Forces.

Thus the Zionist culture that emerged in Jewish Palestine idealized the *New Muscle Jew*, the antithesis of the stereotyped intellectual European Jew. But while both boys and girls participated in Zionist education in the *Yishuv* and both men and women built and developed together their homeland, Zionist culture was unmistakably gendered; for it was largely men who claimed the additional mission of national defense.

Paramilitarism and the cult of toughness

Schools and youth movements also redesigned ceremonies and the celebration of Jewish holidays to fit the Zionist message of bravery, redemption and national liberation. These nationalistic celebrations emphasized nature and homeland, and the revival of old Jewish heroes, in particular the Maccabees and Hasmonites. Because all these heroes were men, the new focus of these holidays contributed greatly to the cult of masculinity that became integral to the emerging Jewish culture in Palestine. Many of the celebrations during the *Yishuv* involved shooting, horseback riding and gymnastics competitions (Alboim-Dror 1986), virtually none of which included women participants. Furthermore, as it had in German youth movements, singing played a crucial role in building attachment to the nation and the homeland and strengthening the commitment to defend both. Ultimately, Israeli songs which recalled warriors' experiences in the battlefield and memories of fallen friends and which told stories about military culture were adopted as if they were youth movement songs, and even became a dominant force in the music repertoire that developed, later, in Israeli popular culture. Thus that largely male warrior culture came to shape youth culture in modern Israel (Shakham 1995).

Even in the early years of Jewish Palestine, the emerging priority of security contributed in crucial ways to the masculine image of the Zionist success story. Once they began arriving in Palestine in large numbers and transforming the land which they saw as unclaimed, Jewish settlers were, inevitably, met with growing resistance by the indigenous Arab population of Palestine. As Arab attacks on Jewish settlements and farms became more violent and more regular, the Jews of the *Yishuv* in the pre-state years resorted to defending themselves, establishing several organizations whose sole focus was protection of the new Jewish communities. These organizations were aided by youth who had been trained in Zionist schools and youth movements for their security mission. But, as Uri Ben Eliezer (1995) argues in *The Emergence of Israeli Militarism 1936–1956*, much of the military activity of both attack and defense was wholly unorganized. Militarism developed in different locations and by different people without much plan and with little coordination (Ben Eliezer 1995: 35).

Among the most important of the Jewish guarding and fighting organizations that were created in the first two or three decades of the century were *Hashomer* ("the guard") established in 1909, the *Hagana* ("the defense") and the *Palmach* ("strike force"), both established in the 1940s. *Hashomer*, in particular, helped to create the warrior associated with Zionism and, hence, contributed to the masculinization of Jewish nationalism. In the words of its own publication:

> "*Hashomer* created a new type of Jew," a brave Jew who is not afraid of danger and who is ready to fight face to face, a Jew who knows the way of life and the manners of the Arabs and the Bedouins and is better than they are in fighting, riding a horse . . . and . . . in his brave spirit.
>
> (*The Hagana* 1968: 16, my translation)

Hashomer was a selective group of men who, like the members of the *Palmach*, the *Hagana* and many other military units that came on their heels, were sworn to secrecy and therefore appeared to be elitist (O. Almog 1997: 168), and indeed they were.[17] *Hashomer* was the first significant Jewish defense force whose members recognized that the mission of security was intimately connected to settlement activity, and that settling on the land in strategically distributed communities had an important security value.[18] The *shomrim* ("guards") became mythologized in the folklore that developed in the *Yishuv* and in the period before statehood. Alexander Zeid, the armed *shomer* on horseback, riding in defense of his people and land, became the epitome of this image of Jewish masculinity, the *New Jew*.

In the late 1930s and early 1940s, as a result of both the ideological program for a strong *New Jew* and increasing Arab attacks on Jewish settlements, paramilitary training became an integral part of the curriculum of both high schools and youth movements in Jewish Palestine. This training of school-age Jewish youth for their military service taught them how to use guns, trained them in survival techniques and instilled in them a sense of responsibility to the community. Students learned Jewish military history, with an emphasis on the battle experiences of contemporary and historical heroes, and developed extensive knowledge of their physical environment. Teenaged members of the youth regiments who came out of these programs – mostly but not exclusively males – joined the *Palmach* and the *Hagana* for their secret missions against both the British and the Arabs, and at times also against other underground Jewish organizations.

In turn, the youth who became leaders, counselors, in the youth movements and who joined elite units became role models in their movement, school and neighborhood. The impact of these men on succeeding generations was demonstrated by the desire of the younger ones to follow in their footsteps. In a letter published in *Siach Shakulim* (1988 [1981]) ("conversations of mourners") Yitzchak Kadmon writes about the inspiring effect on him of his counselor in the Scout movement, Dani Mass, who died in January 1948 during Israel's war of independence:

> Mass is for me a substitute for God's toughness-Israel's God. He is an object of my admiration. A counselor in the Scouts who observes us, kids, from above. . . . Kids gather around him for an activity, all listening: "We can see the revival on the horizon," Dani says quietly. "We will have to sacrifice ourselves, to give all that we have and all that there is within us, so to ensure for us and for those who will come after us . . . a home here on our forefathers' land. I say goodbye to you. I cannot tell you where I am going. It is a glorious place. One day I will call for you to come to me."
>
> I saw him strong . . . wide eyed and with a smile of an angel. We were scared little kids in the dark, boys and girls; we got closer to him and we touched him.
>
> Dani was the figure with whom I wanted to associate. I wanted to imitate him, to hear and see him. Such charisma . . . and in those days I did not yet know what charisma even was.[19]
>
> (my translation)

Dani Mass was a representative of a whole generation: tall, strong, brave young men who were willing to die for their Zionist ideals. They influenced many younger men who came after them and who themselves would become heroes for the generations that came after them, earning places in the "pantheon" of national heroes, which I shall discuss shortly.

From the early days of Zionism several of the new Jewish military and paramilitary units operated underground, and made membership by invitation only (older members recruited new ones). Because of their selectivity and their reputed bravery, because of the hallowed culture of male bonding that was associated with them, these units became legendary in the emerging Jewish Israeli culture and, increasingly, the most desired units for which young men wanted to volunteer. This canonization of the elite military unit continues to date, and further marks the intense connection between militarism, masculinity, heroism and the Jewish nation.

Many of the men who joined the elite units of the *Hagana* and the *Palmach* and, after statehood, the Israeli Defense Forces, were graduates of youth movements and of agricultural farming boarding schools, especially *Kaduri*, *Ben Shemen* and *Mikve Yisrael*, which all prepared youth to be *kibbutz* farmers as well as fighters.[20] The hierarchy in elite units, and the culture of hazing, the intense military training, and the male bonding associated with these units, mimicked the male-culture of *Hashomer* and of the agricultural schools. The routine abuse and cruelty that young men had to endure as a rite of passage in these schools and in the paramilitary units of the *Hagana* and the *Palmach* marked their manliness. These codes of behavior, the rite-of-passage ceremonies and the cult of toughness associated with them further mythologized the male warriors of Israel.

Special units, particularly the tiny, short-lived Unit 101 of the early 1950s and, later, the paratroopers, made specific contributions to the image of the tough man whose dare-devil courage enabled him to do *anything* for his nation. Such units set the standards for acts of extreme courage in the battlefield which became virtually normative for the elite units of the I.D.F. Until the early 1950s the I.D.F. was not involved in attacks on individual communities across the border. But members of Unit 101 and the paratroopers entered enemy territory regularly as they searched for suspected "terrorists" in retaliation against attacks on Jewish communities. These small units' strategies of rounding up and often separating the men, to "make them pay" collectively for disturbing the life of Jewish communities in Israel, along with infiltration and the disparate attacks across the border in the middle of the night, were new at the time and later became normative military operation for I.D.F. missions. Furthermore, infiltration by a handful of Israeli soldiers into enemy territory, blowing up homes and killing suspects (and sometimes their families), round ups and nighttime face-to-face confrontations with the enemy (Morris 1995, Milshtein 1968), became as a result the ultimate tests of bravery. Since then the bar of bravery of the I.D.F. soldier has constantly been pushed up, as members of each generation surpass those of the previous one in what they have been willing to endure in the name of nation-building and national survival. And as the bar of bravery has been going up, so has the bar of masculinity which has mirrored it.

The pantheon of heroes: myth in the making[21]

Men who proved their courage on the battlefields and who were willing to give their life for national survival became cultural legends in Jewish Israel – like, for example, Meir Har-Zion, a "cold-blooded fighting Jew with an armor-plated conscience" (Elon 1971: 232), veteran of many of these infiltration missions and battles about whom Moshe Dayan wrote: "[his] fighting instincts and courage set an example for the entire Israeli Defense Forces" (quoted in Elon 1971: 232). Har-Zion and his fellow fighters have been remembered in modern Israel in a way that echoes the Hellenistic standard of masculine beauty: tall, wide shouldered, beautiful and brave heroes. The process of memorializing these men and retelling stories about their bravery and vision, commitment to the community and willingness to volunteer has mythologized and canonized them collectively, further reinforcing the strong connection between masculinity and Jewish nationalism. The process of canonizing soldiers as heroes has acquired an almost timeless status in Israel's short history, as regardless of the war the mythologizing narrative is exactly the same.

Over the years more than 20,000 Israeli soldiers have died in defense of their country. Their death has been remembered as the price, the sacrifice, that the Jewish nation has had to pay for its independence and freedom (Witztum and Malkinson 1993). Because the price of freedom has been so high, all fallen soldiers are eulogized as heroes.[22] I offer here two representative examples of the kind of narrative that Israeli parents employ to talk about their fallen dead sons, and the kind of heroic memory employed in Israel, both published in *Siach Shakulim* ("Conversations of Mourners") (the translation of both pieces is mine). One mother wrote:

> Ask me: why did he go?
> Went with no return
> I answered: he went with all the heroes

And a father wrote:

> Our son Yisrael died in the battlefield with the last injured man on his back, while taking care of the wounded soldiers under fire. In his actions he set an example. . . . The stories of his heroism are told in the books that appeared after the Six Day War. He can serve as a role model . . . especially for the youth who are about to conscript to the I.D.F.

In the deepest sense, the choice of who gets remembered and what gets memorialized is in every culture, inevitably, ideological, and is crucial to the nationalism that gets constructed. And because remembrance "shapes our links to the past and the ways we remember define us in the present . . . and nurture [our] vision of the future" (Huyssen 1993: 249), memory has been central to the collective consciousness of every nation. National "memories" are perpetually reconstructed as narrative as they are selectively recollected, reinterpreted and retold. Changing over time to fit particular national needs, they rarely reflect *real* events with a high degree of accuracy. Frequently, instead, as many scholars have argued, these narratives – especially the narratives about historical events that shaped the nation – take the form of myth, one of the most essential ingredients of nationalism (Gellner 1996, 1983, Connor 1994, Hobsbawm 1990, Smith 1986). And which stories or events of the past are used in the present depends greatly on the present needs of the nation, since these myths anchor the present in the past and link the past to the present, reviving for the present the glories of the past and marking the uniqueness of the nation.

In Zionism, too, as in other nationalisms, myth and memory have been crucial to the construction of the nation. These myths have emphasized the Jewish struggle for survival both in the Diaspora and in Palestine, as well as the miraculous status of victory won repeatedly in defiance of Jews' numerical inferiority.[23] This notion of *miracle* has been built up in virtually all the myths of Jewish nationalism, further grounding and dramatizing in national mythology the modern Jewish military hero. Significantly, historical military figures like Bar Kochba and Judah Maccabee – rather than important Rabbis like, for example, the Ramban and Rambam, Rashi or the Gaon from Vilna – have served as the heroic exemplars on which Zionism and the Zionist *New Jew* have been constructed. These historical heroes became the first members of the "pantheon" of Zionist heroes; in turn, members of *Hashomer* and others who died after them "miraculously" defending the Jewish colonies have also been inducted into this "pantheon."

The most influential member of the Zionist "pantheon of heroes" to date has been Joseph Trumpeldor, whose heroism and dying words have influenced generations upon generations of Jewish children in Palestine and, later, in Israel. A highly decorated Russian Jewish officer who lost an arm in the Russian-Japanese war, Trumpeldor came to Palestine as a Zionist in 1907, fought against the Ottomans in World War I, and in 1919 became the commander for Northern Galilee. According to legend, Trumpeldor was fatally wounded in the battle of *Tel Hai* but refused to desert his post and be evacuated. When he finally received medical care he is supposed to have said to the physician who treated him: "No matter, it is worthwhile to die for our land" (*ein davar, kedai lamout b'ad artsenu*) (Ben Gurion 1971: 135).[24] Despite the fact that no one but Trumpeldor's doctor (a recent immigrant with limited Hebrew competence) heard them, an improved version of Trumpeldor's words – *it is good to die for our country* – became what is arguably the most influential motto in modern Zionism, recited annually by Jewish school children from the 1920s to the present (Zertal 1994). To die for one's land became the ultimate modern Jewish sacrifice; it gave meaning to the death battle and, as Mosse (1990) argued, it enabled fallen soldiers to continue to have a significant impact on the living.

Rivka Gover, who lost her two sons in the 1948 war, demonstrates this point eloquently: "After all, our sons did not just give us their life, they gave something more important than that. They returned to us our national and human dignity" (1948, printed in *Siach Shakulim* 1988 [1981], my translation). And it was Trumpeldor who came to epitomize that heroic sacrifice, because he stood alone – outnumbered – in defense of his land (Azaryahu 1995). Significantly, a father whose son died in the 1948 war used Trumpeldor's motto to make highly nationalized sense of the loss of his son:

> When we received the message that our son Ilan fell – it was for us, of course, a terrible hit – but nevertheless we were ready for it. After all, we did not raise our son on almonds and raisins alone,[25] we reared him also on *it is good to die for our country.*
>
> (printed in *Siach Shakulim* 1988 [1981], original emphasis, my translation)

Many of the young men who died in Israel's wars, though certainly not all of them, have been memorialized in remembrance albums, at times published by their families and friends, at others by their military unit; and in the case of *Gvilei Esh* (Parchments of Fire) by the newly established government of Israel.[26] In addition to making available to their family and friends the writings of the fallen, their poetry, art or sections of their diaries, these albums often provide biographical data which stress the status of the dead as heroes who died for the sake of the nation. Even the subtitle of *Gvilei Esh* reads (in English translation): "comprising the literary and artistic works of the fallen heroes of the war of liberation in Israel," thus cementing the connections among death, the masculine hero and the Israeli Jewish nation.

Finally, it seems clear that in all arenas of national mythology the hero is always a male, and that the act of heroism is always associated with military might and masculinity. In Hebrew, as in Arabic and in many ancient Semitic languages, both the word "man" (*gever*) and the word "hero" (*gibor*) come from exactly the same three letter root (G-V-R). So do the phrases for "to overcome" (*lehitgaber*), "to strengthen" (*lehagbir*), "masculinity" or "manliness" (*gavriyut*) and "heroism" (*gvurah*) (Hirschfield 1994: 10) as well as one version of the name for God (*gevurah*) (Gal 1984: 190). In a culture like the one that has developed in Israel – as a result of its struggles to survive – national strength and heroism have become deeply coupled with masculinity. Moreover, because all these words are derived from the same root, the hero (*gibor*) and his act of heroism (*gvurah*) become God-like (*gevurah*). Ultimately it may even be the case that because Jewish nationalism has revived the connection of the Jewish people to its biblical homeland and religion, it has felt important to justify this connection, so that *men* become God-like in their military endeavors.

Conclusion

As early as 1899, one of the most famous Hebrew poets of the time, Shaul Tschernichovsky, made an explicit connection to Greek masculine ideals when he wrote about the *New Jew* in his poem "Facing the Statue Apollo":

> Youth-God, sublime and free, the acme of beauty:
> I am a Jew – Your eternal adversary,
> I am the first of my race to return to you!
> My people is old – its God has aged with it . . .
> The God of light, the god of light calls to every sinew in me:
> Life! Life! . . . To every bone, to every vein . . .
> I have come to you, I bow before your image
> Your image – the symbol of the light of life,
> I bow, I kneel to the good and the sublime,
> To what is worshipped in the fullness of life
> To that which is splendid in all creation.
> I bow to life and courage and beauty.
>
> (translated by D. Kuselewitz, 1978)

Apollo was, according to Tschernichovsky, the standard of courage and beauty for the *New Jew*: brave, standing tall, always ready to defend his community with pride; the antithesis of the stereotype of the Diaspora Jew, the success story of Zionism. These images of the tall and proud *New Hebrew*, tiller and defender of the land, appear frequently in Hebrew writings from the late 1890s onward (see Berlowitz 1983). They were also captured and reinforced in posters (see Plates 12.1, 12.2, 12.3 and 12.4) that were created in Palestine by Jewish artists, in the 1930s and the early 1940s as well as after statehood. These images became the blueprint for the construction of Israeli young men, before and after statehood: the masculinist *New Jew* who was always ready to help his people, defend the land and build her.

When national survival is attributed almost exclusively to the heroism of warriors, nationalism and masculinity become inseparable. And when a nation

Plate 12.1 "Our national purpose is protection and security: the response of the Hebrew youth!"

Source: From the poster collection of the Central Zionist Archives, poster number 1052.

Plate 12.2 "And we stood guard day and night" (Nehemia 4:3)

Source: From the poster collection of the Central Zionist Archives, poster number 8.

cultivates its own myths of survival, recalls the struggles of its past, and celebrates its heroes, it perpetuates the intimate tie between *nation* and *male* and continually constructs the image of both its desired nationalism and its desired son. This son, in whose image the younger generation is socialized, is most dear to the nation when he takes up arms and willingly risks his life for the survival of the nation and when he wins military battles on the nation's behalf. The communal goal of military victory connects *masculinity* and *nationalism* in a militarized society like modern Israel.

Nationalism and masculinity, then, actively participate in one another's construction. They are both, in addition, constructed in opposition to an Other. In the Jewish case, in fact, both nationalism and masculinity have been constructed in opposition: first, to the *Ghetto Jew* and, later, to the indigenous Arab population of Palestine. To mobilize the Jewish people, Zionist leaders called up mythologized stories from the past about the Jews' struggles to survive and their unrelenting fears of annihilation ("the whole world is against us"), continually emphasizing the miracle of Jewish survival in the face of numerical inferiority ("the few against the many"). These inevitably masculinist myths – built around victories made possible by the military courage of Jewish men – have shaped the construction of Jewish nationalism so that when we examine it from virtually any angle we find survival and male bravery as twin narratives.

Because the Jewish nation has remained in constant (real or imagined) fear of annihilation – in a state of what Baruch Kimmerling (1993) calls "a cultural militarism" – and has seen itself as engaged in continuous fighting, it has had to rely heavily on its military prowess and on its sons, and to construct its sons as its defenders. Israeli Jewish men have been socialized into their

Plate 12.3 "We look forward, protecting and building"

Source: From the poster collection of the Central Zionist Archives, poster number 1504.

Plate 12.4 "While one hand works the other holds a weapon" (Nehemia 4:11)

Source: From the poster collection of the Central Zionist Archives, poster number 127.

gender roles by the reality of the first 60 years of Zionism in Palestine and by the messages that they have received in the youth movements, the educational system, paramilitary training and, ultimately, in the modern I.D.F. itself.

Zionism and masculinity have become inseparable: not only have they constructed one another but also they continue to be one another's lifeline. Zionism provided the blueprint for the *New Jew*'s gender identity as well as the arena for perfecting the details of his manliness; and masculinity, bravery, and heroism became in turn what the poet Natan Alterman (1973) has called the "silver platter" on which the Jewish state was given. Because both nation and gender are culturally constructed, and because cultures inevitably change through time, we can expect that Jewish Israeli nationalism and gender identity will change as well. Now when the Israeli military no longer needs men as elite fighters, their masculine identity is threatened. They feel that they are "zero" – that they must either be "real men" or they are nothing. But just as the myths on which Zionism is based have been increasingly questioned and the motivation, on the part of young Israeli men to die for the nation is on the decline, we may well find in the future major changes in both Zionism and masculinity. These changes, I believe, will indicate the maturity of the Zionist project.

Acknowledgments

Earlier versions of this chapter were presented in the *Re*-Thinking Marxism Conference (December 6–8, 1996) at Middlebury College. I thank my colleagues Guntram Herb for our many discussions of German nationalism and masculinity, Robert Schine, for sharpening the translation of the plates' captions, and Joanne Jacobson and Bat-Ami Bar On, for ongoing discussions of masculinity and Jewish nationalism. The chapter has benefited from Professor Jacobson's critical eye.

Notes

1 The word "killers" appears several times in the newspaper article. While the article is written in Hebrew it uses the English word – "killers" – but adds the suffix to denote the plural masculine – *killerim*.

2 I distinguish between a nation and a state, and therefore between nationalism and patriotism. While the nation is a glorified ethnic group whose members have an attachment to a specific territory, the state is a political unit. Therefore, I see loyalty to the nation as nationalism and loyalty to the state as patriotism. In the case of Israel, two distinct nationalisms exist in the land: Jewish nationalism and Palestinian nationalism. For Jews in Israel their nationalism and their patriotism overlap and are often interchangeable.

3 See for example Sharoni (1994).

4 Herzl's fondness for Bismarck is evident in several different entries in his diaries: on June 15, 1895 he wrote: "Napoleon was the sick superman, Bismarck was the healthy one . . ." (Herzl 1956: 52). On several different dates he wrote about Bismarck's success at mobilizing Germans into a German nation (see April 22, June 3, 1895).

5 For a more detailed discussion of the marginalized man as "feminized" see for example McClintock (1995) and Enloe (1989).

6 For an excellent discussion of the construction of the "feminized" Jewish man see Boyarin (1997). Boyarin argues that much of the "feminized" identity of Jewish men over the years, from as early as the Talmudic period, was internalized "and that Jewish men identified *themselves* as feminized" (1997: 12). But the problem with Boyarin's assertion is that the activities in which Jewish men engaged and which may have branded them "feminine" in the eyes of the non-Jews were clearly not open to Jewish women at the time and thus could not have contributed to a "feminine" self-identity. This is a clear example, I believe, of how culture specific gender construction is.

7 The *Übermensch* (male ideal) is taken from Nietzsche's moral system as described in his *Also sprach Zarathustra* and critiqued by Nordau in *Degeneration* (1895). The *Übermensch* possesses traits such as "severity, cruelty, pride, courage, contempt

of danger, joy in risk, extreme unscrupulousness" (Nordau 1895: 422), traits that Nordau believed could lead to tyranny. While the *New Jew* should not possess all these specific traits, he should nonetheless be brave, virile and ready to fight.

8 This quote is taken from an open letter that Nordau wrote to Professor Caesar Lombroso, to whom he dedicated his famous book *Degenerates*. According to Sander Gilman (1986), Lombroso, a Jew himself, had initially developed the notion of "degeneration" mostly for prostitutes and criminals, and only later accepted the anti-Semitic notion that Jews were more prone to specific forms of mental illness.

9 In his 1902 essay "Was bedeutet das Turnen für uns Juden?" which appeared in *Jüdische Turnzeitung*, Nordau explained to his Jewish audience that gymnastics would give them the self-confidence which they were lacking. Further he asserted that Jews would make excellent gymnasts because they were smart; it was not enough to be physically fit, he believed, but essential as well to be daring, in complete control of one's muscles and fast and precise – all of which come from the brain.

10 The father of the German Turner Movement was Friedrich Ludwig Jahn, whose *turnen* activities began as early as 1811 in Berlin and were to instill "the love of the fatherland through gymnastics" (Mosse 1975: 128).

11 Internalizing anti-Semitic characteristics of the Jews as feminine and the Zionists' attempts to reverse them brought major criticism upon Herzl and Nordau from Eastern European Jews and from writers such as Karl Kraus (founder of the Viennese literary journal *Die Fackel*) and others, who "equated the attitudes of Zionism, represented by Herzl, with that of the anti-Semites" (Gilman 1986: 235).

12 Judah refers to Judah Maccabee.

13 The German Youth Movement began among schoolboys in Berlin in 1801 and focused initially on allowing boys to roam the countryside without adult supervision. Soon, however, the movement became politicized and the participants (all males) were asked to be part of the German national consciousness (Mosse 1985: 45).

14 Baden-Powell's Boy Scout movement was intended for white boys in whose hands the future of the empire lay. Through participating in the movement white boys would learn, Baden-Powell believed, self-control and become "tall, muscular, eyes straight ahead, body at attention" (Enloe 1989).

15 Members of the Zionist movement were split over territorial solution and not all of them agreed on historical *Eretz Yisrael* as the location for the Jewish homeland. Through the early Zionist Congresses several territorial proposals were discussed, and included locations such as Uganda, Birobidjan and Argentina.

16 According to Lamm (1991), in the first quarter of the twentieth century twenty-five different Zionist youth movements were established in Europe and they differed from one another on the basis of their political ideology.

17 There were also a few women members in *Hashomer*, but they did not participate actively in security missions. Rather, they learned how to use guns and defend themselves, ride horses and participate in the hard life of the colonies.

18 The first Jewish defense force in Palestine was, actually, *Bar Giora* which was formed in 1907 as a secret organization which assumed the guarding and protection of the Jewish colonies from the Arabs and the Bedouin. This was the first organization whose goal was to train Jewish people to use guns for self-defense. It is from this organization that *Hashomer* was born in 1909.

19 *Siach Shakulim* is published by the *Yad Labanim*, an organization dedicated to memorializing the fallen sons. Established after the 1948 war by a mother whose son had died in battle, Yad Labanim has become an official institution in Israel. *Siach Shakulim*, an internal publication of *Yad Labanim* is produced for families

who have lost their loved ones. It includes mostly creative writings by different family members (usually about the fallen), letters to the fallen soldiers and debates about the appropriate commemoration of the sons. Every few years it also publishes letters that have been published before, and this 1948 letter is one of these.

20 A large number of the most decorated and highest ranking officers in the *Palmach* and later in the I.D.F. were graduates of these schools, especially *Kaduri.*

21 The pantheon of Jewish heroes is comprised almost exclusively of men. Although some women did die defending the land their numbers have remained small, and although they contributed to the cult of heroism they have not been made central to it. Two of the more famous women who have been marked as heroes are Hanna Senesh and Chaviva Reich, two paratroopers who joined the British army during World War II; they jumped into Nazi-occupied Hungary on a secret mission and were killed in the line of duty.

22 George Mosse (1990) suggests that with the development of the "cult of the fallen soldier," at least in public, the national gain from the soldiers' death outweighed the personal loss.

23 For a thorough discussion of myths in Zionism see Wistrich and Ohana (1997), Gertz (1995), Zertal (1994) and Zerubavel (1995).

24 The Hebrew word *artsenu* could be translated as either "our land" or "our country." In 1919, when Trumpeldor died, the likelihood of an Israeli state was nothing less than a dream, and therefore the more fitting translation in this context is "our land." But since children have recited this verse in the late 1940s and thereafter, the meaning has most likely become "our country."

25 "Almonds and raisins" refer here to the Yiddish song *rozinkist mit mandelen.* In this specific context, the father emphasizes that he raised his son not only on Yiddishist, Diaspora culture but also on the new Jewish culture of Palestine. *Shlilat Hagola* (the negation of the Diaspora) was an important theme in Zionist ideology – only by rejecting the Diaspora, many believed, could the *New Jew* be born.

26 The first of these commemoration books was the *Yizkor Book, 1911* which was a way to memorialize *Hashomer* members who had died. For an excellent analysis of the debates surrounding this book see Frankel (1996). The first *Yizkor Book* set the standard for memorializing the fallen soldiers in the wars to come, as well as the memorial books for the many Jewish communities in Europe that vanished at the hands of the Nazis.

References cited

Alboim-Dror, R. (1986) *Hebrew Education in Eretz Yisrael: 1884–1914*, Jerusalem: Yad Ben Zvi Institute.

—— (1996) "'He goes and comes back, from within us comes the first Hebrew': on the youth culture of the first *aliyot*," *Alpayim* 12: 104–135 (in Hebrew).

Almog, O. (1997) *The Sabra – A Profile*, Tel Aviv: Am Oved (in Hebrew).

Almog, S. (1984) "From 'muscular Jewry' to the 'religion of labor'," *Zionism* 9: 137–146 (in Hebrew).

Alterman, N. (1973) *HaTur HaShvii*, vol. 1, 1943–1952, Tel Aviv: HaKibbutz Hameuchad, p. 54 (in Hebrew).

Azaryahu, M. (1995) *State Cults: Celebrating Independence and Commemorating the Fallen in Israel, 1948–1956*, Beer Sheva: Ben Gurion University of the Negev (in Hebrew).

Ben Eliezer, U. (1995) *The Emergence of Israeli Militarism 1936–1956*, Tel Aviv: Dvir Publishing House.

Ben Gurion, D. (1971) *Memoirs*, vol. 1, Tel Aviv: Am Oved.

Berkowitz, M. (1993) *Zionist Culture and West European Jewry before the First World War*, Cambridge: Cambridge University Press.
Berlowitz, Y. (1983) "The model of the *New Jew* in the literature of the first *Aliya*," in *Alei Siach* (*Literary Conversations*) 17–18: 54–70.
Bernstein, D. (1987) *The Struggle for Equality: Urban Women Workers in Prestate Israeli Society*, New York: Praeger.
Boyarin, D. (1997) *Unheroic Conduct: The Rise of Heterosexuality and the Invention of the Jewish Man*, Berkeley: University of California Press.
Connor, W. (1994) "When is a nation?" in W. Connor essays *Ethnonationalism: The Quest for Understanding*, Princeton: Princeton University Press, pp. 210–266.
Elon, A. (1971) *The Israelis: Founders and Sons*, New York: Penguin.
Enloe, C. (1989) *Bananas, Beaches and Bases: Making Feminist Sense of International Politics*, Berkeley: University of California Press.
Frankel, J. (1996) "The 'Yizkor' book of 1911 – a note on national myths in the second Aliya," in J. Reinharz and A. Shapira (eds.) *Essential Papers in Zionism*, New York: New York University Press, pp. 422–453.
Gal, R. (1984) *A Portrait of the Israeli Soldier*, New York: Greenwood Press.
Gellner, E., (1983) *Nations and Nationalism*, Ithaca, N.Y.: Cornell University Press.
—— (1996) "The coming of nationalism and its interpretation: the myths of nation and class," in G. Balakrishnan (ed.) *Mapping the Nation*, London: Verso, pp. 98–145.
Gertz, N. (1995) *Captive of a Dream: National Myths in Israeli Culture*, Tel Aviv: Am Oved (in Hebrew).
Gilman, S. (1986) *Jewish Self-Hatred: Anti-Semitism and the Hidden Language of the Jews*, Baltimore, MD: Johns Hopkins University Press.
Guri, C. (1989) "Youth movements as a serving elite," in M. Naor (ed.) *Youth Movements, 1920–1960*, Jerusalem: Yad Ben Zvi, pp. 221–226.
The Hagana (1968), Tel Aviv: Museum of the *Hagana*, no author (in Hebrew).
Herzl, T. (1956) *Diaries*, edited by M. Lowenthal, New York: Dial Press.
Hirschfield, A. (1994) "Men of men: the hero, the man, and heroism," *Mishkafaim* 22: 9–15.
Hoberman, J. (1984) *Sport and Political Ideology*, Austin: University of Texas Press.
Hobsbawm, E. (1990) *Nations before Nationalism: Programme, Myth, Reality*, Cambridge: Cambridge University Press.
Huyssen, A. (1993) "Monument and memory in a postmodern age," *Yale Journal of Criticism* 6, 2: 249–261.
Izraeli, D. (1994) "On the status of the woman in Israel: women in the military," *International Problems: Society and State* 33: 21–26.
Kimmerling, B. (1993) "Is Israel a militarized society?" *Teoria U'vikoret* 4: 123–140 (in Hebrew).
Krüger, M. (1996) "Body culture and nation-building: the history of gymnastics in Germany in the period of its foundation as a nation-state," *International Journal of the History of Sport* 13, 3: 409–417.
Lamm, Z. (1991) *The Zionist Youth Movements in Retrospect*, Tel Aviv: Sifiriat Ha-Poalim (in Hebrew).
Luz, E. (1988) *Parallels Meet: Religion and Nationalism in the Early Zionist Movement (1882–1904)*, Philadelphia, PA: Jewish Publication Society.
McClintock, A. (1995) *Imperial Leather: Race, Gender, and Sexuality in the Colonial Contest*, London: Routledge.

Milshtein, U. (1968) *The Wars of the Paratroopers*, Tel Aviv: Ramdor (in Hebrew).
Morris, B. (1995) *Israel's Border Wars, 1949–1956: Arab Infiltration, Israeli Retaliation and the Countdown to the Suez War*, Tel Aviv: Am Oved (in Hebrew).
Mosse, G. (1975) *The Nationalization of the Masses: Political Symbolism and Mass Movement in Germany from the Napoleonic Wars through the Third Reich*, New York: Howard Fertig.
—— (1985) *Nationalism and Sexuality: Middle Class Morality and Sexual Norms in Modern Europe*, Madison: University of Wisconsin Press.
—— (1990) *Fallen Soldiers: Reshaping Memory of the World Wars*, Oxford: Oxford University Press.
—— (1993) *Confronting the Nation: Jewish and Western Nationalism*, Hanover, N.H.: Brandeis University Press.
—— (1996) *The Image of Man: The Creation of Modern Masculinity*, New York and Oxford: Oxford University Press.
Naor, M. (ed.) (1989) *Youth Movements, 1920–1960*, Jerusalem: Yad Ben Zvi.
Nordau, M. (1895) *Degeneration*, London: William Heinemann.
—— (1900) "Muskeljudentum," *Jüdische Turnzeitung* June: 10–11.
—— (1902) "Was bedeutet das Turnen für uns Juden?" *Jüdische Turnzeitung* July: 109–113.
Pellegrini, A. (1997) "Whiteface performance: 'race,' gender, and Jewish bodies," in J. Boyarin and D. Boyarin (eds.) *Jews and Other Differences: The New Jewish Cultural Studies*, Minneapolis: University of Minnesota Press, pp. 108–149.
Reichel, N. (1997) "'Roots' or 'horizons': the image of the 'ultimate pupil' in Eretz Israel 1889–1933," *Katedra* 83: 55–96 (in Hebrew).
Reinharz, J. (1996) "Ideology and structure in German Zionism, 1880–1930," in J. Reinharz and A. Shapira (eds.) *Essential Papers in Zionism*, New York: New York University Press, pp. 268–297.
Reulecke, J. (1990) "Das Jahr 1902 und die Ursprünge der Männerbund-Ideologie in Deutschland," in G. Völger and K. v. Welck (eds.) *Männer Bande Männer Bünde: zur Rolle des Mannes im Kulturvergleich*, Köln: Rautenstrauch-Joes-Museum Köln, pp. 3–10.
Shachor, S. (1994) "We used to be men, now we are zero," *Yediot Achronot* November 11 (in Hebrew).
Shakham, A. (1995) "Song of the young, songs of the future," in *Shorashim: Studies on the Kibbutz and the Jewish Labor Movement* 9: 175–192 (in Hebrew).
Shapira, A. (1992) *Land and Power: The Zionist Resort to Force, 1881–1948*, London: Oxford University Press.
Sharoni, S. (1994) "Homefront as battlefield: gender, military occupation and violence against women," in T. Mayer (ed.) *Women and the Israeli Occupation: The Politics of Change*, London: Routledge.
Shimoni, G. (1995) *The Zionist Ideology*, Hanover, N.H. and London: Brandeis University Press.
Siach Shakulim (Conversations of Mourners) (1988 [1981]) Tel Aviv: Merkaz Yad Labanim.
Smith, A. (1986) *The Ethnic Origins of Nations*, Oxford: Basil Blackwell.
Tschernichovsky, S. (1978 [1899]) "Facing the statue of Apollo," translated by D. Kuselewitz, Tel Aviv: Eked, p. 91.
Vital, D. (1975) *The Origins of Zionism*, Oxford: Clarendon Press.

United States, and to that extent, the terms of gender and sexual practice within the military must be carefully policed. Second, I suggest that the U.S. military, by mandating individual sacrifice for the common good, fosters a model of transcendent national citizenship that is closely aligned with heterosexual masculinity. Third, I suggest that the military stands as a repository and defender of U.S. national values, including nationally significant gender ideals. Thus within the United States as in other national contexts, concepts of nationhood, gender, and sexuality are closely linked. What is perhaps unique to the United States is the central role that the nation's military establishment plays in linking those concepts, as national political debates about the military roles of women and homosexuals reveal.

Discussion of argument

Whereas other nations possess a range of institutions that might serve to represent common national interests, the United States is a liberal democracy, and thus its commitment to individualism often supersedes its commitment to the common good. Yet to acknowledge that the United States is a liberal democracy is not to imply that it lacks a communitarian component. Rather, as Lane Fenrich (1994) has argued, American political culture locates communitarianism within its military establishment. Indeed, Fenrich has characterized the U.S. military as "*the* representative public institution in the United States." In the absence of other collectivizing institutions, he argues, the military symbolizes the American nation, because it is the one political institution that irrefutably privileges common, *national* interests over the interests of individual citizens (Fenrich 1994).

Just as it stands alone as a symbol of U.S. national community, the military also fosters a model of transcendent national citizenship. While most U.S. political institutions seek to protect the interests of the individual from interference by other individuals or by the state, the military does the opposite: it requires its members to risk their own personal interests for the sake of a national cause. Genevieve Lloyd observes that, within the liberal-democratic political tradition, "transcendent citizenship" is defined as the willingness to sacrifice one's personal well-being for the sake of a common cause. She notes that the transcendent quality of military service is also what separates an idealized version of *masculine* citizenship, most clearly represented by the soldier, from a less valued version of *feminine* citizenship, characterized by lack of resolve and commitment to private concerns (Lloyd 1986).

The military also serves to defend the values that are the very heart and soul of the American nation. Such values range from political ideals like "life, liberty, and pursuit of happiness," to traditional gender and sexual ideals such as heterosexual masculinity and the male-headed home. Which specific values are perceived to lie at the heart of the American nation is a matter of political perspective. What is clear, however, is that *all* participants in

debates about the military status of women and homosexuals believe that nationally significant values are at stake in decisions about the criteria for military membership.

Another dimension of the military's role as defender of national values is the notion that, while the military *defends* the values of the nation, it conforms to a particular, in some cases different, set of values consistent with its national defense mission. According to military experts, values like liberty, equality and justice that might favor expansion of military membership to women and homosexuals are not always appropriate in a military context. Instead, the military fosters its own distinctive values, such as sacrifice, hierarchy and devotion to duty. Military experts argue that these unique values are what make the military "separate and distinct" from – perhaps even superior to – other U.S. political institutions. Indeed, some participants in debates about women in combat and homosexuals in the military clearly regard the military as a *supra-national* institution.

Finally, as feminist political critics frequently observe, the U.S. military traditionally has been an exclusive masculine preserve (Stiehm 1989, Jeffords 1989, Lloyd 1986). Only in recent years have issues like women's exclusion from combat and homosexuals' exclusion from military roles been opened to debate. Susan Jeffords writes that the military and, more specifically, warfare have affirmed heterosexual, masculine prerogatives, even as those prerogatives have confronted multiple challenges from groups within U.S. civil society. She argues that the activity of making war is "the optimal display of masculine collectivity in America" (Jeffords 1989: 73). Consequently, if the military is *the* representative public institution in the United States; if it helps to foster a model of transcendent national citizenship; and if it defends U.S. national values, it does so in ways that reflect its longstanding masculine bias. Recent debates about the military status of women and homosexuals reveal that the model of national community that the military helps to promote is a model of male-bonding. It reinforces the centrality of heterosexual men to American political community, while denying visibility to other social groups.

While national political debate about the gender and sexual politics of the U.S. military takes many forms, I focus primarily on a series of recent Congressional hearings about women's exclusion from combat and the exclusion of gays and lesbians from military roles. The hearings bring together the testimony of national lawmakers, military officials and other witnesses, most of whom either advocate the exclusion of women and homosexuals from military service, or promote the creation of a more inclusive military establishment. Regardless of whether or not they support the extension of military roles to women and homosexuals, virtually all of these witnesses draw explicit links between U.S. military policy and American national identity. As I shall show, their tendency to connect the military's gender and sexual politics with claims about the national welfare reflects the centrality of the military to U.S. national imaginings.

The politics of military membership

The criteria for membership in the U.S. military have long been open to debate. Military service has generally been regarded as an obligation of citizenship, imposed most heavily on young men. But successive groups of ethnic and racial "outsiders" have also recognized that, within the United States, military service is an important *privilege* of citizenship – one that carries with it the political legitimacy necessary to make broader, national claims. In 1948, after many years of agitation, African-American civil rights leaders succeeded in gaining racial integration of the U.S. military. They recognized that full access to military roles was essential to civic equality. And – equally important – they also understood that full access to military service was a crucial means of *national* self-understanding. As one statesman put it, "[B]lack Americans said if we cannot fight and die for our country, then what do we mean in it?"[1]

This question, posed by Black Americans in the wake of World War II, speaks to the military's role as *the* representative public institution in the United States. An earlier generation of African-American civil rights leaders recognized that military participation, and particularly combat, provided individuals and groups with recognition for their contributions to national community. Without such recognition, African Americans meant "nothing" in America. With the recognition that full and equal military participation affords, however, African Americans could take their place alongside other soldiers as respected members of the national community (see Nalty and MacGregor 1981).

The achievement of racial integration in the military forms the backdrop for more recent debates over the admission of women and homosexuals to military roles. While the policy of excluding women from combat roles has been overturned in recent years, most notably since the Persian Gulf War, their exclusion from infantry, artillery and armor positions remains in place. In hearings before Congress, liberal politicians, feminists and some military women have sought to eliminate the remaining restrictions on women in combat, claiming that full access to the military is crucial if women are to be *agents* rather than *objects* in national politics. Advocates of female combatants, like an earlier generation of African-American civil rights leaders, recognize that participation in the military is a crucial means of gaining access to national community. As one witness before a Congressional committee examining the military impact of the E.R.A. stated, "For all women, discrimination in the military has had a profound effect on their status as citizens. . . . [It] interferes with their access to national leadership roles." She elaborated: "Nearly all men are subject to the military call if they are needed in a national emergency. This is seen as a basic responsibility of citizenship – one that is currently denied to half of the citizens of this country."[2]

As this witness's comments reveal, more is at stake for advocates of female combatants than simply enhancing women's political and professional

opportunities. Also at stake are women's opportunities to contribute meaningfully to a common, national cause. Advocates of female combatants recognize that gender inequality in the military prevents women from contributing fully to national community. Moreover, they contend, combat restrictions prevent women from gaining access to other rights and responsibilities that full membership in national community entails.

Gay men and lesbians have similarly sought full and open access to the military as a means of grounding larger political claims. As one gay officer stated before a Congressional committee, "Military service represents one of the most meaningful ways in which an American can serve his or her country."[3] Barred from military service, openly homosexual men and women are denied the most obvious avenue through which American citizens can serve their country. This is particularly true since, in the United States, other opportunities for national service are notably lacking.

Like their opponents, gay and lesbian witnesses and their supporters imbue the military's sexual politics with immense national significance, claiming that only a more inclusive military can adequately represent the equalitarian ideals of the American nation. As one witness stated, "[A]ll people, regardless of race, gender, or sexual orientation, are entitled to the basic freedoms and liberties that are the very foundation of this Nation."[4] Since the military represents the American nation, this witness implied, it must adhere to the values that are this nation's "very foundation." To do otherwise would be to abdicate the military's important role as a representative public institution and defender of American national values.

Military culture

Throughout Congressional debates on women's admission to combat and the admission of homosexuals to military service, both those who favor restricting military membership and those who favor expanding it have emphasized the military's significance as an embodiment of U.S. national community. In testimonial after testimonial, witnesses have lent credence to the notion that the military is an exceptional public institution in the United States. As one Congressman stated, the U.S. military is "special and unique," showing "the face of America." An officer agreed, calling the military "the best institution we have in this country."[5] Another officer claimed that "the Armed Forces today are arguably the most respected institution we have," so that any policy undertaken by the military has a powerful influence on the broader American society.[6]

Yet if feminists and gay rights activists have developed one set of arguments based on the national centrality of the military, their opponents within the military establishment have developed another, quite different set of claims. Opponents of gender and sexual diversity in the military argue that because the military is a representative public institution, it must not permit "unconventional" social and sexual practices to displace traditional American

values. Moreover, they contend, military members must embody the best qualities of American citizenship. Thus membership in the military must be highly selective, since all Americans do not possess the traits necessary to become exemplary soldiers. Finally, witnesses who seek to exclude female combatants and homosexual soldiers present an idealized image of military fraternity. Their descriptions of male bonding help to illuminate the gender and sexual limits both of the military, and of the imagined national community that it helps to represent.

In outlining the military's role as defender of American national values, opponents of gender and sexual diversity in the military have frequently emphasized its commitment to family, heterosexuality and traditional gender roles. One Congressional witness argued against extending combat roles to women on the grounds that women's presence in combat units would render male soldiers unfaithful to their wives. Female combatants would thus threaten the strong family values that characterize military society.[7] Similarly, a soldier opposed to lifting the anti-homosexual ban argued that gay soldiers would alienate others who had a proper respect for marriage and family life. Only the unsavory would remain, and the military would cease to be a defender of traditional American values.[8] In these instances and others, participants in debates about the military status of women and homosexuals drew links between the military's longstanding gender and sexual practices, and the shared values of the nation. By implication, neither female combatants nor gay and lesbian soldiers belong in the military because, in the opinion of these witnesses, their presence would disrupt nationally significant gender values that the military is designed to defend.

Witnesses have also argued that the criteria for membership in the military must be highly selective. They have suggested that the military must not mirror the broader American political community, accepting all who choose to apply. Rather, the military must embody the best qualities of American citizenship, as well as those qualities vital to the national defense. Such considerations favor making a distinction between the "warrior," who represents a superior class of citizens, and other, more ordinary Americans. As one witness commented, "War is a special activity, different and separate from any other pursued by man." He added, "No matter how clearly we see the citizen and the soldier in the same individual, how strongly we conceive of war as the business of the entire nation, the business of war will always remain separate and distinct." This witness drew a distinction between the ordinary "citizen" and the "soldier," whose activities in war are "separate and distinct." War is not "the business of the entire nation," he maintained, but is rather an exceptional and exclusively masculine activity, such that "[w]e must not judge military organizations by the standards we apply to their civilian counterparts."[9]

On the face of things, this argument contradicts the claim that the military represents U.S. national values and embodies U.S. national community. But the contradiction dissolves when we think of the military as a supra-

national institution – an institution that represents only the best values and ideals of the American nation. The appearance of contradiction also diminishes when we distinguish between the American nation, on the one hand, and U.S. civil society, on the other. Whereas most U.S. civic institutions privilege diverse individual interests over any concept of collective good, the military fosters a concept of collective good at the expense of diverse individuals. It provides a coherent definition of U.S. national community precisely by limiting membership as much as possible to heterosexual men, while excluding groups that threaten its coherence, such as female combatants and gay and lesbian soldiers.

Part of what distinguishes military organizations from their civilian counterparts, proponents of military exclusiveness contend, is that such organizations make demands on their members that have no parallel in civilian life. If civil society is based on the rights of the individual, then in the military, collective imperatives displace individual self-interest entirely. As one officer put it, "To even think in . . . terms [of self-interest] as a military man is patently ludicrous and it is counterproductive to the mindset of a warrior." Whereas civilians might think of individual freedoms, he added, a warrior "must think only, only of mission accomplishment and the good of the unit. Never, ever may he think of his own personal well-being."[10] Stated another officer, soldiers "willingly accept . . . abridgment of their freedom of speech, their right to privacy, and control over their living and working conditions."[11] These statements and others suggest that, in the view of many Americans, the military calls upon its members to demonstrate transcendent national citizenship. Because all Americans are not equally suited to forsake self-interest on behalf of a national cause, witnesses claim, the military should restrict the participation of women and homosexuals, while favoring heterosexual men.

But if heterosexual men's unique capacity for transcendent national citizenship is one justification for limiting women's and homosexuals' access to military roles, their exclusive capacity for male bonding is another. Indeed, opponents of gender and sexual diversity in the military frequently invoke idealized images of male bonding. Part of what enables the soldier to efface individual self-interest for the sake of a national cause, many witnesses have argued, is his location within the military fraternity. Such witnesses have contended that men risk their personal welfare not only or even primarily out of commitment to abstract national interests, but also out of a commitment to the tangible embodiment of those national interests, the fraternity of fighting men. Officers' descriptions of the military fraternity reveal much about the gender-exclusive image of national community that the military helps to foster.

Central to officers' descriptions of military fraternity is the insistence that male bonding is crucial to the effectiveness of the military mission. "Combat is a team activity which brings people closer together than any other profession," one opponent of female combatants remarked.[12] An officer called to testify against women's admission to combat stated, "There is such a thing

as male bonding. . . . We are the ones who will be sent forth to do the most daring deed, the most challenging." He added, "That is a healthy thing. That is a warrior's spirit. . . . [T]here is that spirit of we are a band of brothers."[13]

Opponents of sexual and gender diversity in the military and particularly in combat have emphasized the damaging effects that inclusion of women and open homosexuals would have on male bonding and thus on military effectiveness. As one military official stated, the effectiveness of the military unit depends to a large degree not on difference but on *sameness*. He stated: "Men trust each other when they are alike, their values, their similar training, and the same objectives, traditional values given to them by their families before they entered the military." Later, he implied that male bonding might be too fragile to withstand gender and/or sexual diversity. He explained, "[W]e don't want to tamper too much with that warrior spirit and the intangibles that make it up. My concern would be that we could fracture it."[14] Thus the form of community that the military fosters, and that helps to embody American national community, is off limits to groups that cannot perform the heterosexual male bonding that witnesses repeatedly describe.

Other images of fraternity also emerged in hearings on women in combat and homosexuals in the military. To varying degrees, the images suggest the complex homosocial accents of military life. One set of images, used to explain why openly homosexual men and women would be disruptive to the military mission, implies both an emotional and a bodily intimacy among men in combat. Speaking of his own experience in the Korean War, one Marine combat veteran remarked, "You had to develop trust in a buddy system under harsh and unhealthy living conditions." He added, "Black or white made no difference, we were real men facing the same fate." Whereas racial diversity seemed immaterial to this combat veteran – as long as all participants were "real men" – sexual diversity did not, precisely because of the bodily intimacy that combat implied. Not only would the presence of women and homosexuals inject sexual tension into an already intimate combat setting, resulting in private sexual alliances that might undermine unit cohesion, but even more alarmingly, gay soldiers' participation would introduce the threat of AIDS. The witness stated: "In combat men bleed a lot. They bleed a lot on each other."[15] Another witness similarly invoked the threat of AIDS, stating that "In combat . . . blood flows so freely that it is unusual throughout the day not to be wearing someone else's blood."[16]

According to witnesses who oppose diversifying membership in the military, masculine pride, brotherly bonding, and a range of other "intangibles" are among the qualities that might be fractured if the military ceased to be the exclusive province of heterosexual men. Closely associated in witnesses' testimony with American national community, those "intangibles" clearly marginalize women and homosexuals who do not conform to the heterosexual, masculine ideal. Witnesses also insist that the inclusion of women and homosexuals might fracture the military's effectiveness as a defender of

national security, of national values, of the American way of life. Deeply frustrated by those who would make the military "a cauldron of social experiment," proponents of military exclusiveness have reiterated time and again that not only the fraternal bonds of military life but the security of the nation are at stake in debates over women in combat and gay men and lesbians in the military. Female combatants and gay and lesbian soldiers, they contend, pose a serious threat to national security – a threat equal or greater than that posed by any foreign enemy.

Women in combat

If we accept the notion that the military helps to represent U.S. national community, then women's historical exclusion from the military and particularly from combat tells us something about the gendered contours of the American nation. As Anne McClintock observes, "All nationalisms represent relations to political power and to the technologies of violence" (1995: 352). Whereas most nationalisms grant men greater access to the resources of the nation-state, and particularly to its technologies of violence, women's relation to the nation-state is altogether different. Indeed, according to McClintock, women are often "[e]xcluded from direct action as citizens" and "subsumed symbolically into the body politic as its boundary and metaphoric limit" (1995: 353).

American women's status as political objects rather than agents is nowhere more apparent than in the military. While women's military roles have expanded considerably in recent decades, so that 80 percent of all jobs in the armed forces are now open to women and women make up 13 percent of the total armed forces, female soldiers continue to be excluded from the most symbolically central of all military roles: direct ground combat. Significantly, American women also are not subject to conscription; they are not obligated as most men are to defend the nation from peril.

Rather, according to witnesses who oppose female combatants, if men's role is to defend the nation, women's role is to be part and symbol of the nation that is defended. As one writer put it, "Women, at least 'our women,' are not a part of war. Indeed, one of the reasons for fighting is to protect our women and the rest of what is in the image of the world back home."[17] Arguing against female combatants, an analyst for the Heritage Foundation stated that a defining national characteristic is that "Good men protect good women."[18]

Women's status as "boundary and metaphoric limit" of the nation is evident in the rhetorical strategy used by witnesses who would exclude women from combat. They depict women as passive and vulnerable objects, incapable of self-protection, rather than as active and able participants in war. They also claim that the likely consequences of permitting women to serve in combat would be nationally repugnant. "[I]s America ready to see one of its daughters half-naked dragged by a rope through the streets of a foreign capital?"

a Congressman asked.[19] Elaine Donnelly, member of the 1992 Presidential Commission on Women in the Military, cautioned that if women are admitted to combat, "we as a Nation . . . must prepare our soldiers and, indeed, the entire Nation, for the inevitability of deliberate violence against women, especially if they are captured behind enemy lines."[20] Again invoking the national welfare, Donnelly asked, "Do we have to do this? . . . If we do, how will it affect the Nation and how will it affect national security[?]"[21] Among the threats to national security, Donnelly implied, was the ease with which women's bodies and minds could be manipulated. Because women are physically weaker and more susceptible to sexual violence, Donnelly suggested, they are more vulnerable to enemy manipulation than men. Remarking on the likely fate of a captured female combatant, she stated, "[I]t was very easy to imagine total control of her and power over her in a real captive situation."[22]

Clearly, the prospect of female ground combatants disrupts cherished, nationally significant images of femininity. Yet the prospect of women's admission to combat also disrupts cherished images of *masculinity*. The testimony of several Congressional witnesses indicates that women's admission to ground combat would compromise widely held assumptions about the manliness of combat. The statement of one Congressman reveals how gendered images of combat have become. He commented, "[W]hen it comes to repeling out of helicopters or slitting throats at night after scaling up a point to neutralize the German guns on D-Day, I don't see my three daughters or four granddaughters or any women [doing that] . . . do you, gentlemen?"[23] As this witness's comment implies, women are not supposed to engage in military heroics; rather, they are supposed to be the daughters and granddaughters of men who do. Women's status as dependent family members helps to dignify the national role that male defenders play, yet it denies them equal opportunity to express their loyalty to the nation.

Opponents of gender diversity in the military have cited women's physical and psychological inadequacy for combat. Jeff Tuten (1982), a military operations analyst and former infantry officer in the U.S. Army, notes factors ranging from women's upper-body weakness to their lack of urge to kill as justifications for their continued exclusion from combat. Yet one argument in favor of women's admission to combat is that technological change has dramatically altered the nature of combat, obviating arguments that stress women's physical and psychological inadequacy. As Nancy Loring Goldman (1982) observes, "increased reliance on technology has led to an expansion of the military roles that women are defined as capable of performing." She adds, "The line between combat and noncombat becomes more and more difficult to draw" (1982: 16). The technological transformation of combat has undermined many arguments against the inclusion of women in combat, but opponents continue to emphasize the negative consequences of female combatants for unit cohesiveness and male bonding.

Perhaps the central argument against using female combatants is that women's presence would be sexually disruptive and therefore damaging to unit cohesiveness. Reflecting the sexual tension that pervades military culture, one soldier remarked, "It's too dangerous for women to be out on the line. Say you go to war and a woman rips her pants. The man next to her is not going to be concentrating on his job because he is going to be concentrating on the hole in her pants."[24] Stated another soldier, "When all is said and done, they [women] cost more than their worth. The divisiveness, sexuality things – headaches that come with it."[25] As these comments imply, opponents of gender diversity in the military see women as sexually troublesome intruders, rather than as citizens bearing equal rights and responsibilities to the nation.

Indeed, if on one hand non-military women are idealized as vulnerable and defenseless, then a very different set of stereotypes applies to military women, who are seen as sexually unsettling or worse. Sexual harassment in the military is well documented, never more so than in the Tailhook scandal involving high-ranking naval officers in 1991. Commenting on harassment, one female officer stated, "[T]he purpose of harassment . . . is always the same: to humiliate and degrade women, to make them feel they do not belong, and ultimately, to drive them out or keep them out." She added that while sexual harassment exists outside of the military, "It is most prevalent in communities that exclude women as in military combat fields."[26] Harassment has often been said to be the worst in the Marine Corps, arguably the most combative of the U.S. Armed Forces. As one Marine, sympathetic to military women's plight, put it, "I thank God every day that I'm a male Marine in this male Marine Corps. . . . If a woman Marine is a little too friendly, she's a slut. If she doesn't smile at all, she's a dyke. . . . [A] woman Marine in the normal course of a day confronts more stress and more bullshit than a male Marine would in twenty years."[27] As Michelle M. Benecke and Kirsten S. Dodge (1992) have observed, when women are harassed, they often remain silent for fear of being called lesbians. Such fears are well founded, for under the "don't ask, don't tell, don't pursue" policy, allegations have placed the careers of countless female soldiers in jeopardy.

Military women negotiate a difficult path through a morass of negative stereotypes. Meanwhile, their careers are limited by virtue of persistent restrictions on combat. American women's ability to participate in the defense of the nation is arguably greater than it has ever been before, yet "the exclusion of women from the full range of . . . assignments," one witness commented, "hurts . . . women's careers, morale, and acceptance."[28] Another witness stated, "There will always be the connotation that a woman is inferior if she is restricted in categories in which she can serve. . . . [I]f you see people who are restricted, that means they are not quite your equal."[29]

Perhaps the most damning – and nationally significant – image of the female combatant is that of the woman who puts her own personal career aspirations above the collective interests of the nation, willfully insinuating

her way into combat and then proceeding to jeopardize the entire military mission. Opponents of women in combat argue that "such women's clear and primary objective is to create more 'career opportunities' for women, as if the armed forces were just another government-sponsored job program."[30] Capitulating to women who pursue combat assignments not only reflects "bad military judgment," but also is "morally wrong," since "[w]inning in war is often only a matter of inches, and unnecessary distraction or any dilution of that combat effectiveness puts the mission and lives in jeopardy."[31]

The emphasis on female soldiers' selfishness, in contrast to male soldiers' devotion to duty, dramatizes the military's role in grounding a masculine concept of transcendent national citizenship. Political critic Genevieve Lloyd (1986) writes that "the masculinity of war is what it is precisely by leaving the feminine behind. It consists in the capacity to rise above what femaleness symbolically represents: attachment to private concerns, to 'mere life.'" By leaving the feminine behind, Lloyd notes, "the soldier becomes a real man." She concludes: "Womankind is constructed so as to be what has to be transcended in order to be a citizen" (Lloyd 1986: 75).

Gays in the military

If women are one group whose petitions for full and equal membership in the military have fallen on resistant ears in recent decades, then homosexuals are another. The exclusion of both groups speaks to the gender and sexual limits of U.S. military culture, as well as to the gender and sexual limits of U.S. national community. As scholars Garry L. Rolison and Thomas K. Nakayama (1994) observe, the manner in which arguments against homosexuals' full inclusion have been waged reflects the military's preoccupation with preserving the embattled masculinity of heterosexual men. They note that the discourse of exclusion regarding homosexuals is a "defensive discourse" – one that is framed less in terms of the interests of homosexuals themselves, than in terms of the consequences that inclusion of homosexuals would have for heterosexual soldiers. Rolison and Nakayama note that the debate over gays in the military has reached a fever pitch just as anxieties about white, heterosexual manhood are at a high point in the United States: "Now, more than ever before, white males are feeling threatened as their cultural space is being redefined in relation to Others – members of racial/ethnic minorities, women, and gays and lesbians" (1994: 126). They note that "the military . . . claims to be a special arena that should be immune from such social concerns and politics" (1994: 126). They also note that homosexual participation threatens military culture, since in the military, "homosexuality [is] defined in wholly antithetical terms with respect to hegemonic masculinity" (1994: 130).

Among the characteristics that distinguish gay soldiers from heterosexual soldiers, witnesses who favor the exclusion of homosexuals have argued, is that gays are preoccupied with their own selfish concerns; like women, they

are ill equipped to deal with the fact that "[i]n combat training and in war, an individual's desires, interests, or career aspirations are totally subordinated to the accomplishment of the military mission."[32] Instead, one officer stated, "Homosexuals constantly focus on themselves. Their so-called needs, what they want, their entitlements, their rights. They never talk about the good of the unit." He added, "It is this constant focus on themselves, the inability to subjugate or subordinate their own personal desire for the good of the unit, this is an instant indicator of trouble in combat, and frankly, not even in combat."[33] Such comments are a bit ironic, given how preoccupied many supporters of the anti-homosexual ban are with promoting the exclusive interests of heterosexual men.

The fear of homosexual impulsiveness has definite sexual connotations, reflecting the extent to which homophobia pervades military culture. According to RAND investigators, anti-gay stereotypes, including the view that homosexuals are "unable to control their sexual urges and unable to distinguish between those who would and those who would not welcome an advance," are rampant in the military. As one participant in a RAND-sponsored focus group explained, "I'm afraid to be in the showers with them." Stated another, "I'd be afraid to be in a foxhole with a gay person. I don't trust them. I'd be afraid that if I looked the other way, he'd do something."[34]

Particularly intriguing is the way that homosexuals are likened to a foreign enemy. Unlike the heterosexual soldier's trustworthy battlemates, gay military personnel are perceived as threatening outsiders, not really part of the American nation at all. This is evident in the frequently drawn distinction between the civil rights of all Americans and the so-called "special interests" of gay men and lesbians. As one proponent of military exclusiveness remarked, "What is more important, the civil rights of all Americans and the National interests of this great country that makes those civil rights possible . . . or the special interests of a tiny fraction of the population?" The same witness later remarked, "Gays are a danger now, and after . . . lifting the ban, more young Americans will become their victims."[35] Another military officer stated that the military regards gay soldiers as "infectious and life-threatening disease carriers." He made the analogy to the foreign enemy explicit, noting that lifting the ban "would be a threat equal to the enemy threat itself, a great threat upon the health and continuing existence of your own young men."[36]

Also dramatized in debates over homosexuals is an important distinction between the military, which discriminates as a matter of course, and civil society, which is at least nominally committed to equal rights and opportunities for all U.S. citizens. Stated one high-ranking officer, "I do not believe that there should be discrimination against homosexuals in civilian jobs." Yet he continued: "But I strongly believe that the military as an institution is extremely unique, extremely different."[37] Whereas talk of gay rights might be appropriate in civil society, a Congressman remarked, "We are not talking

of civil rights here. We are talking about an effective fighting force in our military and what is conducive to effectiveness and cohesiveness in that fighting force."[38] Stated another officer, "We in the military . . . are very discriminatory. We always have been. We must remain so."[39] The dictates of national security, he argued, require it. Such comments lend credence to the notion that, in the view of some Americans at least, the military is a *supra-national* institution, exempted by virtue of its national defense mission from the anti-discriminatory policies that govern much of American society. Moreover, favorable commentary concerning military discrimination reflects witnesses' perception that the military represents the *best* traits and values of the American nation, in contrast to the vast and questionable diversity of U.S. society as a whole.

Another argument against admitting openly homosexual military personnel is that their presence diminishes the image of the military and makes membership less attractive to the military's dominant and most privileged constituency, heterosexual males. One soldier predicted that if openly homosexual soldiers were allowed to serve, "No one will want to join. . . . Morale will go down." He explained, "We join because of the image, because we do the job right, are macho."[40] Other witnesses have similarly argued that the military will lose its vital national stature if efforts to diversify membership succeed.

Such witnesses' concerns about the macho image of the military serve to tie military ambivalence about women and homosexuals together. The presence of both groups defies the military's image of itself as hypermasculine – as what Susan Jeffords has called "the optimal display of masculine collectivity in America" (Jeffords 1989: 73). But the presence of female and gay combatants does more than defy *masculine* collectivity; it defies *national* collectivity as well. The rhetoric of national security and national defense that surrounds objections to both groups must be taken very seriously. It helps us to understand the centrality of the military as an institution that figures national collectivity – above all, a collectivity of privileged male citizens – in the United States.

Conclusion

The model of national community that I have outlined here is not necessarily the only one available to the American public. Yet as I have sought to demonstrate through an analysis of Congressional testimony, this model is nevertheless influential. Its influence derives from the fact that, as witnesses on both sides of Congressional debate have argued, the U.S. military performs an important representative function. In the absence of other collectivizing institutions, the close-knit fraternity of military life helps to symbolize U.S. national community. Witnesses have also argued that America's traditionally masculine soldiers are exemplars of transcendent national citizenship and defenders of American national values.

13 Gender, sexuality and the military model of U.S. national community

Holly Allen

Considerable national debate has focused on the place of women in the U.S. armed forces. Women's efforts to expand their military roles, and even to gain access to combat assignments, have been a topic of national controversy since the 1970s when efforts to ratify the Equal Rights Amendment (E.R.A.) began. More recently, women's efforts to gain access to military academies like the Citadel and women's contributions to military missions such as Operation Desert Storm have captured national attention. Since the late 1960s, women have expanded their place in the U.S. military, but national debate remains heated about whether they should be admitted to additional combat roles.

Similarly, national controversy has surrounded the question of whether or not homosexuals should be permitted to serve in the U.S. military. Spurred by President Bill Clinton's 1992 campaign promise to lift the ban on gay and lesbian military personnel, debate has raged about the appropriateness of military policies that proscribe soldiers' private sexual lives. Controversy has also surrounded the policy of "don't ask, don't tell, don't pursue" that resulted from Clinton's failed effort to repeal the anti-homosexual ban. Cases of men and women expelled from military service on the grounds of sexual orientation have frequently appeared in the national press, as have challenges to antihomosexual policies by gay rights activists.

In this chapter, I address *why* so much national attention has focused on U.S. military policy and practice regarding women and homosexuals. Why are the issues so hotly contested? Why are they considered newsworthy? What does their centrality to national political debate say about the significance of military institutions to American concepts of citizenship and national community?

I argue that the positions of women and homosexuals in the U.S. military – and the national preoccupation with those positions – reveal a great deal about how national community is imagined in the United States. Among other things, they show that U.S. military conventions are closely linked to a particular, *fraternal* model of U.S. national community. Using Congressional hearings about the military status of women and homosexuals, I suggest, first, that the military is *the* representative public institution in the

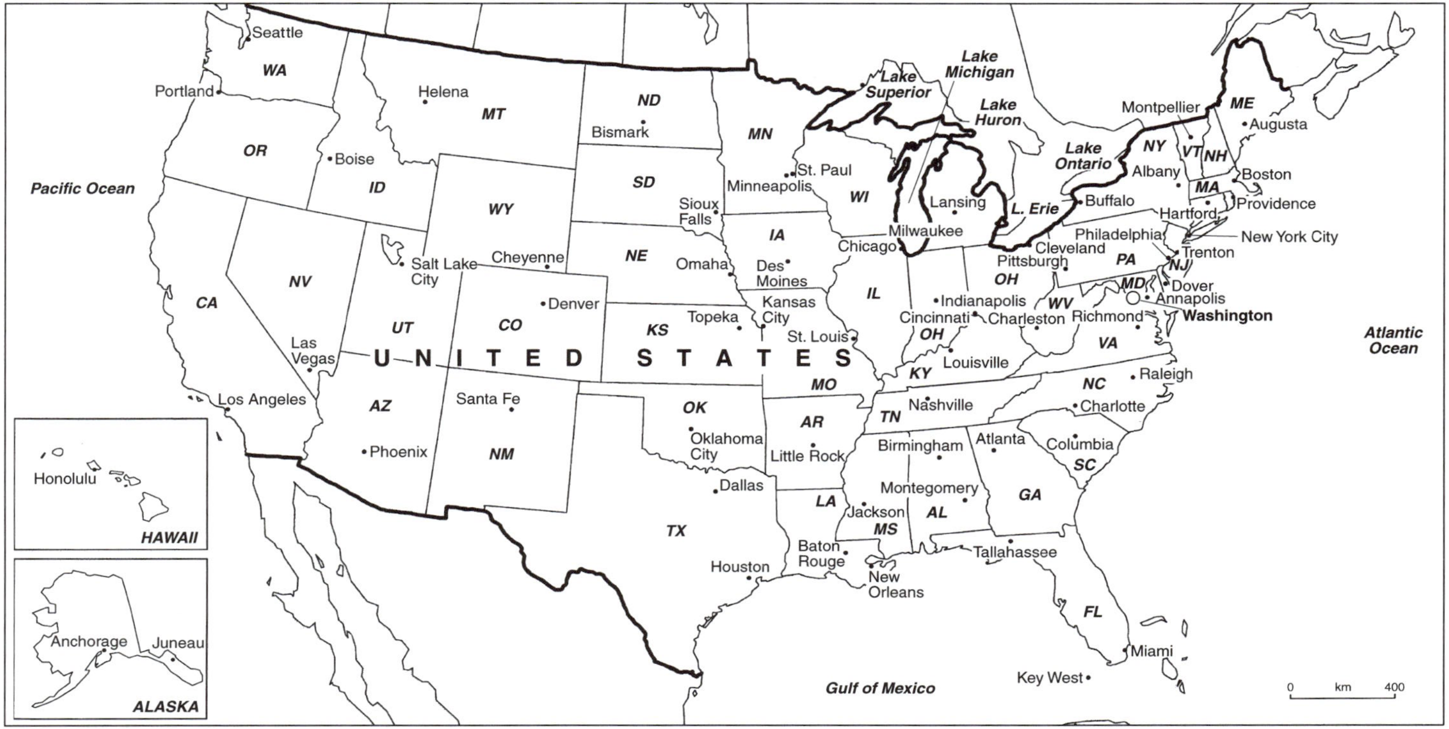

United States of America

—— (1982) *Zionism: The Formative Years*, Oxford: Clarendon Press.

Wistrich, R. and D. Ohana (1997) *Myth, Memory, and Trauma: Transfiguration of Israeli Consciousness*, Jerusalem: Van Leer Jerusalem Institute (in Hebrew).

Witztum, E. and R. Malkinson (1993) "Bereavement and commemoration in Israel: the dual face of national myth," in E. Witztum and R. Malkinson (eds.) *Loss and Bereavement in Jewish Society in Israel*, Jerusalem: Ministry of Defense Publishing House – Cana, pp. 231–255.

Zertal, I. (1994) "The sacrificed and sanctified: the construction of a national martylogy," *Zemanim* 48: 26–45 (in Hebrew).

Zerubavel, Y. (1995) *Recovered Roots: Collective Memory and the Making of Israeli National Tradition*, Chicago and London: University of Chicago Press.

In all of these ways, the military is central to U.S. national imaginings. And it is precisely because of this centrality that advocates for female combatants and gay and lesbian soldiers have sought access to military roles. Like African Americans and other groups before them, women and homosexuals recognize that military service is an important route to membership in U.S. national community. Moreover, they recognize that exclusion from military roles sets a standard of inequality that carries over into other aspects of American political life.

Yet if women and homosexuals have sought full membership in U.S. national community by participating in the military, their efforts have been powerfully resisted by officers, politicians, and others who have sought to preserve the military status quo. Opponents of gender and sexual diversity in the military have argued that the inclusion of women and homosexuals would compromise American national values, particularly the nation's commitment to heterosexuality and traditional gender roles. Moreover, according to opponents, altering current policies would change the military from being a testing ground for transcendent national citizenship into just another "government-sponsored jobs program," since neither women nor homosexuals possess the traits necessary to become exemplary soldiers. Finally, the presence of female combatants and gay and lesbian soldiers would compromise the male bonding that is crucial to military effectiveness. Witnesses maintain that without the strong, fraternal ties that bind American soldiers together, the military would cease to function as the embodiment of U.S. national community.

Yet if the fraternal model of U.S. national community privileges heterosexual men, it also makes strategic use of women and homosexuals. They are important not as equal members of the military fraternity, but rather as marginal groups whose outsider status helps to define the terms of U.S. national belonging. Judith Hicks Stiehm (1982) observes that if men are the "protectors" of the nation, then women are the "protected." Their inability to defend themselves, much less the nation, helps give meaning to the contrasting selflessness, strength and national devotion of male soldiers. Homosexuals perform a similarly negative function in defining the terms of U.S. national belonging. As with women, homosexuals' ostensible selfishness and preoccupation with particular concerns contrasts with the presumptive selflessness and national commitment demonstrated by America's fighting men. Moreover, as Congressional testimony reveals, homosexuals are likened to a foreign enemy. The threat they pose tangibly demonstrates the need for national defense. The military status of both women and homosexuals suggests that, without a politics of exclusion, the fraternal model of U.S. national community would cease to function effectively.

In recent decades, both women and homosexuals have made important advances in U.S. civil society. Yet their continued exclusion from important military roles is highly significant. Only when the military ceases to discriminate on the basis of sexual and gender difference will full national equality be possible. In the meantime, the terms of gender and sexual practice within the military will continue to reveal much about the gender and sexual limits of U.S. national community.

Notes

1 Chair, Policy Implications of Lifting the Ban on Homosexuals in the Military, Hearings before the Committee on Armed Services, House of Representatives, 103rd Congress, 1st Session, 1993, p. 155.
2 Statement of A. C. Hayes, Partner, Law Firm of Csaplar and Bok, The Impact of the Equal Rights Amendment: The Military, Hearings before the U.S. Senate, Subcommittee on the Constitution, Committee on the Judiciary, 1993, p. 303.
3 Testimony before the Senate Armed Services Committee, Hearings on the Military Policy Concerning the Service of Gay Men and Lesbians in the Armed Forces: The Constitutional Invalidity of the Exclusionary Policy, 1993, p. 839.
4 Prepared Statement of Colonel K. Cropsey (Ret.), Policy Implications of Lifting the Ban on Homosexuals in the Military, p. 165.
5 Statements of Committee Chair and Master Chief Jackson, Policy Implications of Lifting the Ban on Homosexuals in the Military, pp. 155–156.
6 Statement of Chair and Brigadier General W. Weise, USMC (Ret.), Policy Implications of Lifting the Ban on Homosexuals in the Military, p. 93.
7 Statement of E. Donnelly, Women in Combat: Hearing before the Military Forces and Personnel Subcommittee of the Committee on Armed Services, House of Representatives, 103rd Congress, 1st Session, 1993, p. 114. For a similar remark concerning homosexuals, see statement prepared by R. H. Knight and D. S. Garcia, Policy Implications of Lifting the Ban on Homosexuals in the Military, p. 142.
8 Statement of Colonel J. Ripley, Policy Implications of Lifting the Ban on Homosexuals in the Military, p. 91. Ripley commented, "[N]ormal, decent Americans will not support this kind of activity. They will prevent their children . . . from joining the military." See also National Defense Research Institute (1993: 233–234).
9 Statement of Prof. E. A. Cohen, Harvard University Department of Government, The Impact of the Equal Rights Amendment: The Military, Hearings before the U.S. Senate, Subcommittee on the Constitution, Committee on the Judiciary, 1993, p. 272.
10 Ripley, Policy Implications of Lifting the Ban on Homosexuals in the Military, pp. 87–88.
11 Jackson, Policy Implications of Lifting the Ban on Homosexuals in the Military, p. 84.
12 Heritage Foundation no. 230, "Should Congress hold hearings before allowing women in combat?" July 27, 1994, Appended to Assignment of Army and Marine Corps Women under the New Definition of Ground Combat, Hearings before the Military Forces and Personnel Subcommittee of the Committee on Armed Services, House of Representatives, 103rd Congress, 2nd Session, 1994, p. 59.
13 Statement of General Mundy, Women in Combat, p. 98.
14 Ripley, Policy Implications of Lifting the Ban on Homosexuals in the Military, pp. 89, 98.
15 Weise, Policy Implications of Lifting the Ban on Homosexuals in the Military, p. 103.
16 Ripley, Policy Implications of Lifting the Ban on Homosexuals in the Military, p. 90.
17 Segal (1982: 273) cited by Cohen, Impact of the Equal Rights Amendment.

18 Heritage Foundation no. 230, "Should Congress hold hearings?" p. 59.
19 Statement of Honorable S. E. Buyer, Representative from Indiana, Assignment of Army and Marine Corps Women under the New Definition of Ground Combat, p. 4.
20 Donnelly, Women in Combat, p. 112.
21 Donnelly, Women in Combat, p. 112.
22 Donnelly, Women in Combat, p. 112.
23 Congressman Dornan, Gender Discrimination in the Military, Hearings before the Military Personnel and Compensation Subcommittee and the Defense Policy Panel of the Committee on Armed Services, House of Representatives, 102nd Congress, 2nd Session, 1992, p. 48.
24 Statement made during RAND-sponsored focus group with military personnel. Cited in National Defense Research Institute (1993: 229).
25 National Defense Research Institute (1993: 229).
26 Statement of Major General J. Holme, USAF (Ret.) and author of *Women in the Military* (Holme 1982), Gender Discrimination in the Military, p. 13.
27 Captain G. Richardson, *The Progressive*, March 1989. Cited in Benecke and Dodge (1992: 167).
28 Holme, Gender Discrimination in the Military, 15.
29 Statement of Admiral Zumweilt, Gender Discrimination in the Military, pp. 38–39.
30 Statement of Elaine Donnelly, President, Center for Military Readiness, Re: The Assignment of Women in or Near Land Combat Units, House Armed Services Subcommittee on Military Forces and Personnel, October 6, 1994, p. 63.
31 The Case against Women in Combat: Executive Summary, House Armed Services Committee, Subcommittee on Military Forces and Personnel, May 12, 1993, p. 68.
32 The Case against Women in Combat, p. 67.
33 Ripley, Policy Implications of Lifting the Ban on Homosexuals in the Military, p. 89.
34 National Defense Research Institute (1993: 230).
35 Weise, Policy Implications of Lifting the Ban on Homosexuals in the Military, p. 98.
36 Ripley, Policy Implications of Lifting the Ban on Homosexuals in the Military, p. 90.
37 Jackson, Policy Implications of Lifting the Ban on Homosexuals in the Military, p. 155.
38 Congressman Bartlett, Policy Implications of Lifting the Ban on Homosexuals in the Military, p. 65.
39 Ripley, Policy Implications of Lifting the Ban on Homosexuals in the Military, p. 90.
40 National Defense Research Institute (1993: 233–234).

References cited

Benecke, M. M. and K. S. Dodge (1992) "Lesbian baiting as sexual harassment: women in the military," in W. J. Blumenfeld (ed.) *Homophobia: How We All Pay the Price*, Boston, MA: Beacon Press.

Fenrich, L. (1994) Roundtable on Gays in the Military, Organization of American Historians Meeting, Atlanta, Georgia, April.

Goldman, N. L. (1982) "Introduction," in N. L. Goldman (ed.) *Female Soldiers: Combatants or Noncombatants? Historical and Contemporary Perspectives*, Westport, CT: Greenwood Press, pp. 3–17.

Heritage Foundation no. 230 (1994) "Should Congress hold hearings before allowing women in combat?" July 27, 1994, Appended to Assignment of Army and Marine Corps Women under the New Definition of Ground Combat, Hearings before the Military Forces and Personnel Subcommittee of the Committee on Armed Services, House of Representatives, 103rd Congress, 2nd Session.

Holme, J. (1982) *Women in the Military: An Unfinished Revolution*, Novato, CA: Presido Press.

Jeffords, S. (1989) *The Remasculinization of America: Gender and the Vietnam War*, Indianapolis: Indiana University Press.

Lloyd, G. (1986) "War, selfhood, and masculinity," in C. Pateman and E. Gross (eds.) *Feminist Challenges: Social and Political Theory*, Boston, MA: Northeastern University Press, pp. 63–76.

McClintock, A. (1995) *Imperial Leather: Race, Gender, and Sexuality in the Colonial Conquest*, New York: Routledge.

Nalty, B. C. and M. J. MacGregor (eds.) (1981) *Blacks in the Military: Essential Documents*, Wilmington, DE: Scholarly Resources.

National Defense Research Institute (1993) *Sexual Orientation and U.S. Military Personnel Policy: Options and Assessment*, Santa Monica, CA: RAND.

Presidential Commission on the Assignment of Women in the Armed Forces (1992) Report to the President, November 15, 1992, Washington, D.C.: U.S. Government Printing Office.

Rolison, G. L. and T. K. Nakayama (1994) "Defensive discourse: blacks and gays in the U.S. military," in W. J. Scott and S. Carson Stanley (eds.) *Gays and Lesbians in the Military: Issues, Concerns, and Contrasts*, New York: Aldine de Gruyter, pp. 121–133.

Segal, M. W. (1982) "The argument for female combatants," in N. L. Goldman (ed.) *Female Soldiers: Combatants or Noncombatants? Historical and Contemporary Perspectives*, Westport, CT: Greenwood Press, pp. 267–290.

Stiehm, J. H. (1982) "The protector, the protected, the defender," *Women's Studies International Forum* 5, 3–4: 367–376.

—— (1989) *Arms and the Enlisted Woman*, Philadelphia, PA: Temple University Press.

—— (1994) "The military ban on homosexuals and the Cyclops effects," in W. J. Scott and S. Carson Stanley (eds.) *Gays and Lesbians in the Military: Issues, Concerns, and Contrasts*, New York: Aldine de Gruyter, pp. 149–162.

Tuten, J. (1982) "The argument against female combatants," in N. L. Goldman (ed.) *Female Soldiers: Combatants or Noncombatants? Historical and Contemporary Perspectives*, Westport, CT: Greenwood Press, pp. 237–265.

United States Congress, House of Representatives, Committee on Armed Services, Subcommittee on Military Personnel (1981) Women in the Military: Hearings before the Military Personnel Subcommittee of the Committee on Armed Services, House of Representatives, 96th Congress, 1st and 2nd Sessions, November 13, 14, 15, 16, 1979 and February 11, 1980, Washington, D.C.: U.S. Government Printing Office.

United States Congress, House of Representatives, Committee on Armed Services, Military Personnel and Compensation Subcommittee (1990) Women in the Military: Hearing before the Military Personnel and Compensation Subcommittee of the Committee on Armed Services, House of Representatives, 101st Congress, 2nd Session, Hearing held March 20, 1990, Washington, D.C.: U.S. Government Printing Office.

United States Congress, House of Representatives, Committee on Armed Services (1993) Policy Implications of Lifting the Ban on Homosexuals in the Military: Hearings before the Committee on Armed Services, House of Representatives, 103rd Congress, 1st Session, Hearings held May 4 and 5, 1993, Washington, D.C.: U.S. Government Printing Office.

United States Congress, House of Representatives, Committee on Armed Services, Military Forces and Personnel Subcommittee (1994) Women in Combat: Hearing before the Military Forces and Personnel Subcommittee of the Committee on Armed Services, House of Representatives, 103rd Congress, 1st Session, Hearing held May 12, 1993, Washington, D.C.: U.S. Government Printing Office.

United States Congress, House of Representatives, Committee on Armed Services, Military Personnel and Compensation Subcommittee (1992) Gender Discrimination in the Military: Hearings before the Military Personnel and Compensation Subcommittee and the Defense Policy Panel of the Committee on Armed Services, House of Representatives, 102nd Congress, 2nd Session, Hearings held July 29 and 30, 1992, Washington, D.C.: U.S. Government Printing Office.

United States Congress, Senate, Committee on the Judiciary (1983) The Impact of the Equal Rights Amendment: Hearings before the Subcommittee on the Constitution of the Committee on the Judiciary, United States Senate, 98th Congress, 1st and 2nd Sessions, Hearings held November 1, 1983, Washington, D.C.: U.S. Government Printing Office.

United States Congress, Senate, Armed Services Committee (1993) Hearings on the Military Policy Concerning the Service of Gay Men and Lesbians in the Armed Forces: The Constitutional Invalidity of the Exclusionary Policy, Senate, 103rd Congress, 1st Session, Hearings held March 31, 1993, Washington, D.C.: U.S. Government Printing Office.

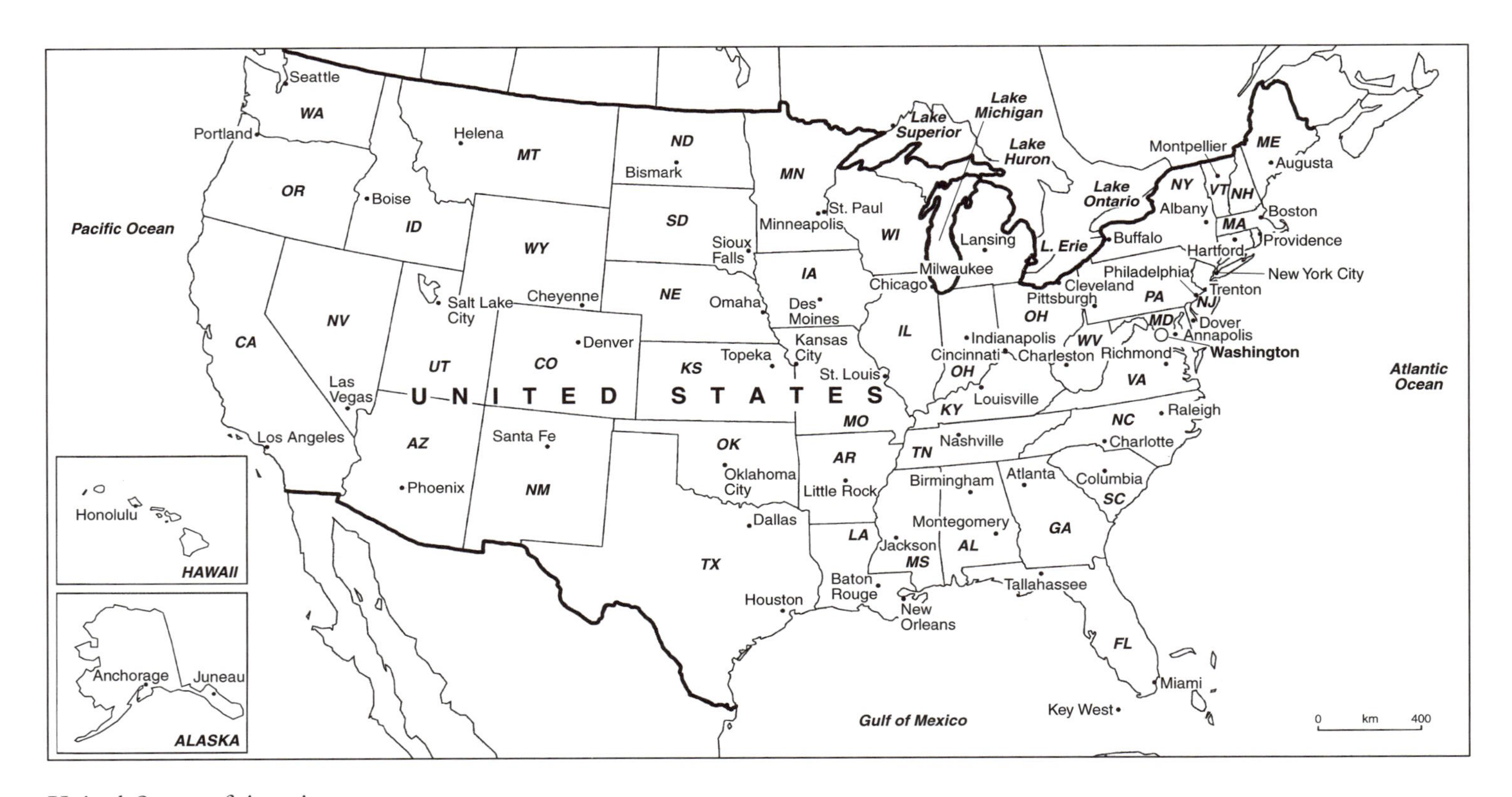

United States of America

14 Angry white men

Right exclusionary nationalism and left identity politics

Andrew Light and William Chaloupka

In a 1996 interview, Colorado State Senator Charles Duke, a leader of the emerging far right wing in U.S. politics, is oddly unsure of his movement's relationship to national identity (Duke 1996).[1] Duke is a political theorist, a former engineer whose model of state legislation to challenge federal authority has been enacted in eleven states and hundreds of counties (Chaloupka 1996: 172). One would expect Duke to have, ready at hand, a view on the right's sense of identity and nationhood. But when asked about the "angry white males" who form the most visible base for new rightist groups, Duke demurs:

> The bulk of Constitutionalists are women, by a long line. They just want the men to be men. There are certainly some angry white males, but I don't even think it's a majority, by any stretch. It's maybe 20 percent. I would think that the bulk of people who get things done are women [*laughs*]. My political campaigns would never have survived without women.
>
> (Duke 1996)

Nonetheless, it takes little prompting to get beyond that thin defensive veneer. Asked whether his constituents see themselves as a victimized group, Duke makes more explicit the far right's focus on male identity in opposition to all other "special interests." Closely connected is a nationalist base to this self-perception:

> I don't hear whining. Many are very aware that we've become a special interest society. Rights have been carved out – the white man thinks it's been carved out of his hide to give it to them. There are a number of people who feel that way. But I don't think it's their driving force. They want their Constitution back, they want to love America for what it stands for.
>
> (Duke 1996)

Duke's handling of the twin topics of nationhood and white male identity is telling. He is less inarticulate when detailing the Constitutionalist principles of sovereignty he has popularized. Nonetheless, the connected themes of white identity, nationalism, and the rightist political response to the left are not far beneath the surface, even for this strategist.

In this chapter, we investigate white identity politics, arguing that left versions of identity politics (e.g. feminism and the politics of racial and sexual orientation) may have invited the far right's formulation of itself around themes of national self-identity. Of course, it is certainly contentious to use terms like "left" and "right" so sweepingly. These terms are no longer rigid designators and can appear to be monolithic. Perhaps it would be better to substitute progressive/reactionary for this distinction, though those terms are also increasingly indeterminate. Surely any full fledged political theory today must be prepared to both call into question and rethink what is meant by right and left on the political spectrum. Nonetheless, in order to survey the issues we wish to take up in this chapter, we shall leave our contribution to the meaning of right and left to another day, and hope that our readers will allow us the use of these terms here.

In making our argument, we shall first briefly describe one way of understanding the terrain of identity politics that has been staked out, primarily by the left. Second, we shall examine how the notion of white male identity has been articulated as a parasitic response to the left's conception of identity as a response to political oppression. Here we shall detail, as much as space allows, the varieties of right-wing national identity, and attempt a useful generalization of the ways in which the adherents to right identity politics have used their embrace of an identity position to subvert the discourse of the left. Finally, we shall turn to an issue quite different from other concerns raised in this collection. Rather than pushing forward an analysis of the relationship between race, sexuality and nationality represented in the existence of right white identity politics, we wish to examine a different topic: what should be the response of leftists confronted with right white identity? As leftists ourselves, we think that an organized political response to the far right is crucial. We have no doubt that such a response is one of the most important tasks for the left today. But as theorists we are also interested in the question of whether a theoretical response is required as well. Not to draw a specious distinction between theory and practice, but is the theoretical apparatus that helps to explain the left's articulation of a politics of identity threatened by the right's embrace of a similar theoretical position? Does the right's subversion of the politics of identity, to serve what leftists perceive as the interests of oppression, constitute a reason to reject identity politics as a viable political stance for the left?

Our approach to this cluster of questions will be to focus on the issue of the theoretical terrain left up for grabs in the right's articulation of an exclusionary political meaning for whiteness and maleness. Has the right succeeded in providing a determinate meaning for what it is to be male and white, namely a meaning tied to a regressive interpretation of American national self-identity? If so, does this move by the right require a theoretical response from the left? In particular, does it require a theoretical response from leftist white men, such as ourselves, who wish to take part in the left political movements dominated by a politics of identity? And how do we avoid the right's claims that our masculinity and race are irrevocably tied to American nationalism? In the last section, our focus will not be on the issue of nationalism *per se*, but, instead, on the theoretical response to a white identity that has been tied to nationalistic themes.

This concluding focus of our chapter points to a prejudice of our approach which should be admitted at the start. This chapter is unabashedly aimed at contributing to a discussion of left political identity, and white male participation in it, for the purposes of unifying the left as a coherent theoretical unit effective as a base for political action. However, we also hope to at least provide a survey of the strong connections being made between white male identity and right-wing nationalism in the U.S.

Identity politics

Very generally, one can describe a politics of identity as a politics where agents ground their political self-conception in some aspect of their own personal identities. Often this identity is defined negatively: one is socially marked by an identity trait and then subject to different forms of oppression as a result of possessing that trait. For example, where a legacy of oppression based on race exists, an identity politics of race can be formed in opposition to that form of oppression, and can help to provide an occasion for racial pride and resistance to that oppression. Different descriptions can be given of the meaning of one's identity in this sense, and of the political implications derivable from that identity. It is rare that one hegemonic view of the political implications of one specific identity will exist, though at times such views may be articulated. But in general, theoreticians of identity politics see the political self as heterogeneous as is the variety of political positions in a given community. While this description may be vague, it is necessarily so given the competing descriptions of identity politics by theorists of and participants in the movements associated with that politics and given the limited aspirations of our discussion here. Certainly, a more thorough discussion of political identity would reveal more nuances among different views on this topic.

We endorse the position that political identities are situated in a variety of different, or heterogeneous, rationalities only with respect to politics. We call this view the argument for differing "political rationalities."[2] The closest approximation of this view is found in the work of the feminist political theorist Iris Marion Young (1990). We only sketch out this view of identity politics here, and leave a more complete account of it for another time. While the apparatus used here for describing an identity politics aids the discussion which will follow, it is not necessary to embrace this view for the substance of the argument concerning white male identity to hold.

By political rationality, we mean a narrower conception of rationality than is commonly articulated. The term refers to the sense in which different politically constituted communities have different standards for what counts as a right or reasonable argument in some political context. It does not simply suggest, for example, that different social groups differ over the right schema for a distribution of goods as part of a model of justice. Political rationalities would represent different conceptions of legitimate criterion for determining whether a model of justice should be distributive. Nothing separates this account from more ordinary metatheoretical claims about political theory unless it is attached to a robust view of political identity. So, if there is a feminist argument against the liberal scheme of justice, it could be true that non-feminists could also reject that same scheme of justice, even for similar reasons. The political rationality of a feminist identity politics is expressed in the idea that the reasons for feminist rejection of liberal justice may emerge out of issues unique to women (for example, out of experiences that are unique to women).

Taken together, the different contemporary leftist groups to which we can ascribe a politics of identity are often referred to as the "new social movements" (Aronowitz 1994: 17).[3] In fact, many leftists believe that identity politics is the wrong framework in which to articulate a left political movement (see, for example, Kellner 1995). But we do not wish to pass judgment here on the question of whether identity politics or the new social movements should or should not be the dominant form of politics today, either in general or specifically on the left. We instead want to investigate how one responds to a political sphere that embraces the existence and political legitimacy of heterogeneous identity formations. As political theorists interested in the pragmatics of doing politics, we think it is important to assume the political legitimacy of the new social movements and to negotiate left solidarity on their terms.

If one accepts identity politics as the political framework in which the new social movements in the developed world will form themselves as anything like a cohesive "left" in the twenty-first century, the right-wing reaction to left identity politics becomes crucial. A provocative question comes to the fore: in what sense did left identity politics invite the "angry white male" pose used to such great effect by the radical right? But to address this question, we first must investigate the current formation of white identity politics. Here we see that the construction of right white male identity politics has focused mainly around the formation of a new American national identity. One of the dangers of this identity is its claim to represent an exclusive interpretation of the political meaning of whiteness and maleness. Understanding white identity will both help us to understand the uses and abuses to which a politics of identity can be put, and, if the abuses can be overcome, give us an idea of the challenge that white right identity poses for white men seeking to embrace a left identity politics.

White male identity

The right has dominated the discourse of identity politics for whites in the U.S., in particular, for white males. This domination cuts two ways. On the one hand, the right's identity position calls attention to how "whiteness" has been created, consolidated, and used to political advantage as part of the process of American nation-building. On the other hand, as left identity politics has become more successful, that success has been refracted in a version of backlash that seeks to normalize hate groups as merely interest groups for whites.

From at least the eighteenth century until quite recently, white identity has formed the basis of a politics of exclusion and cultural hegemony in the U.S. Ronald Takaki (1990) suggests that the formation of the American republic – and, we would add, of American nationalism – was dependent in part on the ability to define slaves and Indians as culturally, morally and mentally inferior to white males.[4] Takaki argues that establishing the advantages of the new Protestant republic over the old decadent English monarchy was in large part dependent on perceiving the superiority of white Americans over imperfect Black and Indian bodies. It was only in the comparison of whites with these minorities that white virtue became self-evident and formulated as a cohesive, unified culture. Hence, the formulation of the American sense of nationhood was dependent on the creation of the myth of white male superiority. White superiority was generalized and universalized into a national system which denied rights and status to non-whites and relegated them to objects of ownership, abuse, discrimination or slaughter (see Takaki 1990: ch. 1, Saxton 1990). Similarly, the virtues of men were advanced over the vices of women in order to justify women's exclusion in the public sphere and repression in the private sphere (see Pateman 1988). Models of cultural imperialism emerged in the period which would persist in American history. In these models, according to Iris Young, "the dominant group constructs the differences which some groups exhibit as lack and negation." These groups "become marked as Other" (Young 1990: 59).

This relationship among the celebration of white male virtue, the active domination of non-whites and the project of nation-building intensified in the nineteenth century in the conquering of the American West. For example, in the political celebrations of the push of the railroad westward, the rhetoric of nation-building at the expense of non-whites was prevalent. This role for the railroad in the industrial revolution was unique to America. For comparison, Wolfgang Schivelbusch (1977) argues, the mechanization of travel in Europe was perceived of as the destruction of a traditional culture and traditional modes of transportation. American railroads, however, were expanding into what was perceived to be worthless wilderness – worthless in part because it was inhabited by non-whites. The opening up of the West represented in some arguments the key to American development because the vast untamed wilderness held the key to American wealth in natural resources. This wealth was something that was needed in order to offset labor and capital shortages in the young Republic (Schivelbusch 1977: 91). But for our argument the most important point is that this technological expression of national expansion was dependent on a pronouncement of white superiority over native Americans. When, for instance, the first transcontinental line was completed in 1869, the editor of the *Cheyenne Leader* proclaimed:

> The iron horse in his restless march to the sea, surprises the aborigines upon their distant hunting grounds and frightens the buffalo from the plains where, for untold ages, his race has gazed in the eternal solitudes. The march of empire no longer proceeds with sagely, measured strides, but has the wings of morning, and flies with the speed of lightning.
>
> (cited in Takaki 1990: 173)

This process of nation-building made possible by the railroads clearly represented the destruction of the Indians and the creation of the national identity of the white republic conceived as a growing empire.

The various local cultures that were the original Indian nations were in large part obliterated with the incursion of the railroad in the West. The train was the great equalizer for settlers, in a certain sense, indiscriminately opening up lands throughout the West for settlement. But of course it was not simply settlement for settlement sake; railroad companies made immense profits through the usurpation of Indian lands and the sale of that land to whites (Takaki 1990: 173–174). More directly, railroads made possible the transportation of buffalo hunters into the area – wiping out a crucial natural resource for the Indians – and further, carried federal troops to “rapidly and effectively respond to Indian resistance.” William Sherman called the railroad a crucial instrument in the battle of “civilization against barbarism” (Takaki 1990: 174–175). By 1872 the Commissioner for Indian Affairs, Francis Amasa Walker, proclaimed that the railroad would directly lead to the extinction of Indians as “the great plough of industrial civilization” (Takaki 1990: 186).

Of course, as barbarians, Indians were considered to be representatives of nature and nature would in turn also be tamed by the rails. And not only tamed of course, but also, as Ronald Takaki points out, sexually conquered, reflecting the male character of the building of the white American nation.

> In his speech at the Chicago Railroad Convention in 1847, William Hall described the emergent railroad as a virile masculine force, plucking out the forests, tearing up and flinging aside the seated hills and making his way into the “body of the continent” with the step of a “bridegroom” going to his chamber. With its locomotive steam ejecting the “pure white jet,” the railroad was penetrating what Henry Nash Smith has called the “virgin lands.”
>
> (Takaki 1990: 150)

Such sexual overtones may seem even stronger when the purpose of railroads was taken by others to include the encouragement of certain moral refinements, including restrictions on sexual mores. In an influential article published in the early days of American rail development, the physician Charles Caldwell (1832) suggested that the railway itself would suppress idleness through the maintenance of a technological order. People would be employed in useful business and because of the increased rate of activity, would not have time for "vicious practices." The railroad would be a mechanism to control passions as people concentrated on its workings rather than the satisfaction of their own desires (Caldwell 1832). Presumably, then, the sexuality of the construction of white male nationality was to be constrained by its own instruments of expansion. Several more recent studies emphasize the connection between the creation of white male identity and the establishment of an American nationalism. We only briefly mention these studies in order to indicate the variety of historical work undertaken on the connection between race, gender and early American national identity.

Michael Rogin shows how mass entertainment forms, including blackface movies, "acted out a racially exclusionary melting pot," even at the same time that such gestures seemed to manifest racial solidarity and concern (Rogin 1996: 8). Noel Ignatiev (1995) focuses on how Irish immigrants differentiated themselves from the African Americans with whom they often shared neighborhoods and job sites:

> The first Congress of the United States voted in 1790 that only "white" persons could be naturalized as citizens. Coming as immigrants rather than as captives or hostages undoubtedly affected the potential racial status of the Irish in America, but it did not settle the issue, since it was by no means obvious who was "white."
>
> (Ignatiev 1995: 41)

On gender, Thomas Dumm notes that "the universal language of violent domination in American culture is that of [male] sexual domination," and that this is one of the ways the American national character is composed (Dumm 1994: 63). In short, various scholars, undertaking a variety of projects, help to show how the formation of the American national character has been dependent on connections to white privilege, and, in particular, white male privilege.

Today, however, many conservatives argue that contemporary right white movements are a reaction to the political legacy of liberalism, which attempted to formulate policies designed to remove the historical scars imposed on these "Others" in the formation of American national identity. Affirmative action, redistribution schemes, etc., were all designed, according to these conservatives, in part against this eighteenth- and nineteenth-century background of minority oppression. The far right goes a step further: in their view whites must formulate a political identity of white racial pride (and prejudice) at the expense of any and all "special interest" minorities. Thus, what we would characterize today as white identity politics is clearly marked by the discourse of neo-fascism, exclusionary blood ties, racial violence and social Darwinism.

Importantly, however, this identity is parasitic on the historical legacy just described. Contemporary formulations of far right identity represent attempts to reclaim the discourse of American nation-building as singularly white and male and to assert that because of this history, white men should hold unequivocal power. But paradoxically, because such discourse is now outside of the mainstream in America, whites who try to revive this form of American nationalism have to organize themselves into special white identity movements.

It is not too much to say that the 1990s became, particularly in the rural northwest, the Me Decade for disillusioned working-class white men. Anecdotally, both of us observed the popularity of Rush Limbaugh on A.M. radio replace the omnipresent country western stations. More gun racks on old pickup trucks carried guns instead of fishing poles. The cultural movement of fear found a form. This weakly defined constituency formed the base for various anti-government groups, including militias, that received a substantial amount of media attention in the U.S. in the mid-1990s. But of course, we need more than anecdotes to connect this perceived trend to the eighteenth- and nineteenth-century history discussed above.

Human rights activists have debated the strength of connections between militias, a diverse phenomenon, and the narrower and much more frightening racist extreme. For the purposes of the present argument, the movement known as "Christian Identity" represents the most visible presence for the term "identity" in rightist politics. The Christian Identity movement (which articulates its position as a theology) predates by many decades the current theories of identity politics briefly described in the first section. It was reportedly created in 1946 by a former Klansman and a racist activist. In the words of one observer of the movement, "Identity teaches that Jesus Christ was an Aryan, not a Jew; the lost tribes of Israel were Anglo-Saxons and other Aryan peoples. . . . The United States may be the promised land but only if Jews are eliminated and non-Christian Aryans subjugated by the true 'white races'" (Bennett 1995: 350). White Christian Identity is thus made coextensive with a renewed form of American nationalism, harking back to the racism represented in nineteenth-century American expansionism.

Politically, Christian Identity is connected (by ideology, affiliation and some direct personal contacts) with a list of familiar fringe racist groups since the late 1970s, including the Order, Richard Butler's northern Idaho-based Aryan Nations, C.S.A. (the Covenant, the Sword and the Arm of the Lord) and Posse Comitatus (Bennett 1995). Members of both C.S.A. and Posse Comitatus committed highly publicized political crimes in the 1980s and 1990s (Bennett 1995: 351–353; see also Flynn and Gerhardt 1989, Stern

1996). Posse Comitatus gained national attention when Gordon W. Kahl, a retired farmer, gun collector and tax protestor, shot and killed two federal marshals who had tried to stop him at a North Dakota roadblock in 1983 (Bennett 1995: 352–353). A group called "the Order" then became national news when a string of robberies, including two armored car robberies in Seattle and Uklah, California, netted over $4 million. Leader Robert Matthews died in a shootout with police on Whidbey Island in 1984. Order members were convicted for the 1984 murder of Denver talk show host Alan Berg. Fugitives fled to the C.S.A. headquarters in the Ozarks, where they were arrested after a four-day siege (Bennett 1995: 348–351). In each case, these are tiny fringe movements, with membership often estimated only in the dozens. But these movements' perceptual impact has been magnified, both because of their sometimes spectacular crimes and confrontations with authority, and because they are so often mentioned as the extreme element of broader rightist movements.

Although Christian Identity represents an important overlay with the politics of identity already described here, a broader and more important development in white male identity politics is the use of identity themes in Klan rhetoric in recent years. In part due to successful lawsuits brought by Morris Dees' Southern Poverty Law Center, the Klan has come under unprecedented pressure in recent years. One wing of the Klan has responded by adopting a moderate, more mainstream, discourse that often closely mirrors identity claims common to the new social movements. As one Klan leader explains, "The KKK does not preach against Negroes. We believe everyone has a right to love their heritage and race" (Bennett 1995: 431–432). Beginning in the 1980s, increasingly extreme forms of racism, predominantly intolerable in American pubic discourse since the civil rights movement, were given a new lease as a form of white political identity.

This translation of hate rhetoric into the terminology of interest group and identity was probably best promoted by David Duke, the former Klan leader who went on to found the National Association for the Advancement of White People, an obvious attempt to translate his racist discourse into the terms of identity politics. Duke achieved considerable national visibility and political success – often to the considerable consternation of the Republican Party which Duke claimed as his own and which wanted to distance itself from Duke's racist legacy (Bennett 1995, Brady 1997: 236). In 1989, Duke, by then a Republican state legislator, ran for Governor of Louisiana. National Republican leaders, including Republican National Committee chair Lee Atwater, denounced Duke and tried to run him out of the party, creating a backlash that probably resulted in Duke's win in the Republican primary. Duke's political success paralleled his public renunciation of bigotry, announcements which human rights advocates ridicule, noting that racist literature continued to be sold by his organization, and even distributed at his campaign headquarters. Yet Duke announced, "I'm just as Republican as Lee Atwater" (Brady 1997: 236). And Duke adopted straightforward identity terminology which harked back to the history of national identity formerly defined as securely white: "White people don't have rights in this country anymore. . . . White people today are the victims of greater racial discrimination than blacks faced anytime in the last one hundred years" (Dees 1996: 60).

Both radical rightists and more moderate conservatives rallied around an ideology that was at once anti-government and nationalistic in the wake of the federal A.T.F. (Bureau of Alcohol, Tobacco, and Firearms) attack and destruction of the Branch Davidian Waco, Texas, complex in 1994 and the violent attempt by federal agents to capture Randy Weaver, who was avoiding arrest warrants at his remote Ruby Ridge, Idaho, cabin.[5] This anti-government position was nationalistic insofar as it represented an attempt to redefine what it means to be an American. According to some rightist accounts, white Americans, the only true nationalists, need to recognize their white history and rebuild their national identity around a rejection of central government authority. This message clearly had an appeal: at its broadest level, the anti-government movement attracted substantial interest and recruited members at a rate never before seen among populist right-wing groups in the U.S.[6] While human rights observers have quietly debated the extent to which the broad militia movement can appropriately be called an extension of earlier racist organizing (Stern 1996: 245), some connections are explicit. For example, Randy Weaver was a long-time adherent to Christian Identity positions, and had attended events at the nearby Aryan Nations compound (Dees 1996: 9–13).

Many in the militia claim that their unique form of nativist culture is under attack. But radicals, including Identity Christians, go well beyond self-defense or racial pride. As Dees (1996) explains, Christian Identity includes an explicitly nationalist justification of white male authority:

> [Christian Identity] maintains that America is the New Jerusalem and that the Constitution was derived from the Bible and given to the white Christian Founding Fathers by God. It contends the U.S. government is nothing more than an expansion of the Christian faith and that the first Ten Amendments of the Constitution (the Bill of Rights) and the Articles of Confederation are the only documents – aside from the Bible – that need to be obeyed. The Israel Message also holds that only white Christian men are true sovereign citizens of the United States. All other Americans, it argues, are merely Fourteenth Amendment "state" citizens, the creation of an illegitimate government.
>
> (Dees 1996: 10–11)

In its most extreme expressions, such as the infamous Turner Diaries, a fictitious account of a future race war popular in the militia movement, this form of identity politics advocates genocide of racial and ethnic minorities as well as summary execution of "mixed blood" race traitors.

The contradiction between, on the one hand, the leftist focus of identity politics currently in vogue in academic circles, and, on the other, the far right identity politics at work in arguments for white supremacy must be addressed. Both left and right identity politics are committed to the importance of personal identity as a foundation for valid political positions. Both could be translated into the language of political rationality outlined earlier. Both see these new orientations as responses to the failure of a liberal political sphere to acknowledge the importance of personal identity and history in political orientation (Young 1990). But the two come to radically different conclusions. Both seek suppression of the power of "majority" culture over perceived persecuted minorities, but for different ends: left identity politics looks to build a ground for emancipation against any hegemonic identity, and celebrates the plurality of identity cultures. Most new social movements embrace one form or another of multiculturalism. The ground for identity recognition is articulated in each case to ensure the flourishing of many different identities: each different cultural identity (at least each identity group that was or has been the mark of majority oppression) has intrinsic worth.

But right identity politics, according to its own adherents, seeks emancipation only for "persecuted" whites. Pushed further, the most extreme of the right white groups, like Christian Identity, seek repression of all others, a romantic retrieval of an earlier American national identity. As we saw above with Duke's rhetoric, the far right argues that whites have lost rights in the U.S., a claim that draws force from the historical background that whites once defined who was to have rights in the U.S. White identity therefore represents a return to a form of American nationalism predicated on the exclusivity of American national character. Multiculturalism (an issue we shall briefly return to at the end of this chapter) becomes the inverse of this argument: American national identity should be plural and diverse.[7]

But behind the observation that the right's nationalism uses an exclusionary base to formulate its version of white identity lurks the specter of an argument that such anti-pluralism is in fact an extension of the general logic of identity politics. Is the right's theoretical move from an articulation of a unique political rationality for white men to an exclusionary nationalism different in kind from the left form of identity politics, or only in degree? The right's success in tying its identity to an exclusive nationalism brings up the worry that any form of identity politics could tie itself to an exclusionary politics, even to an exclusive definition of what it means to be an American. The political rationality of any identity view could easily move from a claim to an exclusive perspective on a set of political issues to a view that dismissed all other perspectives as illegitimate.

The important question then becomes whether the left can successfully separate itself from this extension, and possible abuse, of identity politics, as it manifests itself in the form of white nationalism. And, if such a separation is successful, can other forms of white male identity be separated from this white right historical legacy, or has the right successfully claimed the content of whiteness and maleness in the discourse of exclusionary nationalism?[8]

Rethinking whiteness?

What accounts for this difference between, broadly speaking, right and left forms of identity politics? Why is it the case, for example, that what counts for women's identity politics has a left slant while white males go to the right? If the difference between these two forms of identity could be articulated, then we would have a sound basis on which to separate left and right identities. The anti-multiculturalism of right white nationalists could be expressed as endemic to white nationalism rather than suggestive of a potential within identity politics in general.

The strongest theoretical strategy for separating right and left identity politics would be to argue that each identity formation produces a determinant political orientation (or narrow range of political positions) which are a direct result of the historical development of each identity group. Along these lines we would need an account of how white men, as the historically dominant identity group in America, shaped by a particular nationalist history, have secured their rights and privileges as a direct result of their oppression of other identity groups. This exercise of authority has left some sort of power trace which reemerges in the identity politics of the white right, and which is evidenced in the neo-fascist discourse of the far right. Presumably, then, we could argue that it is whiteness and maleness that are inextricably tied to an exclusionary nationalist discourse and not identity politics in general. But if this argument was accurate then the meaning of whiteness and maleness – and subsequently of all representations of the male gender and the white race – would be shaped by that nationalist legacy.

But we think that such a determining relationship between the history of an identity position and its ideological content is overly simplistic. Even if there is a determinant relationship between the history of an identity group and its politics, this relationship is not necessarily determining. But to successfully argue that the history of an identity position does not necessarily determine its politics we must first be skeptical of any strong expression of the metaphysics of history whereby historical forces necessarily shape the content of individual political views. Such an argument would benefit from a general pragmatist approach to politics. For the pragmatist, politics is made up of individual actors, each with the capacity for free choice, especially in the case of moral choice (see Dewey 1928). Pragmatism views the engagement of any person in politics, not just identity politics, as an act of volition, and not the result of historically determining forces acting on the content of one's identity. People choose to embrace a political position and are free to reject it if they wish. Such a view has the philosophical advantage of theoretical simplicity: it does not require a sophisticated account of how history shapes individual political decisions. On this pragmatist view then, the burden of proof should be on the argument for a necessary relationship between a group's history and its politics. In the case under discussion here, the burden is on the claim to a connection between an exclusionary nationalist history of white men and the current identity formations of white men.

Another reason why the strongly determinist view connecting the history of white nationalism with right white identity formations is wrong emerges out of our earlier claim that identity politics is a form of political rationality. Even before one looks at it through the lens of pragmatism, identity politics assumes that politics is something that is embraced rather than determined. This follows directly from the assumption that heterogeneous political rationalities exist, and from the assertion, common to most if not all interpretations of identity politics, that individuals can embrace several different political identities at once. Additionally, many theorists argue that individuals can adopt different interpretations of the political arguments which are manifest by those individuals' identities. If personal politics were historically determined, how could it be true that individuals can choose to reject or simply reinterpret the relation between their political history and their politics? Identity political theorists certainly do not argue that all women are necessarily feminists simply because they are women. Feminism, according to the political rationality account, is something that women are free to embrace because they are women. One could account for some women's turn away from feminism through a false consciousness analysis, but this would be inconsistent with the non-essentialist discourse at work in much of the identity politics literature, a discourse which distances itself from historically determinist accounts of political identity. So if identity politics is something that is embraced rather than determined, then the historically determining argument for the conservative to neo-fascist exclusionary nationalist arrangements of white identity exaggerate the relationship between the history of an identity and its later ideology.

Now, if it is in principle possible that American white men need not have the political meaning of their race and gender determined by the right-wing nationalist interpretation of identity politics, then history would seem inconsequential for the determination of political identity in general. But the strong historical legacy of whiteness in American national culture cannot simply be ignored in an assessment of white identity. In fact, that strong history infects not only the political interpretation of the meaning of any white identity, but also the understanding of any political project embraced by right white nationalists.

Understanding the ongoing history of the relationship between American exclusionary nationalism and white identity is the key to how left and right identity are to be separated – not theoretically, but rather politically. Clearly the move to the right by many white men, especially toward a politics of exclusion rather than inclusion, is perfectly consistent with the national history of white self-identity that we briefly outlined in the previous section.[9] But not all political issues embraced by right nationalists can be sufficiently described as fully consistent with this politics of exclusion. Leftists should be worried about the breadth of the political ground that is being occupied almost completely by the right through white identity formations and used to serve a particular interpretation of American national identity. Some of that ground has been the traditional terrain of the left. In particular we have in mind the right's calls for resistance to the liberal state, especially to federal interference with individuals and communities. We can not therefore simply endorse an argument that left and right identity can be theoretically separated based on the obvious differences in some of the political claims they make. The right must instead be actively resisted over the political terrain it is coming to occupy. We contend that this project of resistance to the right is more political than theoretical, focusing on issues beyond the articulation of a theoretical difference between left and right forms of identity politics. But examples of the political terrain coming to be occupied by right white nationalists is needed.

For instance, it is fair to say that many on the left were as troubled by the actions of Federal A.T.F. agents in the incident at Waco, Texas, as their counterparts on the right.[10] There is legitimate and reasonable ground upon which to object to the actions of the Department of Justice in this incident, particularly to the tragic events which led to the needless deaths of children at the Davidian compound. While it is easy to dismiss such objections because the targets of the raid were fundamentalist Christian gun-toting followers of a would-be Messiah, one sees an uncomfortable resemblance between the actions at Waco and other incidents in the history of the American left. How far have we come, for example, from the Nixon administration's war on the Black Panthers culminating in the tacit approval and encouragement of the Chicago Police Force's siege of a Panther leader's home in 1969? But to make such objections to Waco publicly is almost impossible in the current political climate, where the critique of the Clinton administration's assault on the Davidian compound comes most strongly from right white nationalists. Especially after the bombing of the Alfred P. Murrah Federal Building in Oklahoma City, which killed 168 people and injured 850 others, the right owns the discourse of the Waco incident.

Timothy McVeigh, convicted of the bombing, has publicly used his objections to the Waco incident as the basis for his actions. Stephen Jones, McVeigh's lead defense attorney, has gone so far as to use Waco and Ruby Ridge as part of the defense's plea to avoid the death penalty, suggesting that McVeigh's reaction to Waco made the bombing in Oklahoma a political act. According to the *New York Times*, Jones argued that "many people died in the sieges at Waco and at Ruby Ridge, Idaho . . . and these events led Mr. McVeigh to conclude that 'the Federal Government was the enemy of this country'" (Thomas 1997: A14). Thus, to attempt to reclaim the incident at Waco from the right's use of it as a defense for their extremist actions – as a basis for any kind of critique of the liberal state in general, or the Clinton administration in particular – risks association with American neo-fascism.

Breaking the hold of this political hegemony should be high on the agenda for the identity politics of leftists. Unless they do make this break, objections to any form of centralized authority will continue to be associated with the exclusionary nationalism of the right. McVeigh's actions were clearly targeted against the Federal government, but also, as Jones said, in defense of the people of "this country." The Oklahoma incident brings home the lesson that the right's exclusionary nationalism is not a nationalism of the country as it now stands, but of a view of the U.S. as the right thinks it should be – an image informed by the blatant white supremacy of earlier American national culture. It is striking to note that McVeigh's concerns over Waco were presented as supposedly common among Americans. According to defense lawyer Richard Burr, McVeigh's actions belie a concern about the power of the Federal government that all of us should, or perhaps do, share: "That we have not expressed that concern before this tragedy means that we all bear some responsibility for Oklahoma City" (Thomas 1997: A1). According to Stephen Jones: "His [McVeigh's] fears [about the Waco incident] are not alone." And again from Burr, "Mr. McVeigh's beliefs 'did not arise out of thin air,' and they were not 'the delusions of a madman'" (Thomas 1997: A14). What do such statements mean? Most plausibly that even the motivation for the most extreme actions of the far right should be common to most Americans, namely, a justifiable worry about the power of the Federal government. By extension of this assertion that many, if not most, Americans share McVeigh's concerns about centralized federal power, the right's interpretation of American national culture as anti-federal is supposed to serve as the proper basis for a true description of American national identity.

But this explication of the proper content of American national culture as anti-federalist could be expressed from the left as well, or more specifically, non-leftists could accuse the left of holding a similar position. And if the far right's anti-federalism is thought of as connected to apologetics for terror bombing, then any leftist who articulates an anti-federal position becomes guilty by association of harboring similar extreme tendencies. Therefore, if leftist identity groups wish to separate themselves from the right's nationalism, without giving up the ability to legitimately criticize abuses of federal power, they must actively articulate the difference between their version of a communal political identity from that provided by the right. More specifically, they must show how their conception of individual and group identity does not lend itself to a homogenous conception of American culture as so anti-federalist as to justify the same extremes evidenced in the far right. This is not an easy political puzzle. The "right" or "left" coding of recent political spectacles is certainly overdetermined. It is yet possible, however, that this quandary can be resolved on political, rather than theoretical, grounds. A unified, concentrated attack on the claims of right white nationalists by a more cohesive left could be effective in this regard. Perhaps the focus of such a response by leftists could be a rearticulation of multiculturalism, as an alternative to nationalism, as the political center of progressive political identity. We return to this point below.

But even if we can understand that leftists have a clear political interest in challenging right-wing white identity, the question remains: would a revitalized left identity politics in opposition to white nationalism require any major theoretical changes in regard to the meaning of whiteness or maleness? For example, to retain white male participation in left political coalitions of identity groups, leftists may need to retheorize white identity making it more politically suitable for leftist political activity. The left would need to reconstitute the meaning of whiteness away from the rightist nationalist legacy of the past. It seems then that we are back in a theoretical muddle.

But is the articulation of a new meaning for whiteness a serious problem? Possibly. Many authors have recognized that those identity groups that have historically been the objects of white cultural imperialism in America represent the best basis for reconfiguring identity as a foundation for progressive political movements (see e.g. Aronowitz 1992, Young 1990, Mouffe 1990, Laclau 1989, Laclau and Mouffe 1985). Following this line of thought, Iris Young calls for a "revolution in subjectivity," against cultural imperialism:

> The dissolution of cultural imperialism thus requires a cultural revolution which also entails a revolution in . . . subjectivity. Rather than seeking a wholeness of the self, we who are the subjects of this plural and complex society should affirm the otherness within ourselves, acknowledging that as subjects we are heterogeneous and multiple in our affiliations and desires.
>
> (Young 1990: 124)

Young's claim that we should embrace a pluralist account of the self has much appeal as the basis for a challenge to some form of essentialist feminism, for example, in favor of a more fluid feminism which acknowledges heterogeneous identities. Such a feminism would avoid the early problems of feminist insensitivity to the multiple jeopardies of poor Black women, for whom sexism was only one of several sites of oppression, and not necessarily always the most important. But when it comes to white males in the new social movements, this revolution in subjectivity cannot begin and end with a simple embrace of the multiple identity positions within white males. Given the predominantly anti-pluralist and anti-multicultural legacy of white male identity framework as it has been defined by the white right, a more complicated and effective process of negotiating a change in white male identity would be needed.

Of first importance for leftist white men in the project of rejecting the exclusionary nationalist character which has been assigned by the right to their race and gender would have to be the conscious rejection of white male privilege. The blindness that white men exhibit to their own privilege is astonishing, and in this context appalling.[11] Peggy McIntosh (1992) has argued that whites and men are taught not to recognize the character of their privilege and are institutionally sustained in their myopia. Marilyn Frye (1983) argues that the blindness of white men to their privilege excludes whiteness and maleness as reasons for social failure.

> If a man has little or no material or political power, or achieves little of what he wants to achieve, his being male is no part of the explanation. Being male is something he has going *for* him, even if race or class or age or disability is going against him.
>
> (Frye 1983: 16)

Of course, the far right has come a long way in subverting the likely truth of Frye's claim that white men have their gender and race going for them, by arguing that white men are now the victims *by virtue* of their identity. White men, according to this explanation (as exemplified in the quote from Charles Duke which opened this chapter), are victims of a liberal state which punishes them not for their past privilege, but for their superior abilities. Even more mainstream conservatives, participating in Republican caucuses rather than militia groups, have embraced something like this position in their rejection of affirmative action as a form of white male punishment. Moderates' charges of "reverse racism" are evocative of just the same sort of oppression which has sparked the call to arms of other identity political groups. And as we have seen, such arguments are offered as a defense of an older national American culture. These are indeed troubled waters. To make such arguments involves a willful misreading of the historical role of white male privilege articulated above, painstakingly demonstrated by historians like Takaki (1990, 1993), Alexander Saxton (1990) and David Roediger (1992). If white male identity was to be reconfigured to another form of identity it would have to include a challenge to white males' denial of their own political history. One cannot simply wish away the historical connections between white history and white politics; these two must instead be actively separated.

A leftist redefinition of white identity would necessarily include a clear understanding of the history of white male privilege, acknowledging the role of that privilege in everyday life. An active challenge has to be made to those who would ignore, or continue, the myth of white male superiority. Whites must, as Charles Mills has said, become "*race traitors*" – they must embrace the legacy of those whites before them who were (and sometimes still are) accused by white supremacists of betraying their racial telos.[12] But while a politics of rejection of privilege could form the basis for an active politics of critique of contemporary social structures, it does not provide a proactive vision for moving beyond this negative stage. What positive programs can white men embrace? Only those which suppress white male privilege? Such an answer would be insufficient, as many different political configurations could accomplish this goal, some of them authoritarian in framework. Something else must accompany the call for the active rejection of white male privilege. And since, following Young, we would be trying to flesh out a politics which embraces a pluralistic self, the next place to turn is toward other aspects of identity which may be embraced by white males. More accurately, some other forms of identity must be secured into which white male identity may be placed in acts of political coalition formation.

Of course, there are reasons to be skeptical of such a project – including the implausibility of the political successes of existing attempts, such as is found in the contemporary "men's movement," to reform white identity. Such movements run into problems because they court essentialism implying that male identity has a determinant meaning. More importantly, to claim that white right identity politics requires us to recapture white male identity as an identity position is perhaps to award right-wing whites more credibility than we think they ultimately deserve. Left identity politics must not be driven to reformulate itself theoretically out of regard for what the right whites are doing with white male national identity. There is no clear need to theoretically reclaim whiteness, even though thinking through this possibility reveals a clear need for white men in the new social movements to acknowledge and confront their legacy of white male privilege. This rejection of white male privilege can be accomplished in the process of creating a stronger left political front without worrying exclusively about the theoretical status of whiteness.

The discomfort that Charles Duke exhibited in talking about the white male identity position is telling. Except for the extremely marginal example of Christian Identity, right-wingers have not theorized identity politics nearly as thoroughly as they have such questions as gun ownership and federal authority. The challenge that leftists face is political, not theoretical; practical, not existential. Left identity theorists should not turn backflips to articulate their theoretical differences from the right and to mollify worries that their interpretation of American politics could come to represent a warped form of identity nationalism, as has emerged on the right. But in order to allow left identity theorists and activists to ignore the theoretical challenge of the right identity movement for the left's use of the politics of identity, the rest of the left (e.g. more traditional socialists) must accept the identity basis of their colleagues in the new social movements for the purposes of forming a more unified left political front. On that political front we are cautiously optimistic. In the wake of the Oklahoma City bombing, the strength of the radical right appears diminished, though at this point we have no clear empirical evidence to back up this claim. The trial of Timothy McVeigh turned out to be a quick affair, offering no opportunity for reinvigorating the right. The political limits of the right's position are apparently as severe as its theoretical problems.

Conclusion: toward a multicultural nationalism

We have argued that the white right has merged a vision of exclusionary nationalism with an appropriation of the foundations of an identity politics often thought to be the purview of the left. The establishment of this white nationalist identity position deserves a response from the left, if only because the exclusivity of the right's position leaves left identity politics susceptible to being charged with a similar form of exclusion. One answer to the right would be for leftist white men to try to reclaim the political meaning of their gender and race, disassociating themselves from the right white identity movement's claim that maleness and whiteness are necessarily tied into an exclusionary national history and culture. We think that while leftist white men must reject their white male privilege in American society, they need not be bothered with the theoretical task of rethinking the political meaning of white male identity. Instead, we argue that the left, broadly construed, should concern itself more with resisting the right's takeover of certain political claims as a matter of practical political action. In particular, the left needs to resist the right white's growing hegemony over the discourse of resistance to centralized authority. Even more important, the left must resist the right's effort to define what it means to be a "real" American, such as was seen in the explanation for Timothy McVeigh's actions by his defense attorney. The right has tried to define American national identity in opposition to its own perception of the federal government.

Such a definition of the meaning of American national identity must not go unchallenged. In resisting the right's political stance, perhaps the formation of a competing view of national identity can also help in developing a clearer left political position for identity politics. As support for diversity and multiculturalism has developed, the left can argue that a diverse national identity – a pluralist vision of the American nation – makes sense in many ways, for example by countering hatred, fostering internationalism and generally encouraging a national self-image of inclusivity. Just as the right makes claims about the historical basis of white supremacy, the left can claim that inclusive pluralism is part of the American national character and has a strong and identifiable history (see Takaki 1993, Hollinger 1995). Supporting such concerns involves everyone – including white men – in a set of political claims that are widely defensible in terms of identity politics, if not necessarily from an identity position: namely the rejection of white male privilege, and resistance to a resurgence of white exclusionary nationalism.

Pushing this reflection on the potential of multiculturalism as the basis for a new national identity a bit further, we believe that left identity theorists and activists have much at stake in reducing the essentialism or dogmatism sometimes associated with multiculturalism, especially if multiculturalism is to be used to counter the right white view of American national culture. An essentialist multiculturalism would hold that there is a necessary content to the character of any particular culture that is determined by that country's history. But such an essentialism serves only the white right's claim that exclusivity and white supremacy are necessarily part of America's culture because the essentialist position argues that there can in principle be only one exclusive content to the character of any culture.

Certainly, by pressing for actual inclusion, as opposed to simply the symbolic representation of inclusion found in tokenism, multiculturalists could put pressure on institutions which have historically supported white hegemony. But in this sense, multiculturalism could help to generate the defensive, backlash stance taken by the right against movements for inclusion. But articulating strong, self-confident political identities is at the center of the multiculturalist project, whether or not promoting such identities stirs up resentment on the right. This tension strikes us as a more or less permanent paradox built into identity politics – the push for resistance to the old American national identity causes a backlash by whites wanting, in turn, to retrieve that national culture.

A multicultural response to the right white's interpretation of American national culture must not be deterministic. It must not theorize identity in essentialist terms, but must be more pragmatic, seeing the embrace of any political identity by any person to be an act of volition or personal choice. If identity is seen as something we embrace rather than something that is determined, then it will be harder to tie any particular interpretation of an identity position to an exclusionary nationalism. The creation of a more pluralistic, multicultural vision of an American national culture than that offered by the right, like the embrace of any personal or communal political identity, must also proceed as an act of conscious, communal, and democratic deliberation.

Notes

1 Duke briefly achieved mainstream visibility when the F.B.I. called him in to negotiate in the Montana Freemen standoff in 1996.

2 This articulation of identity politics as a claim for differing "political rationalities" is taken from Andrew Light's manuscript in preparation, *Marx, Post-Marxism, and Identity Politics.*

3 Of course, many other theorists also use this term in the literature, and so we take it to be an adequate description of how to refer to these groups as a whole.

4 For some details on how the idea of wilderness was also used in this construction of social identity, see Light (1995).

5 For extensive reviews of Waco and Ruby Ridge, see Dees (1996: chs 1 and 4) and Stern (1996: 19–64).

6 An ABC News/*Washington Post* poll reported that "Thirteen percent of Americans support private armed militias. . . . Twelve percent are afraid of the government, nine percent are angry at it, nine percent say violence against it can be justified and six percent call it their 'enemy.'" "Distrust of Government," ABC News/*Washington Post* Poll, May 17, 1995. Posted to America On-line (keyword ABC/NEWS/POLLS), transcript ab5np036–7. Stern (1996: 13) reports that a few months before the Oklahoma bombing, human rights observers called the militia movement the fastest growing radical right phenomenon they had ever seen.

7 There are certainly other forms of identity politics which also embrace repression of all others. We shall not discuss here forms of extreme Black Nationalism, for example, that reject multiculturalism. We imagine though that our critique of such groups would be similar to our critique of white supremacists. But importantly a critique of non-white nationalists would also be different given the forms of oppression that non-white groups have historically faced.

8 For another interpretation of the political content of right white identity see Pfeil (1995). Pfeil claims, as we do later, that the right white identity movements do not, and cannot, exclusively claim the content of white maleness. Pfeil further argues that racism and sexism are not an intrinsic part of the identities of the men who are outside of the core far right groups, and that the real causes of the creation of the far right militant identities lies in the exacerbated class antagonisms of the late 1980s and early 1990s. We are willing to agree that the subjects of Pfeil's largely anecdotal reflections – white men at the periphery of the far right groups – have a largely economic motivation for their flirtation with these movements. Still, given our account so far, we would strongly disagree with a claim that the core of these movements do not tie their militant anti-federal nationalism to a robust sexism and racism. It is unclear, however, if Pfeil believes the core members of these groups do not hold an ideology which is intrinsically sexist and racist.

9 David Roediger (1992) has even argued that this has been an extraordinarily important part of the history of the white working class.

10 We are indebted to Moishe Postone for pointing out this argument. Some may object to the following analysis that the left was as vocal as the right in criticizing the Clinton administration's handling of the Waco incident. But the point we are making here is still valid: the right now owns the discourse of radical anti-federalism.

11 In a public forum sponsored at UC Riverside shortly after the Rodney King Riots in Los Angeles, one of us (Light) made an argument like this as the official representative of "white people" on a multiracial panel. We are indebted however to Valerie Broin for pointing out the importance of the priority of self-conscious disempowerment as a prerequisite for reconfiguring white male identity, in a public response she gave to an earlier version of part of this chapter. The references to the existing literature on this point were provided by Professor Broin.

12 Consistent with this argument, Mills (1994) called for the active pursuit by white scholars of new work on the nature and function of white prejudice, in his closing address, "Taking race seriously," to the Radical Philosophy Association meeting at Drake University in Des Moines, Iowa, November 5, 1994.

References cited

Aronowitz, S. (1992) *The Politics of Identity: Class, Culture, Social Movements*, New York: Routledge.

—— 1994) "The situation of the left in the United States," *Socialist Review* 23, 3: 7–42.

Bennett, D. H. (1995) *The Party of Fear: The American Far Right from Nativism to the Militia Movement*, New York: Vintage.

Brady, J. (1997) *Bad Boy: The Life and Politics of Lee Atwater*, Reading, MA: Addison-Wesley.

Caldwell, C. (1832) "Thoughts on the moral and other indirect influences of railroads," *New England Magazine* 2, April: 292–297.

Chaloupka, W. (1996) "The county supremacy and militia movements: federalism as an issue on the radical right," *Publius* 26, 3: 161–176.
Dees, M. with J. Corcoran (1996) *Gathering Storm: America's Militia Threat*, New York: HarperCollins.
Dewey, J. (1928) "Philosophies of freedom," in H. M. Kallen (ed.) *Freedom in the Modern World*, New York: Coward-McCann.
Duke, C. (1996) Interview with William Chaloupka, conducted at Duke's home in Monument, CO, August 16.
Dumm, T. L. (1994) *United States*, Ithaca, N.Y.: Cornell University Press.
Flynn, K. and G. Gerhardt (1989) *The Silent Brotherhood: Inside America's Racist Underground*, New York: Free Press.
Frye, M. (1983) *The Politics of Reality: Essays in Feminist Theory*, Trumansburg, N.Y.: The Crossing Press.
Hollinger, D. (1995) *Postethnic America*, New York: Basic Books.
Ignatiev, N. (1995) *How the Irish Became White*, New York: Routledge.
Kellner, D. (1995) *Media Culture: Cultural Studies, Identity and Politics between the Modern and the Postmodern*, New York: Routledge.
Laclau, E. (1989) *Reflections on the Revolution of Our Time*, London: Verso.
Laclau, E. and Mouffe, C. (1985) *Hegemony and Socialist Strategy*, London: Verso.
Light, A. (1995) "Urban wilderness," in D. Rothenberg (ed.) *Wild Ideas*, Minneapolis: University of Minnesota Press, pp. 195–211.
McIntosh, P. (1992) "White privilege and male privilege," in M. Anderson and P. Collins (eds.) *Race, Class, and Gender: An Anthology*, Belmont, CA: Wadsworth.
Mills, C. (1994) "Taking race seriously," in B. Stone (ed.) *Proceedings of the Radical Philosophy Association*, Vol. I, no publisher, pp. 291–293.
Mouffe, C. (ed.) (1990) *Dimensions of Radical Democracy*, London: Verso.
Pateman, P. (1988) *The Sexual Contract*, Stanford, CA: Stanford University Press.
Pfeil, F. (1995) "Sympathy for the devil: notes on some white guys in the ridiculous class war," *New Left Review* 213, September/October: 115–124.
Roediger, D. (1992) *The Wages of Whiteness: Race and the Making of the American Working Class*, London: Verso.
Rogin, M. (1996) *Blackface, White Noise: Jewish Immigrants in the Hollywood Melting Pot*, Berkeley: University of California Press.
Saxton, A. (1990) *The Rise and Fall of the White Republic: Class Politics and Mass Culture in Nineteenth-Century America*, London: Verso.
Schivelbusch, W. (1977) *The Railway Journey*, Berkeley: University of California Press.
Stern, K. S. (1996) *A Force upon the Plain: The American Militia Movement and the Politics of Hate*, New York: Simon & Schuster.
Takaki, R. (1990) *Iron Cages: Race and Culture in nineteenth-Century America*, Oxford: Oxford University Press.
—— (1993) *A Different Mirror: A History of Multicultural America*, Boston, MA: Little, Brown.
Thomas, J. (1997) "Jury deliberates fate of McVeigh," *New York Times*, June 13: A1, A14.
Young, I. M. (1990) *Justice and the Politics of Difference*, Princeton, N.J.: Princeton University Press.

Index